Mathematical and Numerical Modelling in Electrical Engineering

MATHEMATICAL MODELLING:
Theory and Applications

VOLUME 1

This series is aimed at publishing work dealing with the definition, development and application of fundamental theory and methodology, computational and algorithmic implementations and comprehensive empirical studies in mathematical modelling. Work on new mathematics inspired by the construction of mathematical models, combining theory and experiment and furthering the understanding of the systems being modelled are particularly welcomed.

Manuscripts to be considered for publication lie within the following, non-exhaustive list of areas: mathematical modelling in engineering, industrial mathematics, control theory, operations research, decision theory, economic modelling, mathematical programming, mathematical system theory, geophysical sciences, climate modelling, environmental processes, mathematical modelling in psychology, political science, sociology and behavioural sciences, mathematical biology, mathematical ecology, image processing, computer vision, artificial intelligence, fuzzy systems, and approximate reasoning, genetic algorithms, neural networks, expert systems, pattern recognition, clustering, chaos and fractals.

Original monographs, comprehensive surveys as well as edited collections will be considered for publication.

Mathematical and Numerical Modelling in Electrical Engineering Theory and Applications

by

Michal Křížek
Academy of Sciences, Prague, Czech Republic

and

Pekka Neittaanmäki
University of Jyväskylä, Jyväskylä, Finland

Springer-Science+Business Media, B.V.

A C.I.P. Catalogue record for this book is available from the Library of Congress

ISBN 978-90-481-4755-7 ISBN 978-94-015-8672-6 (eBook)
DOI 10.1007/978-94-015-8672-6

Printed on acid-free paper

"For the mathematician, not calculation, but clear thinking is characteristic: the ability to disregard inessential things."

PÉTER RÓZSA
Math. Intelligencer **12** (1990), p. 60

Contents

Glossary of symbols

$C, \widehat{C}, C_i, \ldots$	generic constants (different at each occurrence)
$\mathbf{R}^d$	d-dimensional Euclidean space
$\cdot$	scalar product in $\mathbf{R}^d$
(a, b)	open interval in $\mathbf{R}^1$
$[a, b]$	closed interval in $\mathbf{R}^1$
Ω	problem domain (bounded connected open set in $\mathbf{R}^d$)
$\text{meas}_d\,\Omega$	d-dimensional Lebesgue measure of Ω
$\overline{\Omega}$	closure of Ω
$\partial\Omega$	boundary of Ω
n	outward unit normal to $\partial\Omega$
$\mathcal{L}$	set of all bounded domains with a Lipschitz continuous boundary
dist	distance
$v\|_K$	restriction of function v to set K
$n \cdot w$	normal component of vector function w on $\partial\Omega$
$n \wedge w$	tangential component of vector function w on $\partial\Omega$
v^+, v^-	positive and negative parts of v
$\partial_n v$	normal derivative of v on $\partial\Omega$
$D^{\mathbf{i}} v$	$\mathbf{i}$th generalized derivative of v ($\mathbf{i}$ multi-index)
$\partial_j v$	first generalized derivative of v ($j \in \{1, \ldots, d\}$)
Δ	Laplace operator
grad	gradient
div	divergence
rot	rotation of vector function for $d = 2, 3$
curl	rotation of scalar (vector) function for $d = 2$ ($d = 3$)
Ker	kernel
span	linear span
ln	natural logarithm
e	Euler number
$\det A$	determinant of matrix A
A^T	transposed matrix to A
A^C	complex conjugate matrix to A
A^H	conjugate transposed matrix to A
Re	real part
Im	imaginary part

i	imaginary unit
$i,\ j,\ k$	integer indices (subscripts)
δ_{ij}	Kronecker's symbol: $\delta_{ij} = 1$ for $i = j$, $\delta_{ij} = 0$ otherwise
$\mathbb{C}$	complex plane
$\exists$	there exist(s)
$\forall$	for all
a.a.	almost all
a.e.	almost everywhere
$P_k(\Omega)$	space of polynomials of degree at most k defined on Ω
$Q_1(\Omega)$	space of d-linear polynomials defined on Ω
$L^p(\Omega)$, $p \in [1, \infty)$	Lebesgue space of measurable functions v defined on Ω for which $\int_\Omega \lvert v(x)\rvert^p\, dx$ is finite
$L^\infty(\Omega)$	Lebesgue space of measurable essentially bounded functions defined on Ω
$C(\overline{\Omega})$	space of continuous functions defined on $\overline{\Omega}$
$C^k(\overline{\Omega})$	space of functions whose classical derivatives up to order k belong to $C(\overline{\Omega})$
$C^\infty(\Omega)$	space of infinitely differentiable functions in Ω
$C_0^\infty(\Omega)$	space of infinitely differentiable functions with compact support in Ω
$H^k(\Omega)$	Sobolev space of functions whose generalized derivatives up to order k belong to $L^2(\Omega)$
$H_0^1(\Omega)$	space of functions from $H^1(\Omega)$ whose traces vanish on $\partial\Omega$
$H^{1/2}(\partial\Omega)$	space of traces of all functions from $H^1(\Omega)$
$W_p^k(\Omega)$	Sobolev space of functions whose generalized derivatives up to order k belong to $L^p(\Omega)$
$H(\mathrm{div}^0)$	space of divergence-free functions
$H(\mathrm{rot}^0)$	space of rotation-free functions
V	Banach space of test functions
V^*	space of linear continuous functionals on V
$\rightharpoonup$	weak convergence
$\rightarrow$	strong convergence
$\langle b, v\rangle$	scalar product or value of functional $b \in V^*$ at point $v \in V$
v	test function
u	classical or weak (variational, generalized) solution
$\lVert \cdot \rVert_V$	norm in V
$(\cdot\,,\cdot)_V$	scalar product in V
$\lVert \cdot \rVert_{k,\Omega}$	norm in $H^k(\Omega)$
$\lvert \cdot \rvert_{k,\Omega}$	seminorm in $H^k(\Omega)$

$(\cdot\,,\cdot)_{k,\Omega}$	scalar product in $H^k(\Omega)$
$\Vert\cdot\Vert_{k,p,\Omega}$	norm in $W_p^k(\Omega)$
$\langle\cdot\,,\cdot\rangle_{0,\partial\Omega}$	scalar product in $L^2(\partial\Omega)$
$\vert\cdot\vert$	absolute value
$\Vert\cdot\Vert$	Euclidean norm
$(\cdot\,,\cdot)$	Euclidean scalar product
$a(\cdot\,,\cdot)$	bilinear or sesquilinear form
$F(\cdot)$	linear form
J	potential functional
J'	Gâteaux differential of J
K	element
$\hat{K}$	reference element
P_K	space of shape functions
Σ_K	set of degrees of freedom
(K, P_K, Σ_K)	finite element
$\dim P_K$	dimension of P_K
h_K	diameter of K (diam K)
h	discretization parameter
$\mathcal{T}_h$	triangulation (partition, decomposition)
V_h	finite element space
$\pi_h v$	V_h-interpolant of v
$\operatorname{supp} v$	support of function v
$\{v^i\}_{i=1}^N$	basis functions in V_h
$\mathcal{O}(\cdot)$	Landau's symbol: $f(\alpha) = \mathcal{O}(g(\alpha))$, if $\vert f(\alpha)\vert \le C\vert g(\alpha)\vert$ as $\alpha \to 0$ or $\alpha \to \infty$
$\oplus$	direct sum
$\emptyset$	empty set
$x \in A$	element x belongs to set A
$x \notin A$	element x does not belong to set A
$\{x \in A\vert\mathcal{P}(x)\}$	set of all elements x from A which possess property $\mathcal{P}(x)$
$A \subset B$	A is subset of set B
$A \cap B$	intersection of sets A and B
$A \cup B$	union of sets A and B
$A \backslash B$	subtraction of B from A
$f\colon A \to B$	function f from A to B
$x \mapsto f(x)$	function which assigns value $f(x)$ to x
$\square$	Halmos symbol

Foreword

Mathematical modeling plays an essential role in science and engineering. Costly and time consuming experiments (if they can be done at all) are replaced by computational analysis. In industry, commercial codes are widely used. They are flexible and can be adjusted for solving specific problems of interest. Solving large problems with tens or hundreds of thousands unknowns becomes routine.

The aim of analysis is to predict the behavior of the engineering and physical reality usually within the constraints of cost and time. Today, human cost and time are more important than computer cost. This trend will continue in the future.

Agreement between computational results and reality is related to two factors, namely mathematical formulation of the problems and the accuracy of the numerical solution. The accuracy has to be understood in the context of the aim of the analysis. A small error in an inappropriate norm does not necessarily mean that the computed results are usable for practical purposes.

Analysing the same engineering problem by different methods could sometimes lead to results which are significantly different. This could be caused by various reasons, e.g., different models in mathematical formulations were used or the numerical solution does not approximate the data of interest well. This can happen especially when various modern adaptive codes with *a posteriori* error estimation are used. Obviously, to understand the reasons for such discrepancy is very important. Decisions in engineering are still made by humans and not by computers, although computers are the main tools. It is necessary to realize that the computer always provides data, color graphs or movies, correct or incorrect. Hence to understand the basic mathematical background of the modelling is of major importance.

This book presents in a balanced way, the main features of modelling in electrical engineering. It clearly addresses without many prerequisites, mathematical and computational aspects. It presents mathematical ideas, and tools and approaches which are essential in these problems, illustrating these ideas with examples. Although no book of this type can make the reader an expert with great experience in the modelling of electrical engineering, it is an excellent book to learn the basic ideas and tools. It contributes significantly to a way of creatively assessing whether the numbers provided by computers are reliable as the basis for crucial engineering decisions. Hence it is a very valuable book.

Ivo Babuška

1. Introduction

At the present time many areas of science and engineering are becoming more mathematical. In this process, numerical methods play an important role. They are employed even in those areas which up to recently defied any application of mathematical tools. Nowadays the *finite element method* (FEM) is considered to be one of the most effective numerical methods for the solution of problems in mathematical physics, i.e., problems described by differential, integro-differential or integral equations, systems of these equations, variational inequalities, systems of differential-algebraic equations, etc. Typical examples are, for instance, a partial differential equation describing magnetic or electric potential, the Schrödinger equation, the heat conduction equation, the Maxwell's system of equations, the Einstein equations, ...

The FEM has been developed during the last fifty years. The discovery of the FEM is usually attributed to Prof. Richard Courant in 1943 (see [Courant, p. 21]). Nevertheless, in [Ciarlet, Lions, p. 5] there are some older references to finite element-like methods. The first monograph on the FEM is probably that of Synge [Synge] in 1957. There we can already find the original proof of the approximation properties of continuous and piecewise linear functions over a triangulated plane domain. The notion *element* was introduced in the 1950's by aerospace engineers performing elasticity computations. They divided a continuum into smaller pieces – called elements. The notion *finite element* was introduced by mathematicians later, in the 1960's. From that time the theory of the FEM has also been rigorously investigated. It represents one of the most significant developments in the history of numerical analysis. References to the growing literature on the FEM are collected, e.g., in [Mackerle], [Noor].

The main advantage of the FEM is that it enables us to simulate many physical processes on a computer. Thus it can substitute for the creation of expensive technical models (prototypes) or for the performance of complicated measurements. For instance, in designing and manufacturing electrical devices we can compute their electromagnetic, temperature or stress fields by the FEM.

A great deal of progress has been made in FEM software. The whole computational process can be essentially automated; including the following steps:

1. preprocessing of input data,
2. generation of triangulations,
3. assembling FE-matrices,
4. solving discrete problems,
5. postprocessing of output data,
6. a posteriori error estimates,
7. graphical illustration of results.

Real-life technical problems are mostly solved by standard program packages, e.g., ABAQUS, ALGOR, ANSYS, COSMOS, ELLPACK, I-DEAS FEM, INTERLIB, LUSAS, MARC, MODULEF, MSC/PROBE, MSC/NASTRAN, NISA, NONSAP, PAFEC, PATRAN, PLTMG, STRIPE, STRUDL, SYSTUS, TPS10, etc.

However, the user usually does not know which method was chosen by the package to solve the problem, i.e., which numerical quadrature was employed, which method was used to approximate possible nonlinearities and to solve the discrete problem etc.

This book discusses the theoretical background to these questions. It deals with FE-analysis of some important, mostly nonlinear, problems arising in electrical engineering. Of course, we are not able to cover all technically interesting cases. Nevertheless we introduce several typical techniques for solving nonlinear engineering problems which can be then modified to examine other problems not included in this book. The book can be regarded as a continuation of our book [Křížek, Neittaanmäki, 1990], where other nonlinear problems (obstacle problem, minimal surface problem, contact problem, Dirichlet–Signorini problem, etc.) were treated.

Note that there is no general numerical method capable of solving an arbitrary nonlinear problem and, therefore, every class of problems has to be investigated individually (cf. [Lions], [Zeidler]). The mathematical investigation of a nonlinear problem consists of several steps, including, as a rule, the proof of the existence (and, if applicable, also uniqueness) of the true and the approximate solution, the treatment of convergence questions, continuous dependence of solutions upon data, numerical solution of the approximate problem which often means solving large sparse linear or nonlinear algebraic systems, etc. Thus the approximate solution of partial differential equations still represents a great challenge to numerical analysis.

The aim of this book is to show engineers why it is useful to deal with Hilbert spaces, imbedding theorems, weak convergence, monotone operators, compact sets, etc., when solving real-life technical problems. We also hope that mathematicians will see the importance and necessity of dealing with material anisotropy, inhomogenity, nonlinearity and complicated geometrical configurations of electrical devices, which we do not meet when solving academic examples with the Laplace operator on square or ball domains.

This book has arisen as a result of many years of experience in solving actual engineering problems (for the companies ČKD–Elektrotechnika Prague and ŠKODA Pilsen) and our lectures on the FEM at the University of Jyväskylä and the Mathematical Institute of the Czech Academy of Sciences. The reader is assumed to be familiar with linear algebra, real analysis and basic ideas of the FEM (see, e.g., [Golub, Van Loan], [Gelbaum, Olmsted], [Johnson]).

We would like to express our sincere gratitude to Prof. I. Hlaváček for his encouragement and fruitful discussions. Furthermore, we are indebted to our colleagues J. Chleboun, S. Korotov, M. Lehtinen, L. Liu, Z. Milka, P. Monk, M. Pospíšek, M. Práger, V. Preiningerová, K. Segeth, D. Stará, J. Toivanen and E. Vitásek for many valuable suggestions and their help in preparation of this book.

We wish to thank also Mrs. Anja Roikonen and Ms. Marja-Leena Rantalainen for their skillful typing of the manuscript in TEX. The work on the book was partly supported by the grant No. 201/94/1067 of the Grant Agency of the Czech Republic. Finally we gratefully acknowledge the Czech and Finnish Academy for their support.

We hope you enjoy this book!

The authors

2. Mathematical modelling of physical phenomena

2.1. General remarks

The word "physics" is derived from the Greek "fysis" which means nature. Physics investigates fundamental natural phenomena and thus physical knowledge has a very general character. This is also the reason why physics penetrates other areas including electrical engineering.

Physical phenomena are direct consequences of four basic interactions (gravity, weak, electromagnetic and strong interaction). Each physical phenomenon can be described by means of appropriate mathematical tools, using, e.g., the theory of differential equations, the theory of groups, the theory of probability, etc. Throughout this book we shall deal with phenomena which arise in electrical engineering and which can be modelled by differential and algebraic equations (or inequalities). Note that such mathematical models always represent a certain idealisation of some physical reality. Therefore, we may never identify any physical phenomenon with the corresponding mathematical model. This sometimes happens and then leads to misunderstanding or wrong interpretation of results obtained.

The origin of each equation in the model should be known. This was pointed out already by H. Poincaré, i.e., we always have to know whether the equation under consideration is

- a definition (convention, notation),
- an assumption (hypothesis, presupposition),
- an assertion (corollary, deduction),
- an equation obtained by an experiment (measurement, observation).

We should have in mind several important features in the mathematical modelling of some physical problem. Firstly, one should realize that no physical relation (except definitions) holds entirely exactly, since the universe is composed of elementary particles and their behaviour can be predicted only with some probability (due to the Heisenberg principle of uncertainty and the Born interpretation of wave functions). That is why each physical equation can only approximately describe reality. Any model problem thus includes several simplifications. The most common simplification is a "continuous approach", where discrete quantities of the universe are replaced by continuous ones. This idea allows us to employ the theory of differential or integral equations.

Secondly, we must never overestimate mathematics itself. We must not forget that mathematics is only a "game" (even though this game has precisely defined notions and rules). Mathematical models exist only in our minds. They are mainly invented to provide a better understanding of the universe.

In verifying the reliability of the model problem, it is important to check the existence of its solution. Only if the solution exists it is reasonable to look for its analytical expression or to employ some computational method to get its approximation. Consider, for instance, the initial problem

$$y(0) = 0 \quad \text{and} \quad y'(x) = \begin{cases} 1 & \text{for } y(x) \le 0, \\ -1 & \text{for } y(x) > 0. \end{cases}$$

Its solution does not exist, since the right-hand side is not a Darboux function, (i.e., its range is not an interval). However, any numerical method applied to this initial value problem gives some numerical results, which are, of course, absurd.

Model problems (based on various physical laws) treated in this book are described classically by means of partial differential equations with some initial or boundary conditions. It is known that they need not be solvable (e.g., if coefficients of the equation have jump discontinuities). To ensure the existence of a solution we shall investigate the so-called weak (variational) formulation of these problems, which has an integral form. Note that the integral form of any physical law is more natural than its differential form. Moreover, the weak formulation enables us to use powerful tools of functional analysis and to apply the FEM. Contrary to classical methods (such as the collocation method, finite difference method,...), usually no additional assumptions on the smoothness of the true solution are required.

In checking the reliability of the model, it is also important to investigate the uniqueness (respectively multiplicity) of the solution. This also plays an important role in proposing a numerical method for finding the solution. Especially, when the model corresponds to some nonlinear relation, the uniqueness of the solution need not be guaranteed. One of the most famous example of nonuniqueness is the snap-back phenomenon in thyristor technology (see [Markowich, p. 43]). In this case several different states are possible for the same prescribed voltages. In [Gilbarg, Trudinger, p. 209] or [Meyers, p. 177], other examples of nonuniqueness can be found. It is sometimes difficult to distinguish which solution has a physical interpretation and which not.

Once the existence (and perhaps the uniqueness) of the true solution of a mathematical model is known, it is reasonable to investigate the continuous dependence of the solutions on data and approximation properties of the model, i.e., whether the model describes the physical phenomenon appropriately. This important step is usually checked experimentally by suitable measurements. Note that mathematical models treated in this book yield deterministic solutions whereas no physical phenomenon is strictly deterministic. This fact need not trouble us, since for practical purposes it is always sufficient to find some approximate solution. A general computational scheme is illustrated in Figure 2.1.

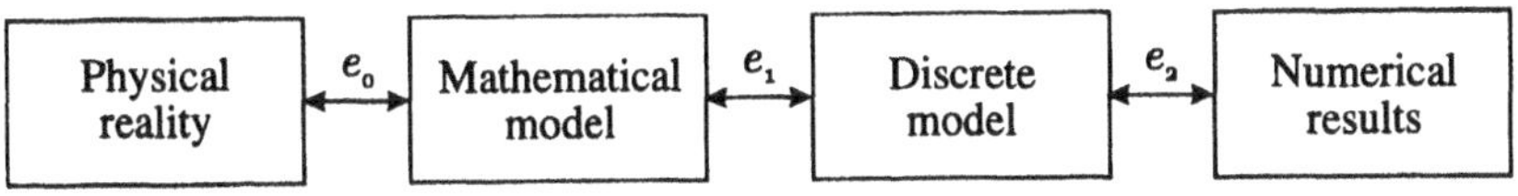

Figure 2.1.

Since no equation describes physical reality exactly, we make an error e_0 (modelling error). Mathematical models are usually expressed as infinite dimensional problems. They are approximated by finite dimensional problems, which yields the error e_1 (discretization error). This error may also include the error of numerical integration, error of approximation of the boundary of the examined region, etc. Finally, the error e_2 arises during a computation of the discrete model. It contains, of course, round-off errors, but may include other errors (like iteration error, etc.). Let us point out that numerical analysis investigates only the errors e_1 and e_2, but not the error e_0.

The difference between the true solution of a mathematical model and its numerical approximation can be sometimes estimated using a posteriori error estimates (cf. Section 4.3). This step is very important, since real-life problems are often nonlinear, large, ill-conditioned and nonstable. They can have multiple solutions, singularities, etc. If we do not perform a reliable a posteriori error analysis, we do not know, in fact, how close the numerical solution is to the exact one (see [Babuška, 1985], [Babuška, Miller]).

2.2. Some models described by differential equations

One of the greatest discoveries in physics are the *Maxwell equations* describing the electromagnetic field,

$$
\begin{aligned}
\operatorname{rot}\mathcal{H} &= \mathcal{J} + \frac{\partial \mathcal{D}}{\partial t}, && \text{(Ampere's law)}\\
\operatorname{rot}\mathcal{E} &= -\frac{\partial \mathcal{B}}{\partial t}, && \text{(Faraday's law)}\\
\operatorname{div}\mathcal{D} &= \rho, && \text{(Poisson's law)}\\
\operatorname{div}\mathcal{B} &= 0,\\
\mathcal{D} &= \varepsilon\mathcal{E},\\
\mathcal{B} &= \mu\mathcal{H}.
\end{aligned}
\tag{2.1}
$$

Recall that $\mathcal{H}$ is the magnetic field (magnetic intensity), $\mathcal{E}$ is the electric field, $\mathcal{D}$ is the electric flux density, $\mathcal{B}$ is the magnetic flux density (magnetic induction), $\mathcal{J}$ is the current density, ρ is the charge density, μ is the permeability, ε is the permittivity, $\partial(\cdot)/\partial t$ denotes time derivative, div and rot are first order differential operators which are classically defined for $v = (v_1, v_2, v_3)^T$ as follows

$$\operatorname{div} v = \sum_{i=1}^{3} \partial_i v_i, \tag{2.2}$$

$$\operatorname{rot} v = (\partial_2 v_3 - \partial_3 v_2, \partial_3 v_1 - \partial_1 v_3, \partial_1 v_2 - \partial_2 v_1)^T, \tag{2.3}$$

where $\partial_i v_j = \partial v_j / \partial x_i$ for $i = 1, 2, 3$, and $(\cdot)^T$ denotes the transposition of a vector. The script letters in (2.1) stand for three-dimensional vectors depending upon three space variables x_1, x_2, x_3 and time t, e.g., $\mathcal{B} = \mathcal{B}(x_1, x_2, x_3, t)$. The function ρ is scalar, but ε and μ can be scalar or 3×3 symmetric matrix functions (in the case of anisotropic media). Moreover, they may depend on other quantities (not only on time and space variables). For instance, $\mu = \mu(\mathcal{B})$, which means that the relation $\mathcal{B} = \mu\mathcal{H}$ could be nonlinear (cf. Section 7.2).

From (2.2) and (2.3) we see that the system (2.1) contains 8 partial differential equations and 6 algebraic equations. They are studied, e.g., in the books [Feynman, Leighton, Sands], [Haus, Melcher], [Solymar], [Steele]. Since $\operatorname{div}\operatorname{rot} \equiv 0$, the equations in (2.1) are not independent. The mathematical theory of electromagnetic fields deals with the solution of these equations. To this end some initial and boundary conditions are usually prescribed.

Suppose now that all functions in the system (2.1) are time independent. We will indicate this fact by the use of the italic letters, e.g., $B(x_1, x_2, x_3) = \mathcal{B}(x_1, x_2, x_3, t)$. Then (2.1) splits into two smaller mutually independent systems

$$\begin{aligned} \operatorname{rot} E &= 0, \\ \operatorname{div} D &= \rho, \\ D &= \varepsilon E, \end{aligned} \tag{2.4}$$

and

$$\begin{aligned} \operatorname{rot} H &= J, \\ \operatorname{div} B &= 0, \\ B &= \mu H. \end{aligned} \tag{2.5}$$

Notice that (2.4) contains the electric quantities E and D whereas (2.5) contains the magnetic quantities H and B. We shall investigate systems (2.4) and (2.5) in Chapters 3, 6 and 7. The time-harmonic Maxwell equations will be examined in Chapters 12 and 13.

Next let us introduce another important equation arising in mathematical modelling of many physical phenomena, the continuity equation. Let S be a physical quantity like mass, charge, heat, etc. Then we have

$$\begin{aligned} \frac{\partial}{\partial t}\int_\Omega (\text{volume density of } S)\, dx = &\int_\Omega (\text{volume density of sources of } S)\, dx \\ &- \int_{\partial\Omega} n^T (\text{surface flux density of } S)\, ds, \end{aligned} \tag{2.6}$$

where Ω is a domain with the boundary $\partial\Omega$ and the outward unit normal n (see Figure 2.2). The left-hand side of (2.6) represents the time variation of S. The first integral on the right-hand side accounts for sources of the quantity S in Ω and the second integral accounts for the flux of S through the boundary $\partial\Omega$.

Consider the above equation in the case that S is heat and denote by Q the volume density of heat, by f the volume density of heat sources and by q the heat flux. The functions Q and f are scalar, whereas q is a vector function. Note that

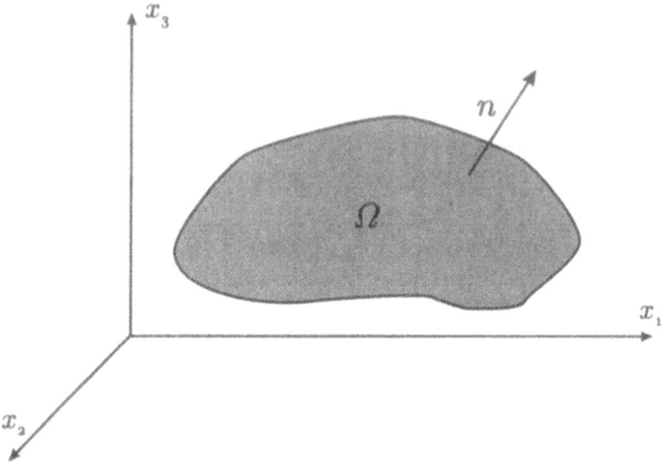

Figure 2.2.

the heat sources can be caused by electric currents, alternating electromagnetic fields, chemical reactions, etc. Now the well-known Green's formula

$$\int_\Omega \operatorname{div} q \, dx = \int_{\partial\Omega} n^T q \, ds$$

and (2.6) imply that

$$\int_\Omega \frac{\partial Q}{\partial t} \, dx = \frac{\partial}{\partial t} \int_\Omega Q \, dx = \int_\Omega f \, dx - \int_{\partial\Omega} n^T q \, ds = \int_\Omega f \, dx - \int_\Omega \operatorname{div} q \, dx.$$

Since this relation holds for any volume Ω, we obtain the differential equation

$$\frac{\partial Q}{\partial t} = f - \operatorname{div} q, \tag{2.7}$$

which is called the *continuity equation*. This equation in the stationary case, i.e.,

$$\operatorname{div} q = f,$$

is also called the *equilibrium equation*.

The volume density of heat Q satisfies the relation

$$Q = c\rho u, \tag{2.8}$$

where ρ is the specific mass, c is the specific heat capacity and u the temperature. These functions are scalar and are supposed here to be independent of time. According to the *Fourier law*, the heat flux is proportional to the gradient of the temperature, i.e.,

$$q = -\mathcal{A} \operatorname{grad} u. \tag{2.9}$$

Here $\mathcal{A}$ is a scalar function in the case of isotropic media, but it is a 3×3 symmetric matrix function for anisotropic media. It may also depend on the temperature u (see Chapter 9). In this case the Fourier law is nonlinear. Substituting (2.8) and (2.9) into (2.7), we arrive at the following partial differential equation of parabolic type which describes nonstationary heat conduction

$$c\rho\frac{\partial u}{\partial t} - \operatorname{div}(\mathcal{A}\operatorname{grad} u) = f. \tag{2.10}$$

In Chapter 11, we prescribe some initial and boundary conditions for this equation.

If all functions occurring in (2.10) are time independent, we get

$$-\operatorname{div}(\mathcal{A}\operatorname{grad} u) = f. \tag{2.11}$$

This equation is of elliptic type and describes stationary (steady-state) heat conduction. In Chapters 4, 8, and 9, we prescribe several different boundary conditions for (2.11).

Now, let us consider the solution $u = u(x,t)$ of (2.10) for $x \in (-\infty,\infty)$, $c = \rho = \mathcal{A} = 1$ and $f = 0$. In Figure 2.3, we see the initial temperature for $t = 0$ and the corresponding temperature for some arbitrarily small $t > 0$. We observe that the speed of heat conduction modelled by the equation (2.10) is thus infinite, even though u is extremely small for $|x| \gg 0$. To overcome this difficulty, we can consider the Cattaneo–Maxwell law instead of (2.9)

$$\frac{\partial q}{\partial t} = -\tau^{-1}(\mathcal{A}\operatorname{grad} u + q), \tag{2.12}$$

where τ is a given constant (e.g., $\tau \approx 10^{-11} - 10^{-9}$ [s] for metallic materials). This relation is more exact than the Fourier law (2.9). Differentiating (2.7) with respect to t, we get by (2.8), (2.12) and again by (2.7), the equation of hyperbolic type

$$\tau\frac{\partial}{\partial t}\left(c\rho\frac{\partial u}{\partial t}\right) + c\rho\frac{\partial u}{\partial t} - \operatorname{div}(\mathcal{A}\operatorname{grad} u) = f + \tau\frac{\partial f}{\partial t}. \tag{2.13}$$

In this case, it can be shown that the speed of heat conduction is finite. We see that (2.13) for $\tau \to 0$ become (2.10). However, we shall not deal with the equation (2.13) in this book (for its numerical solution see, e.g., [Baker], [Johnson], [Rektorys, 1982]).

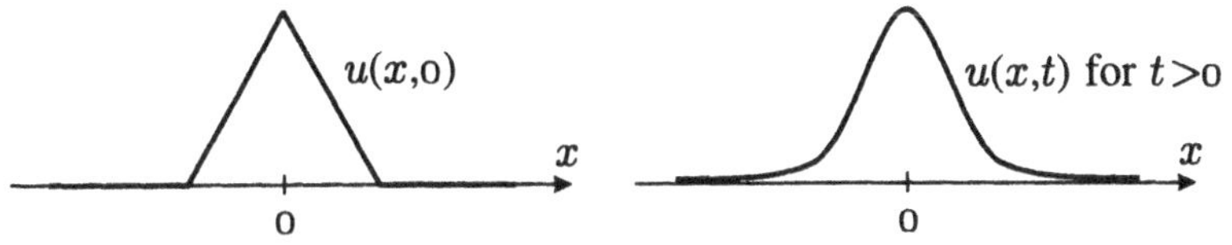

Figure 2.3.

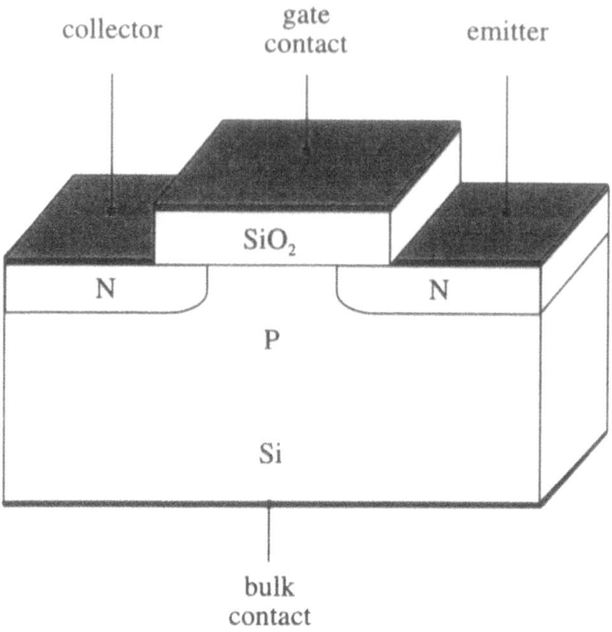

Figure 2.4.

Finally, we briefly introduce the semiconductor equations for the MOS-transistor (Metal-Oxide-Semiconductor), see Figure 2.4. Very fast progress in semiconductor technology has brought an increasing interest in the mathematical and also numerical modelling of semiconductor devices. The full dynamic model is described by a system of 10 partial differential equations in the three-dimensional space and time. Eight of them are the Maxwell equations (2.1) and the remaining two are the continuity equations (cf. (2.7)) for the current densities of holes and electrons.

Let us confine ourselves to the stationary case only (cf. (2.4) and (2.5)). Moreover, if we neglect the influence of the magnetic field, then the behaviour of the MOS-transistor can be described by a system of three partial differential equations. The first one is for the electrostatic potential u_0,

$$-\operatorname{div}(\varepsilon \operatorname{grad} u_0) = q(F + P - N), \tag{2.14}$$

and the second and third equations express the continuity of the current densities J_P and J_N of holes and electrons, respectively,

$$-\operatorname{div} J_P = qR, \tag{2.15}$$

$$-\operatorname{div} J_N = -qR, \tag{2.16}$$

where J_P and J_N depend on the concentration P of holes and the concentration N of electrons in the following way

$$J_P = q\mu_P P \operatorname{grad} u_P, \tag{2.17}$$

$$J_N = -q\mu_N N \operatorname{grad} u_N. \tag{2.18}$$

Here and above q is the elementary charge, F is proportional to the net density of ionized impurities, ε is the permittivity, μ_P and μ_N are the mobilities of holes and electrons, u_P and u_N are the electrochemical potentials of the holes and electrons, respectively, the function R characterizes the difference between the rate at which hole-electron carrier pair recombine and the rate at which they are generated in the semiconductor. Note that $E = -\operatorname{grad} u_0$ is the electric field for which $\operatorname{rot} E = 0$ (cf. (2.4) and (2.14)).

There are several models describing statistics of carrier concentrations. If, for instance, Boltzmann's statistics is supposed to model the semiconductor device, then (see [Markowich, p. 25], [Zlámal, 1986])

$$P = N_I \mathrm{e}^{(u_P - u_0)/U}, \quad N = N_I \mathrm{e}^{(u_N + u_0)/U}, \tag{2.19}$$

where N_I is the intrinsic carrier density and U is the thermal voltage. Now from (2.14)–(2.19) we get the following nonlinear system for the unknowns u_0, u_P and u_N,

$$\begin{aligned} -\operatorname{div}(\varepsilon \operatorname{grad} u_0) &= q(F + N_I \mathrm{e}^{(u_P - u_0)/U} - N_I \mathrm{e}^{(u_N + u_0)/U}), \\ -\operatorname{div}(\mu_P N_I \mathrm{e}^{(u_P - u_0)/U} \operatorname{grad} u_P) &= R, \\ -\operatorname{div}(\mu_N N_I \mathrm{e}^{(u_N + u_0)/U} \operatorname{grad} u_N) &= R. \end{aligned} \tag{2.20}$$

In Chapter 10, we will examine a simplified form of the system (2.20) with some mixed boundary conditions.

3. Mathematical background

3.1. Basic definitions and theorems from functional analysis

The aim of this section is to recall briefly those definitions and statements from functional analysis which will be used frequently throughout the book.

A set V is called a *linear vector space*, if it has the following properties: If $v, w \in V$ are arbitrary and α is a real (or complex) number then $v + w$ and αv also belong to V; moreover, these operations obey the usual rules of algebra, i.e.:

$$\begin{aligned} v + w &= w + v, & v + (w + z) &= (v + w) + z, \\ \alpha(v + w) &= \alpha v + \alpha w, & (\alpha + \beta)v &= \alpha v + \beta v, \\ \alpha(\beta v) &= (\alpha\beta)v, & 1 \cdot v &= v, \end{aligned}$$
$$\text{if } v + w = v + z \quad \text{then } w = z.$$

Let V be a linear vector space. A *norm* on V is a real function denoted by $\|\cdot\|_V$, which satisfies the following conditions:

$$\|v + w\|_V \leq \|v\|_V + \|w\|_V \quad \text{(triangle inequality)}, \tag{3.1}$$
$$\|\alpha v\|_V = |\alpha|\|v\|_V, \tag{3.2}$$
$$\|v\|_V \neq 0 \quad \text{if } v \neq 0, \tag{3.3}$$

for any $v, w \in V$ and any scalar α (by a scalar we mean real or complex number). It follows easily from (3.1) and (3.2) that a norm is a nonnegative function on V, i.e., $\|v\|_V \geq 0$ for any $v \in V$. A linear vector space, which is equipped with a norm, is called a *normed linear vector space.*

Recall that a *Banach space* V is a normed linear vector space which is complete, that is, for each Cauchy sequence $\{v_j\}_{j=1}^{\infty} \subset V$ there exists an element $v \in V$ such that

$$\|v_j - v\|_V \to 0 \quad \text{as } j \to \infty.$$

Suppose that V, U are two Banach spaces. We say that T is a *linear operator* from V to U if

$$T(v + w) = T(v) + T(w)$$

and

$$T(\alpha v) = \alpha T(v)$$

for any $v, w \in V$ and any scalar α. A linear operator $T\colon V \to U$ is thus continuous if there exists a constant $C > 0$ such that

$$\|Tv\|_U \leq C\|v\|_V \quad \forall v \in V. \tag{3.4}$$

Throughout the book the symbols C, C_1, C_2 are reserved for *generic constants* which are positive. They are not necessarily the same at each occurrence and do not depend on the functions being estimated (or on a discretization parameter which will be introduced later).

Theorem 3.1. *(Theorem on Isomorphism) If V and U are Banach spaces and T is a one-to-one linear continuous operator from V onto U, then the inverse T^{-1} is linear and continuous, too.*

For the proof see, e.g., [Hewitt, Stromberg, p. 216]. This theorem is a direct consequence of the well-known closed graph theorem.

Let V be a linear vector space and let α^C denotes the complex conjugate number to α. A *scalar (inner) product* on $V \times V$ is a complex function denoted by $(\cdot\,,\cdot)_V$ which satisfies the following conditions:

$$(v+y,w)_V = (v,w)_V + (y,w)_V, \tag{3.5}$$
$$(\alpha v,w)_V = \alpha(v,w)_V, \tag{3.6}$$
$$(v,w)_V = ((w,v)_V)^C \text{ and } (v,v)_V \geq 0, \tag{3.7}$$
$$(v,v)_V \neq 0 \text{ if } v \neq 0, \tag{3.8}$$

for any $v, w, y \in V$ and any scalar α. If V is a real space then $(\cdot\,,\cdot)_V$ is a real function and the symbol $(\cdot)^C$ in (3.7) can be omitted.

For the so-called *induced norm*

$$\|v\|_V = \sqrt{(v,v)_V}, \quad v \in V, \tag{3.9}$$

we have the well-known *Cauchy–Schwarz inequality*

$$|(v,w)_V| \leq \|v\|_V \|w\|_V \quad \forall v,w \in V. \tag{3.10}$$

Moreover,

$$\|v\|_V = \sup_{w \neq 0} \frac{|(v,w)_V|}{\|w\|_V}.$$

A linear operator whose values are scalars is called a *linear form (linear functional)*. A Banach space with a scalar product is called a *Hilbert space*.

Theorem 3.2. *(Riesz theorem) Let V be a Hilbert space. Then for any linear continuous form F defined on V there exists exactly one element $u \in V$ such that*

$$F(v) = (v,u)_V \quad \forall v \in V. \tag{3.11}$$

For the proof see [Taylor, p. 245].

Recall that a scalar mapping $a(\cdot\,,\cdot)$ defined on $V \times V$, where V is a linear vector space, is said to be a *sesquilinear form*, if for any fixed $v \in V$ the mappings $a(\cdot\,,v)$ and $(a(v,\cdot))^C$ are linear. If moreover $a(\cdot\,,\cdot)$ is only real-valued then $a(\cdot\,,\cdot)$ is called a *bilinear form*. A sesquilinear form $a(\cdot\,,\cdot)$ is said to be *continuous* if there exists a constant $C_1 > 0$ such that

$$|a(v,w)| \leq C_1 \|v\|_V \|w\|_V \quad \forall v,w \in V. \tag{3.12}$$

Lemma 3.3. *(Lax–Milgram lemma) Let V be a Hilbert space and let $a(\cdot\,,\cdot)$ be a continuous sesquilinear form for which there exists a constant $C_2 > 0$ such that*

$$|a(v,v)| \geq C_2 \|v\|_V^2 \quad \forall v \in V \text{ (V-ellipticity condition).} \tag{3.13}$$

Then for any linear continuous form F defined on V there exists exactly one element $u \in V$ such that

$$a(v,u) = F(v) \quad \forall v \in V. \tag{3.14}$$

For the proof see [Nečas, 1967, p. 38]. Actually, the Lax–Milgram lemma says that $\sqrt{|a(\cdot\,,\cdot)|}$ is a norm in V equivalent to the $\|\cdot\|_V$-norm.

A sesquilinear form $a(\cdot\,,\cdot)$ is said to be *Hermitian* if

$$a(v,w) = (a(w,v))^C \quad \forall v,w \in V. \tag{3.15}$$

Theorem 3.4. *Let the assumptions of Lemma 3.3 be satisfied and let a sesquilinear form $a(\cdot\,,\cdot)$ be Hermitian. Then the problem (3.14) is equivalent to the problem: Find $u \in V$ such that*

$$J(u) = \inf_{v \in V} J(v), \tag{3.16}$$

where J is a quadratic functional given by

$$J(v) = \frac{1}{2} a(v,v) - \mathrm{Re}(F(v)), \quad v \in V. \tag{3.17}$$

Proof. From (3.15) we see that the functional (3.17) is real-valued. Let u be the solution of (3.14). Then by (3.17), (3.15) and the sesquilinearity of $a(\cdot\,,\cdot)$, we arrive at

$$\begin{aligned} 2J(v) &= a(v,v) - F(v) - (F(v))^C = a(v,v) - a(v,u) - a(u,v) \\ &= a(v-u, v-u) - a(u,u) \quad \forall v \in V. \end{aligned} \tag{3.18}$$

Now, form (3.13) we see that J attains its minimum just for $v = u$ as the last term $-a(u,u)$ in (3.18) is constant.

Conversely, from (3.18) and the V-ellipticity of $a(\cdot\,,\cdot)$ we observe that just the element u minimizes J, where u is the solution of (3.14). Hence, both the problems (3.14) and (3.16) are equivalent. □

Remark 3.5. The functional (3.17) is called the *energy functional.* Note that in the real case, the relation (3.15) reduces to $a(v,w) = a(w,v)$ and the corresponding bilinear form is then said to be *symmetric.*

3.2. The spaces $L^p(\Omega)$ and $C^k(\overline{\Omega})$

Let $\mathbf{R}^d$ stand for the *Euclidean space* equipped with the norm

$$\|x\| = \left(\sum_{j=1}^{d} x_j^2\right)^{1/2}, \quad x = (x_1, ..., x_d)^T \in \mathbf{R}^d. \tag{3.19}$$

The *Lebesgue space* of real or complex-valued functions defined over an open set $\Omega \subset \mathbf{R}^d$, which are integrable with the power $p \in [1, \infty)$, is denoted by $L^p(\Omega)$ and equipped with the norm

$$\|v\|_{0,p,\Omega} = \left(\int_\Omega |v|^p \, dx\right)^{1/p}, \quad v \in L^p(\Omega). \tag{3.20}$$

When $p = 2$, we write shortly $\|\cdot\|_{0,\Omega} = \|\cdot\|_{0,2,\Omega}$ and define the scalar (inner) product as follows

$$(v, w)_{0,\Omega} = \int_\Omega v w^C \, dx.$$

Recall the well-known Hölder inequality

$$\left|\int_\Omega vw \, dx\right| \leq \|v\|_{0,p,\Omega} \|w\|_{0,q,\Omega}, \quad v \in L^p(\Omega), \; w \in L^q(\Omega),$$

which holds for any $p, q \in (1, \infty)$ satisfying the equality

$$\frac{1}{p} + \frac{1}{q} = 1.$$

The Lebesgue space of measurable essentially bounded functions over Ω is denoted by $L^\infty(\Omega)$ and equipped with the norm

$$\|v\|_{0,\infty,\Omega} = \operatorname*{ess\,sup}_{x \in \Omega} |v(x)|.$$

For any $p, q \in [1, \infty]$, $p \leq q$, the following algebraic imbedding

$$L^q(\Omega) \subset L^p(\Omega)$$

holds. Moreover, we have also the topological imbedding; namely, there exists a constant $C > 0$ such that

$$\|v\|_{0,p,\Omega} \leq C\|v\|_{0,q,\Omega} \quad \forall v \in L^q(\Omega).$$

We will not specify the term "real" or "complex" in further definitions of function spaces, since these definitions are essentially the same for both cases. Spaces of complex-valued functions will be used especially in Sections 3.5, 12.1–12.3 and 13.2.

By $\overline{\Omega}$ we denote the closure of Ω and by $\partial\Omega$ the boundary of Ω. Then clearly

$$\overline{\Omega} = \Omega \cup \partial\Omega, \quad \partial\Omega = \overline{\Omega} \cap (\mathbf{R}^d \setminus \Omega).$$

Recall that a *domain* Ω is an open and connected set in $\mathbf{R}^d$. Throughout the book, the symbol $d \in \{1, 2, ...\}$ is solely reserved for the dimension of Ω.

If Ω is a bounded domain then the space of continuous functions over $\overline{\Omega}$ is denoted by $C(\overline{\Omega})$ and equipped with the norm

$$\|v\|_{C(\overline{\Omega})} = \max_{x \in \overline{\Omega}} |v(x)|. \tag{3.21}$$

Obviously we have

$$\|v\|_{C(\overline{\Omega})} = \|v\|_{0,\infty,\Omega} \quad \forall v \in C(\overline{\Omega}).$$

The symbol $C^k(\overline{\Omega})$, $k \in \{0, 1, ...\}$, stands for the space of functions whose classical derivatives up to order k belong to $C(\overline{\Omega})$. Moreover, we set

$$C^\infty(\overline{\Omega}) = \bigcap_{k=1}^{\infty} C^k(\overline{\Omega}),$$

i.e., $C^\infty(\overline{\Omega})$ is the space of infinitely differentiable functions over $\overline{\Omega}$. Finally by $C_0^\infty(\Omega)$ we denote the space of infinitely differentiable functions with a compact support in Ω (see Figure 3.1), that is

$$C_0^\infty(\Omega) = \left\{v \in C^\infty(\overline{\Omega}) \mid \operatorname{supp} v \subset \Omega\right\},$$

where

$$\operatorname{supp} v = \overline{\{x \in \Omega \mid v(x) \neq 0\}}. \tag{3.22}$$

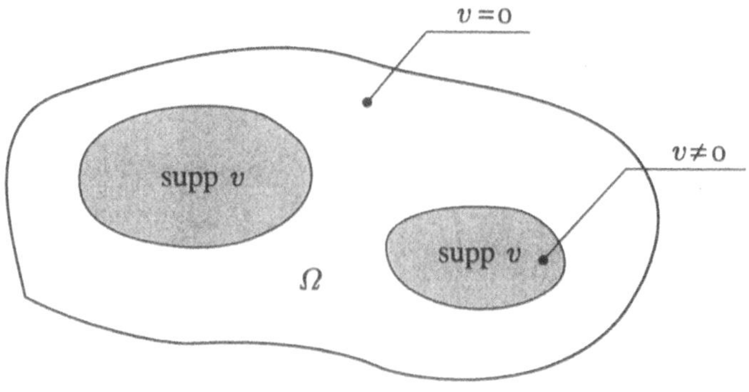

Figure 3.1.

3.3. Sobolev spaces $H^k(\Omega)$ and $W_p^k(\Omega)$

Weak solutions of many problems in electromagnetism and other fields are looked for in Sobolev spaces. First we briefly recall some basic properties of the Sobolev space $H^k(\Omega)$, $k = \{0, 1, 2, ...\}$, i.e., classes of real or complex functions defined on a domain $\Omega \subset \mathbf{R}^d$ whose generalized derivatives up to the kth order belong to $L^2(\Omega)$. We shall consider Sobolev spaces defined only on bounded domains with a Lipschitz continuous boundary which is a sufficiently wide class of domains for practical purposes. The next definition is due to [Mazja, p. 22].

Definition 3.6. *A bounded set $\Omega \subset \mathbf{R}^d$ is said to have a Lipschitz continuous boundary if for any $z \in \partial\Omega$ there exists a neighbourhood $U = U(z)$ such that the set $U \cap \Omega$ can be expressed, in some Cartesian coordinate system $(x_1, ..., x_d)$, by the inequality $x_d < F(x_1, ..., x_{d-1})$, where F is a Lipschitz continuous function.*

Denote by $\mathcal{L}$ the set of all bounded domains with a Lipschitz continuous boundary.

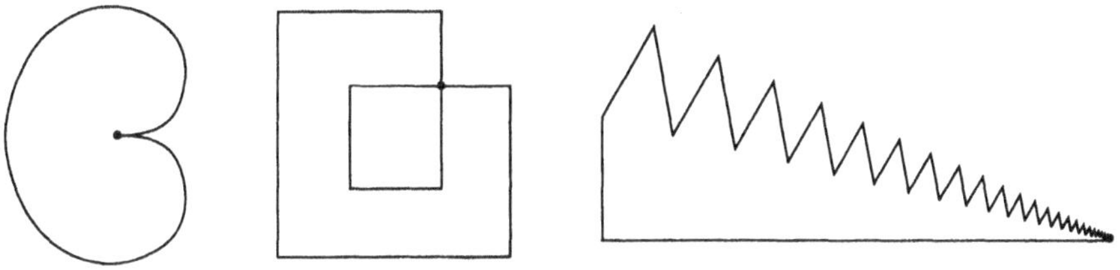

Figure 3.2.

Figure 3.2 shows some domains whose boundaries are not Lipschitz continuous ones (black dots on the boundaries indicate where the Lipschitz continuity is violated). We see that the assumption $\Omega \in \mathcal{L}$ is not too restrictive for applications on bounded domains. Moreover, if $\Omega \in \mathcal{L}$ then by [Nečas, 1967, p. 88] the outward normal exists almost everywhere (a.e.) on $\partial\Omega$. This is the reason why we will confine ourselves to domains with a Lipschitz continuous boundary, since later we shall deal with, e.g., the normal derivative, normal component of flux etc. A further reason is that there are several definitions of the Sobolev spaces and in the case $\Omega \notin \mathcal{L}$ the Sobolev spaces need not be uniquely defined. Also, some important theorems such as Theorems 3.8–3.11, 3.14 need not be valid when $\Omega \notin \mathcal{L}$ (see, e.g., [Amick, p. 83, p. 92]). Let us point out that there are other (nonequivalent) definitions of the Lipschitz continuity of $\partial\Omega$. For instance, we say that Ω has a Lipschitz boundary if for any $x \in \partial\Omega$ there exists a neighbourhood U and a Lipschitz continuous one-to-one mapping $F\colon U \to (0,1)^d$ such that the inverse F^{-1} is also Lipschitz continuous.

So from now on let $\Omega \in \mathcal{L}$. For any $v \in C^\infty(\overline{\Omega})$ and the so-called *multi-index* $m = (m_1, ..., m_d)$ we define the *mth classical derivative* as follows

$$D^m v = \frac{\partial^{|m|} v}{\partial x_1^{m_1} \cdots \partial x_d^{m_d}},$$

where $m_1, ..., m_d$ are non-negative integers and

$$|m| = m_1 + ... + m_d.$$

A function $v \in L^2(\Omega)$ is said to have the mth generalized derivative in $L^2(\Omega)$ if there exists a function $z \in L^2(\Omega)$ such that

$$\int_\Omega zw \, dx = (-1)^{|m|} \int_\Omega vD^m w \, dx \quad \forall w \in C_0^\infty(\Omega).$$

Then the function z is called the *mth generalized derivative* of v and we set $D^m v = z$. One may easily check (see, e.g., [Křížek, Neittaanmäki, 1990, p. 7]) that $D^m v$ is well-defined. Note that the generalized derivative is sometimes looked for in the larger space $L^{1,\mathrm{loc}}(\Omega)$.

For the generalized derivatives we can derive an interesting property which is not valid for the classical derivatives. Consider, for instance, the function

$$v(x) = \operatorname{sign} x_1 + \operatorname{sign} x_2 \quad \text{for } x = (x_1, x_2) \in \Omega,$$

where $\Omega = (-1,1) \times (-1,1)$. We will show now that the second mixed derivative $\partial_{12} v = D^{(1,1)} v$ exists in $L^2(\Omega)$, but the first derivatives $\partial_1 v = D^{(1,0)} v$ and $\partial_2 v = D^{(0,1)} v$ do not exist in $L^2(\Omega)$.

First we prove that $\partial_{12} v \in L^2(\Omega)$. Let $w \in C_0^\infty(\Omega)$ be arbitrary. Then by the Fubini theorem

$$\begin{aligned} \int_\Omega v \partial_{12} w \, dx &= \int_{-1}^1 \int_{-1}^1 (\operatorname{sign} x_1 + \operatorname{sign} x_2) \partial_{12} w(x) \, dx_1 \, dx_2 \\ &= \int_{-1}^1 \operatorname{sign} x_1 \left(\int_{-1}^1 \partial_2(\partial_1 w(x)) \, dx_2 \right) dx_1 \\ &\quad + \int_{-1}^1 \operatorname{sign} x_2 \left(\int_{-1}^1 \partial_1(\partial_2 w(x)) \, dx_1 \right) dx_2. \end{aligned}$$

However,

$$\int_{-1}^1 \partial_2(\partial_1 w(x_1, x_2)) \, dx_2 = \partial_1 w(x_1, 1) - \partial_1 w(x_1, -1) = 0 \quad \forall x_1 \in (-1,1),$$

since obviously $\partial_1 w \in C_0^\infty(\Omega)$. Analogously, we find that

$$\int_{-1}^1 \partial_1(\partial_2 w(x_1, x_2)) \, dx_1 = 0 \quad \forall x_2 \in (-1,1).$$

Hence, $\partial_{12} v$ exists in $L^2(\Omega)$ and, moreover, $\partial_{12} v \equiv 0$ in Ω.

Second we prove that $\partial_1 v$ does not belong to $L^2(\Omega)$. Assume, on the contrary, that there exists $z \in L^2(\Omega)$ for which

$$\int_\Omega zw\,dx = -\int_\Omega v\partial_1 w\,dx \quad \forall w \in C_0^\infty(\Omega).$$

Let $w \in C_0^\infty(\Omega)$ be arbitrary and let us consider the integral

$$\begin{aligned}\int_\Omega v\partial_1 w\,dx &= \int_{-1}^1 \left(\int_{-1}^1 \operatorname{sign} x_1 \partial_1 w(x)\,dx_1\right) dx_2 \\ &\quad + \int_{-1}^1 \operatorname{sign} x_2 \left(\int_{-1}^1 \partial_1 w(x)\,dx_1\right) dx_2 \equiv I_1 + I_2.\end{aligned}$$

The second integral I_2 is zero, because

$$\int_{-1}^1 \partial_1 w(x_1, x_2)\,dx_1 = w(1, x_2) - w(-1, x_2) = 0 \quad \forall x_2 \in (-1, 1).$$

For the first integral I_1 we get

$$\begin{aligned}I_1 &= \int_{-1}^1 \left(\int_0^1 \partial_1 w(x_1, x_2)\,dx_1 - \int_{-1}^0 \partial_1 w(x_1, x_2)\,dx_1\right) dx_2 \\ &= -2\int_{-1}^1 w(0, x_2)\,dx_2.\end{aligned}$$

Hence,

$$\int_\Omega zw\,dx = 2\int_{-1}^1 w(0, x_2)\,dx_2.$$

Setting $y(x_1, x_2) = x_1 w(x_1, x_2)$ for $(x_1, x_2) \in \Omega$, we see that $y \in C_0^\infty(\Omega)$ and thus y can be used as a test function, i.e.,

$$\int_\Omega z(x) x_1 w(x)\,dx = \int_\Omega zy\,dx = 2\int_{-1}^1 y(0, x_2)\,dx_2 = 0.$$

This implies that $z(x)x_1 = 0$ for a.a. $x \in \Omega$, since $w \in C_0^\infty(\Omega)$ was arbitrary. Therefore, $z \equiv 0$ in $L^2(\Omega)$ and we always have

$$\int_{-1}^1 w(0, x_2)\,dx_2 = 0.$$

However, this relation cannot hold for any $w \in C_0^\infty(\Omega)$, what is a contradiction.

In a similar way we can show that $\partial_2 v$ does not belong to $L^2(\Omega)$. Consequently, the jump function v has no first generalized derivatives. The existence of the first generalized derivatives of discontinuous functions is discussed in Remark 3.13.

Now for $k = 0, 1, \ldots$, the *Sobolev spaces* $H^k(\Omega)$ are defined as

$$H^k(\Omega) = \{v \in L^2(\Omega) \mid D^m v \in L^2(\Omega), \quad |m| \le k\}.$$

It can be verified that $H^k(\Omega)$, equipped with the scalar product

$$(v, w)_{k,\Omega} = \sum_{|m| \le k} \int_\Omega D^m v (D^m w)^C \, dx, \quad v, w \in H^k(\Omega), \tag{3.23}$$

is a Hilbert space, where the symbol $(\cdot)^C$ denotes the complex conjugate number. Let us further introduce the induced norm

$$\|v\|_{k,\Omega} = \left(\sum_{|m| \le k} \int_\Omega |D^m v|^2 \, dx \right)^{1/2}, \quad v \in H^k(\Omega), \tag{3.24}$$

and seminorm

$$|v|_{k,\Omega} = \left(\sum_{|m| = k} \int_\Omega |D^m v|^2 \, dx \right)^{1/2}, \quad v \in H^k(\Omega). \tag{3.25}$$

For simplicity the same symbols will be used also for vector functions, we shall write, e.g.,

$$\|v\|_{k,\Omega} = \left(\sum_{l=1}^{q} \|v_l\|_{k,\Omega}^2 \right)^{1/2} \quad \text{for } v = (v_1, \ldots, v_q)^T \in (H^k(\Omega))^q.$$

Moreover, the subscript Ω will be often omitted, i.e.,

$$(\cdot\,,\cdot)_k = (\cdot\,,\cdot)_{k,\Omega}, \quad \|\cdot\|_k = \|\cdot\|_{k,\Omega}, \quad |\cdot|_k = |\cdot|_{k,\Omega}.$$

In particular, we have

$$\|v\|_1^2 = \|v\|_0^2 + \|\operatorname{grad} v\|_0^2.$$

Clearly,

$$\|v\|_{k-1} \le \|v\|_k \quad \forall v \in H^k(\Omega), \ k = 1, 2, \ldots$$

and

$$L^2(\Omega) = H^0(\Omega) \supset H^1(\Omega) \supset H^2(\Omega) \supset \cdots.$$

Note that each classical derivative is also the generalized derivative, and thus we have

$$C^k(\overline{\Omega}) \subset H^k(\Omega), \quad k = 0, 1, \ldots.$$

Definition 3.7. *A set $\Gamma \subset \partial\Omega$ is said to be a (relatively) open set in $\partial\Omega$ if for any $x \in \Gamma$ there exists an open ball $B \subset \mathbb{R}^d$ containing x such that $B \cap \partial\Omega \subset \Gamma$.*

The *Lebesgue space* of square integrable function over an open set $\Gamma \subset \partial\Omega$ is denoted by $L^2(\Gamma)$ and equipped with the standard norm

$$\|v\|_{0,\Gamma} = \left(\int_\Gamma |v|^2 \, ds \right)^{1/2}, \quad v \in L^2(\Gamma).$$

Further we recall some important properties of Sobolev spaces.

Theorem 3.8. *(Trace theorem) Let $\Omega \in \mathcal{L}$. Then there exists exactly one linear continuous operator $\gamma\colon H^1(\Omega) \to L^2(\partial\Omega)$ such that*

$$\gamma v = v\Big|_{\partial\Omega} \quad \forall v \in C^\infty(\overline{\Omega}).$$

The function γv for $v \in H^1(\Omega)$ is called the *trace of* v and we denote it, for simplicity, by $v\Big|_{\partial\Omega}$. Theorem 3.8, in fact, says that there exists a constant $C > 0$ such that

$$\|v\|_{0,\partial\Omega} \le C \|v\|_{1,\Omega} \quad \forall v \in H^1(\Omega). \tag{3.26}$$

The trace theorem enables us to define the space

$$H_0^1(\Omega) = \left\{ v \in H^1(\Omega) \mid v = 0 \text{ on } \partial\Omega \right\}.$$

Note that the spaces $H_0^1(\Omega)$ and $H^1(\Omega)$ can also be defined as the completion of $C_0^\infty(\Omega)$ and $C^\infty(\overline{\Omega})$, respectively, under the norm $\|\cdot\|_{1,\Omega}$, i.e.,

$$H_0^1(\Omega) = \overline{C_0^\infty(\Omega)}, \quad H^1(\Omega) = \overline{C^\infty(\overline{\Omega})}. \tag{3.27}$$

Denote by $H^{1/2}(\partial\Omega)$ the space of traces of all functions from $H^1(\Omega)$. Then due to [Nečas, 1967, p. 87] we have the following density relation

$$L^2(\partial\Omega) = \overline{H^{1/2}(\partial\Omega)}, \tag{3.28}$$

where the closure is taken under the $\|\cdot\|_{0,\partial\Omega}$–norm.

Theorem 3.9. *(Rellich theorem) Let $\Omega \in \mathcal{L}$. Then the identity mapping from $H^1(\Omega)$ to $L^2(\Omega)$ is compact (i.e., any bounded sequence in $H^1(\Omega)$ contains a subsequence converging in $L^2(\Omega)$).*

Theorem 3.10. *(Green's theorem) Let $\Omega \in \mathcal{L}$. Then for each $i \in \{1, ..., d\}$*

$$\int_\Omega w \partial_i v \, dx + \int_\Omega v \partial_i w \, dx = \int_{\partial\Omega} n_i v w \, ds \quad \forall v, w \in H^1(\Omega), \tag{3.29}$$

where n_i are the components of the outward unit normal to $\partial\Omega$ and $\partial_i v = \dfrac{\partial v}{\partial x_i}$.

Theorem 3.11. *(Sobolev imbedding theorem) Let $\Omega \in \mathcal{L}$ and let k be an integer such that $2k > d$. Then*

$$H^k(\Omega) \subset C(\overline{\Omega}) \tag{3.30}$$

and there exists a constant $C > 0$ such that

$$\|v\|_k \geq C\|v\|_{C(\overline{\Omega})} \quad \forall v \in H^k(\Omega). \tag{3.31}$$

Proofs of Theorems 3.8–3.11 can be found, e.g., in [Adams], [Kufner, John, Fučík], [Nečas, 1967].

Remark 3.12. The continuity expressed by the inclusion (3.30) may be still improved. For instance if $d = 1$ and $v \in H^1((0,1))$ is arbitrary then by (3.30), (3.10) applied to $L^2((0,1))$ and (3.25), we can write

$$|v(z) - v(y)| = \left|\int_y^z v'(x)\,dx\right| \leq |z-y|^{1/2}\left(\int_y^z |v'(x)|^2\,dx\right)^{1/2} \leq |v|_1|z-y|^{1/2}, \quad \forall y, z \in (0,1),$$

which means that v is a Hölder continuous function with exponent $\frac{1}{2}$.

Remark 3.13. In multi-dimensional cases functions from $H^1(\Omega)$ need not be continuous. Consider only the important case $d = 2$ and let $\Omega \in \mathcal{L}$. We shall briefly investigate if the following three kinds of discontinuities can occur for functions from $H^1(\Omega)$.

1. Suppose that, e.g., the line segment $\Gamma = [-1,1] \times \{0\}$ is contained in $\overline{\Omega}$, i.e., $\Gamma \subset \overline{\Omega}$. According to [Nečas, 1967, p. 81], the norm associated to the space $H^{1/2}(\Gamma)$ is defined by

$$\|v\|_{\frac{1}{2},\Gamma} = \left(\|v\|_{0,\Gamma}^2 + \int_{-1}^1\int_{-1}^1 \frac{|v(x)-v(y)|^2}{|x-y|^2}\,dx\,dy\right)^{\frac{1}{2}}, \quad v \in H^{\frac{1}{2}}(\Gamma).$$

For the jump function

$$s = \begin{cases} 0 & \text{on } [-1,0) \times \{0\}, \\ 1 & \text{on } [0,1] \times \{0\}, \end{cases}$$

defined on Γ, we find that

$$\int_{-1}^1\int_{-1}^1 \frac{|s(x)-s(y)|^2}{|x-y|^2}\,dx\,dy = 4\int_{-1}^0\int_0^1 \frac{dx\,dy}{(x-y)^2} = 4\int_{-1}^0\left(\frac{1}{y-1} - \frac{1}{y}\right)dy = \infty,$$

i.e., s cannot be the trace over Γ of a function from $H^1(\Omega)$. Therefore, for any $g \in C(\Gamma)$ and any real $\alpha \neq 0$ the jump function $g + \alpha s$ (see Figure 3.3) also cannot

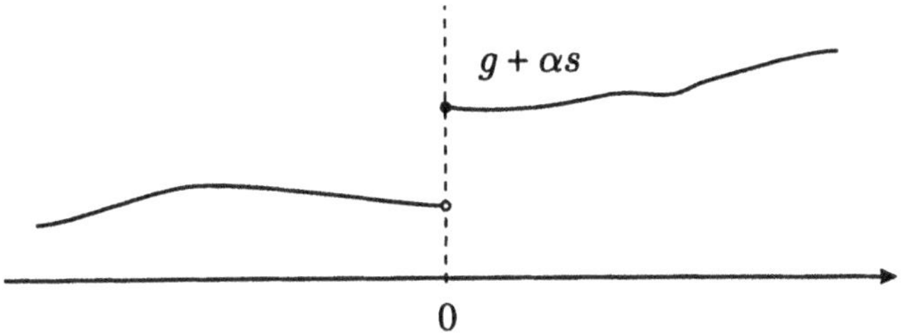

Figure 3.3.

be the trace over Γ of some function from $H^1(\Omega)$. We observe that zero is a point of discontinuity of the first kind. The above approach can be modified also to a curved segment Γ (see [Nečas, 1967, p. 75]).

2. For any $z \in \overline{\Omega}$ and any $R \in (0, \frac{1}{2})$ set

$$U_R(z) = \{x \in \mathbf{R}^2 \mid \|x - z\| < R\}$$

and let

$$v(x; z, R) = \begin{cases} (-\ln \|x - z\|)^{\frac{1}{4}} - (-\ln R)^{\frac{1}{4}} & \text{for } x \in U_R(z), \\ 0 & \text{otherwise.} \end{cases}$$

The function $v(\cdot\,; z, R)$ is continuous in the whole of $\overline{\Omega}$ except the point z, where there is a weakly singular point (a pole) – see Figure 3.4. Such a point is sometimes called a point of discontinuity of the second kind. Applying polar coordinates, it may be shown (cf. [Křížek, Neittaanmäki, 1990, p. 10]) that for any fixed z and R we have $v \in H^1(\Omega)$ and

$$\|v\|_1^2 \leq 20R^2\sqrt{-\ln R} + 2/\sqrt{-\ln R}.$$

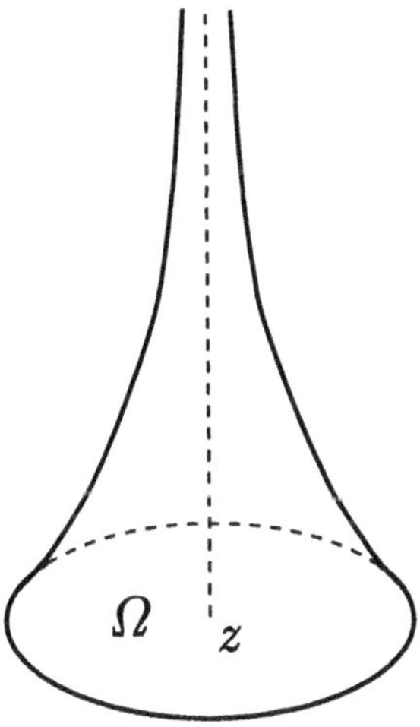

Figure 3.4.

3. Using the above function $v \in H^1(\Omega)$, we will construct now a bounded function $w \in H^1(\Omega)$ which has a strongly singular point. For any $k \in \{1, 2, ...\}$ set

$$\begin{aligned} R_k &= \mathrm{e}^{-k^4}, \\ z_{k,1} &= 2\sum_{j=1}^{k-1} R_j, \quad z_{1,1} = 0, \\ z_k &= (z_{k,1}, 0)^{\mathrm{T}}, \\ v_k(x) &= v(x; z_k, R_k), \\ w_k(x) &= \min_{x\in\overline{\Omega}}(v_k(x), 1). \end{aligned}$$

The sequence $\{z_{k,1}\}$ clearly converges to a finite number. Assume that $U_{R_k}(z_k) \subset \Omega$ for any $k \in \{1, 2, ...\}$ and put

$$w = \sum_{k=1}^{\infty} w_k.$$

Since the supports $U_{R_k}(z_k)$ of w_k are disjoint (see Figure 3.5), we have

$$\begin{aligned} \|w\|_1^2 &= \sum_{k=1}^{\infty} \|w_k\|_1^2 \le \sum_{k=1}^{\infty} \|v_k\|_1^2 \le \sum_{k=1}^{\infty} \left(20R_k^2\sqrt{-\ln R_k} + 2/\sqrt{-\ln R_k}\right) \\ &= 20\sum_{k=1}^{\infty} \mathrm{e}^{-k^8} k^2 + 2\sum_{k=1}^{\infty} k^{-2} < \infty. \end{aligned}$$

Hence $w \in H^1(\Omega)$ and we see that w is bounded, $0 \le w(x) \le 1$ for all $x \in \overline{\Omega}$. The function w is continuous in the whole of $\overline{\Omega}$ except the point $z = (\lim_{k\to\infty} z_{k,1}, 0)^T$, where is a strongly singular point (see Figure 3.6).

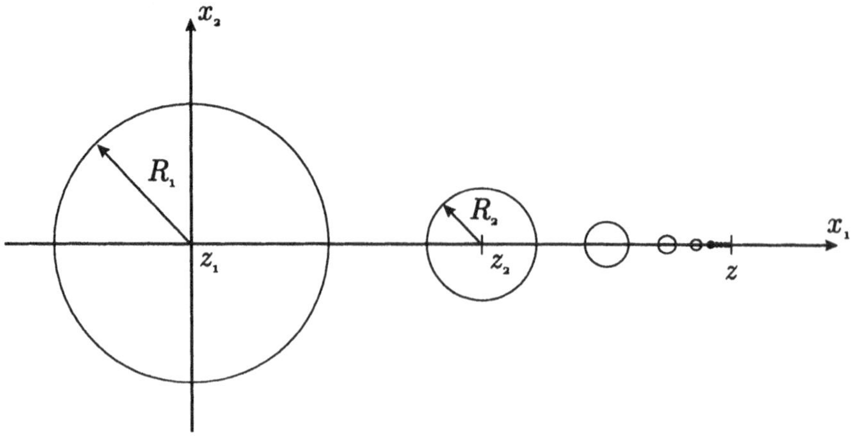

Figure 3.5.

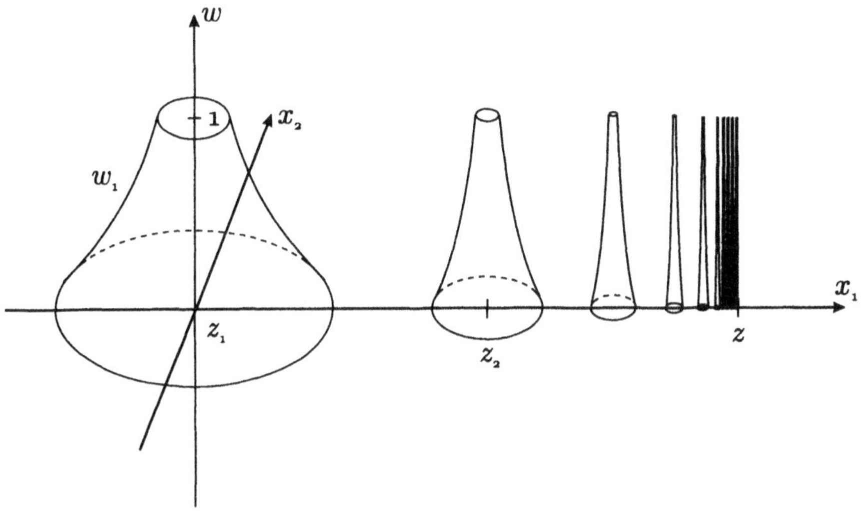

Figure 3.6.

Using the function w, it is not difficult to construct a bounded function from $H^1(\Omega)$ which has a strongly singular point at each point $(x_1, x_2)^T \in \overline{\Omega}$, where x_1 and x_2 are rational numbers.

Theorem 3.14. *(Generalized Poincaré's inequality) Let $\Omega \in \mathcal{L}$ and let $\omega \neq \emptyset$ be an open set either in Ω or in $\partial\Omega_0$, where $\Omega_0 \subseteq \Omega$, $\Omega_0 \in \mathcal{L}$. Then there exists a constant $C > 0$ such that*

$$\|v\|_1 \leq C \left(|v|_1^2 + |\int_\omega v \, d\omega|^2 \right)^{1/2} \quad \forall v \in H^1(\Omega), \tag{3.32}$$

where $d\omega$ stands for dx or ds.

Proof. We slightly modify the proof of [Nečas, 1967, p. 18], where Poincaré's inequality is proved for $\omega = \Omega$, (for a general real case see also [Hlaváček, Nečas, p. 309]). Assume (3.32) is not true. Then there exists a sequence $\{v_j\}_{j=1}^\infty \subset H^1(\Omega)$ such that

$$\|v_j\|_1^2 > j \left(|v_j|_1^2 + |\int_\omega v_j \, d\omega|^2 \right), \tag{3.33}$$

and we may assume that v_j are chosen so that

$$\|v_j\|_1 = 1, \quad j = 1, 2, \ldots. \tag{3.34}$$

The imbedding $H^1(\Omega) \subset L^2(\Omega)$ is compact by the Rellich Theorem 3.9. Hence, there exist $z \in L^2(\Omega)$ and a subsequence of $\{v_j\}$, still denoted by $\{v_j\}$, such that

$$\|v_j - z\|_0 \to 0 \quad \text{as } j \to \infty. \tag{3.35}$$

According to (3.33) and (3.34), we come to

$$|v_j|_1 \to 0 \quad \text{as } j \to \infty, \tag{3.36}$$

and thus, by (3.35) and (3.36), $\{v_j\}$ is Cauchy sequence in $H^1(\Omega)$, and again by (3.35),

$$\|v_j - z\|_1 \to 0 \quad \text{as } j \to \infty. \tag{3.37}$$

From here and (3.36) we have $|z|_1 = 0$, i.e., z is a constant function a.e. in Ω. Let the symbol $\operatorname{meas}\omega$ stand for the d or $(d-1)$-dimensional measure of $\omega \subset \Omega$ or $\omega \subset \partial\Omega_0$, respectively. Using the Cauchy–Schwarz inequality (3.10) in $L^2(\omega)$ and (3.37), we find that

$$\left|\int_\omega (v_j - z)\, d\omega\right| \le \left(\operatorname{meas}\omega \int_\omega |v_j - z|^2\, d\omega\right)^{1/2} \le C\|v_j - z\|_1 \to 0 \quad \text{as } j \to \infty. \tag{3.38}$$

Here the second inequality follows either from

$$\|v\|_{0,\omega} \le \|v\|_{0,\Omega} \le \|v\|_{1,\Omega} \quad \forall v \in H^1(\Omega) \tag{3.39}$$

if ω is an open set in Ω, or from (see (3.26))

$$\|v\|_{0,\omega} \le \|v\|_{0,\partial\Omega_0} \le C\|v\|_{1,\Omega_0} \le C\|v\|_{1,\Omega} \quad \forall v \in H^1(\Omega), \tag{3.40}$$

if ω is an open set in $\partial\Omega_0$. Moreover, from (3.33)

$$\int_\omega v_j\, d\omega \to 0 \quad \text{as } j \to \infty. \tag{3.41}$$

The constant function z is thus zero by (3.38) and (3.41). This contradicts (3.34) and (3.37). □

Corollary 3.15. *Let the assumptions of Theorem 3.14 be satisfied. By the Cauchy–Schwarz inequality (see (3.10))*

$$\left|\int_\omega v\, d\omega\right|^2 \le \operatorname{meas}\omega \int_\omega |v|^2\, d\omega \quad \forall v \in H^1(\Omega),$$

and (3.32), we have

$$\|v\|_1 \le C(|v|_1^2 + \|v\|_{0,\omega}^2)^{1/2} \quad \forall v \in H^1(\Omega). \tag{3.42}$$

This inequality is usually called *Friedrichs' inequality* when $\omega \subset \partial\Omega$. In particular,

$$\|v\|_1 \le C|v|_1 \quad \forall v \in H_0^1(\Omega). \tag{3.43}$$

Finally in this section we briefly introduce the Sobolev spaces $W_p^k(\Omega)$ of functions whose generalized derivatives up to order $k \in \{0, 1, ...\}$ belong to $L^p(\Omega)$, $p \in [1, \infty]$. For $p < \infty$ they are equipped with the norm

$$\|v\|_{k,p,\Omega} = \left(\sum_{|m| \le k} \int_\Omega |D^m v|^p \, dx \right)^{\frac{1}{p}}, \quad v \in W_p^k(\Omega),$$

and seminorm

$$|v|_{k,p,\Omega} = \left(\sum_{|m| = k} \int_\Omega |D^m v|^p \, dx \right)^{\frac{1}{p}}, \quad v \in W_p^k(\Omega),$$

and for $p = \infty$,

$$\|v\|_{k,\infty,\Omega} = \max_{|m| \le k} \|D^m v\|_{0,\infty,\Omega}, \quad |v|_{k,\infty,\Omega} = \max_{|m| = k} \|D^m v\|_{0,\infty,\Omega}.$$

We will again write

$$\|\cdot\|_{k,p} = \|\cdot\|_{k,p,\Omega}, \quad |\cdot|_{k,p} = |\cdot|_{k,p,\Omega}.$$

The Sobolev space $W_p^k(\Omega)$ is the completion of $C^\infty(\overline{\Omega})$ in the $\|\cdot\|_{k,p}$-norm. Sobolev spaces are Banach spaces and for $p = 2$ they become Hilbert spaces, i.e.,

$$H^k(\Omega) = W_2^k(\Omega).$$

The Sobolev imbedding theorem for $p \in [1, \infty]$ and $k \in \{1, 2, ...\}$ takes the form

$$W_p^k(\Omega) \subset C(\overline{\Omega}) \quad \text{if } pk > d.$$

Note that the condition $pk > d$ is not necessary but only sufficient. For instance, we have $W_1^2(\Omega) \subset C(\overline{\Omega})$ for $d = 2$ (see [Kufner, John, Fučík, p. 300]), i.e., the condition $pk > d$ is not satisfied. For more details about Sobolev spaces $W_p^k(\Omega)$ we refer to [Adams], [Mazja], [Nečas, 1967].

3.4. Classical and variational formulation of a 2nd order elliptic problem

In this section all functions are real-valued (for the complex case see Sections 3.5, 12.1–12.3, and 13.2). We introduce a weak (variational) formulation of the following classical boundary value problem with mixed boundary conditions, which describes, for example, an electric or magnetic potential of a stationary linear problems.

Find $u \in C^2(\overline{\Omega})$ such that

$$-\operatorname{div}(\mathcal{A}\ \operatorname{grad} u) = f \quad \text{in } \Omega, \tag{3.44}$$
$$u = \overline{u} \quad \text{on } \Gamma_1, \tag{3.45}$$
$$n^{\mathrm{T}}\mathcal{A}\ \operatorname{grad} u = g \quad \text{on } \Gamma_2, \tag{3.46}$$

where $\Omega \in \mathcal{L}$, n is the outward unit normal, $\mathcal{A} = (a_{ij}) \in (C^1(\overline{\Omega}))^{d\times d}$ is a symmetric matrix which is uniformly positive definite (i.e., there exists a constant $C > 0$ such that

$$\xi^T\mathcal{A}(x)\xi \geq C\|\xi\|^2 \quad \forall \xi \in \mathbf{R}^d \quad \forall x \in \overline{\Omega}), \tag{3.47}$$

$f \in C(\overline{\Omega})$, $\overline{u} \in C^2(\overline{\Omega})$, $g \in C(\overline{\Gamma_2})$, and Γ_1, Γ_2 are relatively open sets with respect to the topology on $\partial\Omega$,

$$\overline{\Gamma}_1 \cup \overline{\Gamma}_2 = \partial\Omega \tag{3.48}$$

and

$$\operatorname{meas}_{d-1}(\overline{\Gamma}_1 \cap \overline{\Gamma}_2) = 0$$

(i.e., Γ_1 and Γ_2 are disjoint). To apply the finite element method later, we, moreover, assume from now on that Γ_1 and Γ_2 have a finite number of components.

The function $u \in C^2(\overline{\Omega})$ satisfying (3.44)–(3.46) is called the *classical solution.* (Note that the classical solution is sometimes looked for in the space $C^1(\overline{\Omega}) \cap C^2(\Omega)$, i.e., $u \in C^1(\overline{\Omega})$ and $u|_{\Omega} \in C^2(\Omega)$.) The conditions (3.45) and (3.46) are called, respectively, *Dirichlet* and *Neumann boundary conditions.* The problem (3.44)–(3.46) is referred as the *Dirichlet* and *Neumann problem*, respectively, if $\Gamma_2 = \emptyset$ and $\Gamma_1 = \emptyset$. For the so-called Newton boundary conditions, see Chapter 9 or [Křížek, Neittaanmäki, 1990].

The matrix $\mathcal{A}$ describes physical properties of the medium Ω. If $\mathcal{A}$ is independent of $x \in \Omega$ we call the medium *homogeneous* (otherwise inhomogeneous). If $a_{ii}(x) = a_{11}(x)$ for $i = 2, ..., d$, and $a_{ij}(x) = 0$ for $i \neq j$, the medium is said to be *isotropic* (otherwise anisotropic). If $\mathcal{A}$ is the identity matrix, the equation (3.44) is called the *Poisson equation.* If moreover $f = 0$, then (3.44) is called the *Laplace equation* and the associated operator

$$\Delta u = \operatorname{div} \operatorname{grad} u = \partial_{11}u + ... + \partial_{dd}u \tag{3.49}$$

is called the *Laplace operator*, where

$$\partial_{ii}u = \frac{\partial^2 u}{\partial x_i^2}.$$

In practical problems the coefficients a_{ij} are often not continuous (they are, e.g., piecewise constant). Then the classical divergence operator in (3.44) cannot be employed. That is why we introduce a weak formulation of the problem (3.44)–(3.46), which enables us to consider also nonsmooth functions a_{ij} or f, g. So let us rewrite the equation (3.44) as follows

$$-\sum_{i,j=1}^{d} \partial_i(a_{ij}\partial_j u) = f \tag{3.50}$$

and introduce the space of test functions

$$V = \left\{ v \in H^1(\Omega) \mid v = 0 \text{ on } \Gamma_1 \right\}, \tag{3.51}$$

which is well-defined due to the Trace Theorem 3.8. Multiplying (3.50) by an arbitrary test function $v \in V$ and then integrating over Ω, we obtain

$$-\int_\Omega \sum_{i,j} \partial_i(a_{ij}\partial_j u)v \, dx = \int_\Omega fv \, dx. \tag{3.52}$$

Applying now Green's Theorem 3.10 for $w = w_i = \sum_j a_{ij}\partial_j u$, we find that

$$\int_\Omega \sum_{i,j} \partial_i v a_{ij} \partial_j u \, dx = \int_\Omega fv \, dx + \int_{\partial\Omega} \sum_{i,j} n_i a_{ij} \partial_j uv \, ds. \tag{3.53}$$

Since $v = 0$ on Γ_1, we obtain by (3.46) that

$$\int_\Omega (\operatorname{grad} v)^T \mathcal{A} \operatorname{grad} u \, dx = \int_\Omega fv \, dx + \int_{\Gamma_2} gv \, ds. \tag{3.54}$$

Defining now the bilinear form

$$a(v, w) = \int_\Omega (\operatorname{grad} v)^T \mathcal{A} \operatorname{grad} w \, dx, \quad v, w \in H^1(\Omega), \tag{3.55}$$

and the linear form

$$F(v) = \int_\Omega fv \, dx + \int_{\Gamma_2} gv \, ds, \quad v \in V, \tag{3.56}$$

we see that any classical solution u of (3.44)–(3.46) (if it exists) satisfies the equation

$$a(v, u) = F(v) \quad \forall v \in V. \tag{3.57}$$

From now on we shall assume that $a_{ij} \in L^\infty(\Omega)$, $f \in L^2(\Omega)$, $g \in L^2(\Gamma_2)$ and $\overline{u} \in H^1(\Omega)$. Moreover let $a_{ij}(x) = a_{ji}(x)$ and let (3.47) hold only for a.e. $x \in \Omega$. We see then that the integrals in (3.55) and (3.56) are still well-defined.

Referring to Section 2.1, we recall that an integral form of a physical law is more natural than its differential form. From formula (3.51) we begin to use implicitly the concept of completion. Instead of looking for $u \in C^2(\overline{\Omega})$ satisfying the differential equation (3.44) with the boundary conditions (3.45)–(3.46), we shall solve the integral form equation (3.57). Its solution will be searched in the Hilbert space $H^1(\Omega)$, which is, of course, complete and bigger than $C^2(\overline{\Omega})$. This allows us to employ some useful results from functional analysis (e.g., the Lax–Milgram Lemma 3.3) and apply the finite element method. First we will deal with the case $\Gamma_1 \neq \emptyset$.

Definition 3.16. *Let $\Omega \in \mathcal{L}$ and $\Gamma_1 \neq \emptyset$. A function $u \in H^1(\Omega)$ is called a weak (or generalized) solution of the given problem (3.44)–(3.46) if*

$$u - \overline{u} \in V \tag{3.58}$$

and

$$a(v, u) = F(v) \quad \forall v \in V. \tag{3.59}$$

For convenience the weak solution is denoted by u as the classical solution. The condition (3.45) is replaced by (3.58) using (3.51). In the next theorem we show that the weak solution exists uniquely. Note that the classical solution $u \in C^2(\overline{\Omega})$ of (3.44)–(3.46) need not exist, in general. However, if it exists then, by (3.57), it is also the weak solution.

Theorem 3.17. *Let $\Omega \in \mathcal{L}$ and $\Gamma_1 \neq \emptyset$. Then there exists exactly one weak solution $u \in H^1(\Omega)$ of the problem (3.58)–(3.59).*

Proof. Setting

$$u^0 = u - \overline{u}, \tag{3.60}$$

the problem (3.58)–(3.59) may be equivalently rewritten as follows: Find $u^0 \in V$ such that

$$a(v, u^0) = F(v) - a(v, \overline{u}) \quad \forall v \in V. \tag{3.61}$$

The bilinear form $a(\cdot\,,\cdot)$ is evidently continuous (cf. (3.12)),

$$|a(v,w)| \leq \|\operatorname{grad} v\|_0 \sum_{i,j=1}^{d} \|a_{ij}\|_{L^\infty(\Omega)} \|\operatorname{grad} w\|_0 \leq C\|v\|_1\|w\|_1, \quad \forall v, w \in V. \tag{3.62}$$

Taking $\omega = \Gamma_1$ in Friedrichs' inequality (3.42), we see by (3.47) that $a(\cdot\,,\cdot)$ is V-elliptic (cf. (3.13))

$$a(v,v) \geq C \int_\Omega (\operatorname{grad} v)^T \operatorname{grad} v \, dx = C|v|_1^2 \geq C'\|v\|_1^2 \quad \forall v \in V. \tag{3.63}$$

Since $v \mapsto F(v) - a(v, \overline{u})$ is a linear continuous form, the existence of a unique $u^0 \in V$ satisfying (3.61) follows now from the Lax–Milgram Lemma 3.3. □

According to [Gilbarg, Trudinger, p. 176], the weak solution u of the problem (3.44)–(3.46) belongs to $H^2(\Omega)$ provided the coefficients a_{ij} are Lipschitz continuous, $\overline{u} \in H^2(\Omega)$, $f \in L^2(\Omega)$ and $\Gamma_1 = \partial\Omega$ is smooth. If $\mathcal{A}$ is the identity matrix, $\overline{u} = 0$, $f \in L^2(\Omega)$, $\Gamma_1 = \partial\Omega$ and Ω convex (not necessarily smooth), then also $u \in H^2(\Omega)$ (see [Kadlec]). For other higher regularity results for the weak solution, we refer to [Agmon], [Dauge], [Douglis, Nirenberg], [Grisvard], [Nečas, 1967, 1983], etc.

Consider now the case $\Gamma_1 = \emptyset$, i.e., the Neumann problem. In this case the test function $v \equiv 1$ belongs to $V = H^1(\Omega)$ and thus by (3.56), (3.57) and (3.55), we find that

$$\int_\Omega f\,dx + \int_{\partial\Omega} g\,ds = 0. \tag{3.64}$$

This is the so-called *equilibrium condition* which is necessary for solvability of the Neumann problem. We observe that if u is the classical solution then $u + constant$ is also the classical solution. According to (3.55) and (3.47), we see that there exists a constant $C > 0$ such that

$$a(v,v) \geq C|v|_1^2 \quad \forall v \in H^1(\Omega), \tag{3.65}$$

and clearly

$$a(v,w) = 0 \quad \text{for } v \equiv 1 \text{ and any } w \in H^1(\Omega). \tag{3.66}$$

In particular, $a(1,1) = 0$, i.e., the bilinear form $a(\cdot\,,\cdot)$ is not $H^1(\Omega)$-elliptic and thus we cannot employ the Lax–Milgram Lemma 3.3 as in the proof of Theorem 3.17. There are several ways to avoid this difficulty.

A variational formulation of the Neumann problem can be given (see, e.g., [Ciarlet, p. 33]) in the quotient space $H^1(\Omega)/P_0(\Omega)$, where $P_0(\Omega)$ is the space of constant functions over Ω. However, this space is not suitable for finite element approximations. Another way is to restrict the space $V = H^1(\Omega)$ of test functions as follows

$$W = \left\{ w \in H^1(\Omega) \mid \int_\omega w\,d\omega = 0 \right\}, \tag{3.67}$$

where ω is described in Theorem 3.14. Then by (3.65) and (3.32), the bilinear form $a(\cdot\,,\cdot)$ is W-elliptic,

$$a(w,w) \geq C|w|_1^2 \geq C'\|w\|_1^2 \quad \forall w \in W. \tag{3.68}$$

A variational solution of the Neumann problem can then be found using the space W. We return to this space of test functions in Remark 6.1. Note that we must not fix w in (3.67) only at one point for $d > 1$, since by Remark 3.13 functions from $H^1(\Omega)$ need not be continuous.

Here we will present a weak formulation of the Neumann problem with the space of test functions

$$V = H^1(\Omega)$$

and a modified bilinear form

$$\tilde{a}(v,w) = a(v,w) + \lambda \int_\omega v\,d\omega \int_\omega w\,d\omega, \quad v,w \in H^1(\Omega), \tag{3.69}$$

where $a(\cdot\,,\cdot)$ is given by (3.55), ω is described in Theorem 3.14 and $\lambda > 0$ is a fixed number.

Definition 3.18. *Let $\Omega \in \mathcal{L}$ and $\Gamma_1 = \emptyset$. A function $u \in H^1(\Omega)$ is called a weak (or generalized) solution of the Neumann problem (3.44)+(3.46) if*

$$\tilde{a}(v,u) = F(v) \quad \forall v \in H^1(\Omega), \tag{3.70}$$

where

$$F(v) = \int_\Omega fv\,dx + \int_{\partial\Omega} gv\,ds, \quad v \in H^1(\Omega), \tag{3.71}$$

and where $f \in L^2(\Omega)$ and $g \in L^2(\partial\Omega)$ satisfy (3.64).

Theorem 3.19. *Let $\Omega \in \mathcal{L}$ and $\Gamma_1 = \emptyset$. Then there exists exactly one weak solution $u \in H^1(\Omega)$ of the problem (3.70). This solution u satisfies the condition*

$$\int_\omega u\,d\omega = 0. \tag{3.72}$$

Proof. By (3.69) and (3.63) and the generalized Poincaré's inequality (3.32), we see that the bilinear form $\tilde{a}(\cdot,\cdot)$ is $H^1(\Omega)$-elliptic,

$$\tilde{a}(v,v) = a(v,v) + \lambda\left(\int_\omega v\,d\omega\right)^2 \geq C|v|_1^2 + \lambda\left(\int_\omega v\,d\omega\right)^2 \geq C'\|v\|_1^2 \quad \forall v \in H^1(\Omega).$$

From (3.62) and the Cauchy–Schwarz inequality (3.10) we further find that $\tilde{a}(\cdot,\cdot)$ is continuous

$$\begin{aligned}
|\tilde{a}(v,w)| &\leq |a(v,w)| + \lambda\left|\int_\omega v\,d\omega\right|\left|\int_\omega w\,d\omega\right| \\
&\leq C\|v\|_1\|w\|_1 + \lambda\,\mathrm{meas}\,\omega\left(\int_\omega v^2\,d\omega\right)^{1/2}\left(\int_\omega w^2\,d\omega\right)^{1/2} \\
&\leq C'\|v\|_1\|w\|_1 \quad \forall v,w \in H^1(\Omega),
\end{aligned}$$

where the last inequality follows from (3.39) or (3.40). Since the linear form (3.71) is obviously continuous, the existence of a unique $u \in H^1(\Omega)$ satisfying (3.70) follows from the Lax–Milgram Lemma 3.3.

Finally, according to (3.69), (3.66), (3.70), (3.71) and (3.64), we find that

$$\lambda\,\mathrm{meas}\,\omega\int_\omega u\,d\omega = \tilde{a}(1,u) = F(1) = \int_\Omega f\,dx + \int_{\partial\Omega} g\,ds = 0,$$

which yields the required relation (3.72). □

Since we have used the modified bilinear form $\tilde{a}(\cdot,\cdot)$ in the weak formulation (3.70), it is not clear whether a smooth weak solution is the classical solution of the Neumann problem and conversely.

Theorem 3.20. *Let $\Omega \in \mathcal{L}$, $\Gamma_1 = \emptyset$, $a_{ij} \in C^1(\overline{\Omega})$ for $i,j = 1,...,d$, $f \in C(\overline{\Omega})$ and let $g \in C(\partial\Omega)$. Let the weak solution u of the problem (3.70) belong to $C^2(\overline{\Omega})$. Then u is the classical solution of the problem (3.44)+(3.46). Conversely, let $u \in C^2(\overline{\Omega})$ be the classical solution of the problem (3.44)+(3.46). Then u is the weak solution of the problem (3.70) provided (3.72) holds.*

Proof. So let $u \in C^2(\overline{\Omega})$ be the solution of (3.70). Then (3.72) holds and we find that

$$\tilde{a}(v,u) = a(v,u) = F(v) \quad \forall v \in H^1(\Omega). \tag{3.73}$$

We show now that

$$w \equiv -\sum_{i,j} \partial_i(a_{ij}\partial_j u) - f \in C(\overline{\Omega})$$

is zero. By (3.73) and Green's Theorem 3.10 we have

$$\begin{aligned} 0 = a(v,u) - F(v) &= \int_\Omega \left(\sum_{i,j} \partial_i v a_{ij} \partial_j u - fv \right) dx \\ &= \int_\Omega \left(-\sum_{i,j} \partial_i(a_{ij}\partial_j u) - f \right) v\, dx = \int_\Omega wv\, dx \quad \forall v \in H_0^1(\Omega). \end{aligned} \tag{3.74}$$

Since $H_0^1(\Omega)$ is dense in $L^2(\Omega)$ there exists a sequence $\{v^i\}_{i=1}^\infty \subset H_0^1(\Omega)$ such that

$$\|v^i - w\|_0 \to 0 \quad \text{as } i \to \infty. \tag{3.75}$$

Now from (3.74), (3.75) and the Cauchy–Schwarz inequality (3.10) we find that

$$\begin{aligned} \int_\Omega w^2\, dx &= \int_\Omega wv^i\, dx + \int_\Omega w(w - v^i)\, dx = \int_\Omega w(w - v^i)\, dx \\ &\le \|w\|_0 \|w - v^i\|_0 \to 0 \quad \text{as } i \to \infty. \end{aligned} \tag{3.76}$$

Hence $\int_\Omega w^2\, dx = 0$ and thus

$$w = 0 \text{ in } \Omega, \tag{3.77}$$

i.e., (3.44) holds.

To prove (3.46), we set

$$z \equiv \sum_{i,j} n_i a_{ij} \partial_j u - g \in L^2(\partial\Omega).$$

According to (3.73), Green's Theorem 3.10 and (3.77), we get

$$\begin{aligned} 0 = a(v,u) - F(v) &= \int_\Omega \left(\sum_{i,j} \partial_i v a_{ij} \partial_j u - fv \right) dx - \int_{\partial\Omega} gv\, ds \\ &= \int_{\partial\Omega} n_i a_{ij} \partial_j uv\, ds - \int_{\partial\Omega} gv\, ds = \int_{\partial\Omega} zv\, ds \quad \forall v \in H^1(\Omega). \end{aligned} \tag{3.78}$$

Due to the density (3.28) there exists a sequence $\{v^i\}_{i=1}^{\infty} \subset H^1(\Omega)$ such that

$$\|v^i - z\|_{0,\partial\Omega} \to 0 \quad \text{as } i \to \infty. \tag{3.79}$$

Using now (3.78) and (3.79), we can proceed as in (3.76) to obtain that $z = 0$ in $L^2(\partial\Omega)$, i.e., (3.46) holds.

Conversely, if $u \in C^2(\overline{\Omega})$ is the classical solution of (3.44)+(3.46) satisfying (3.72). Then clearly $u \in H^1(\Omega)$ and by (3.57) and (3.69) we see that u also satisfies the following relation

$$\tilde{a}(v,u) = a(v,u) = F(v) \quad \forall v \in H^1(\Omega).$$

□

Remark 3.21. For the time being we have not used the assumption that the coefficient matrix $\mathcal{A} = (a_{ij})$ is symmetric. In this case the Neumann problem (3.70) may be equivalently rewritten, by Theorem 3.4, as follows: Find $u \in H^1(\Omega)$ which minimizes the quadratic functional

$$J(v) = \frac{1}{2}\tilde{a}(v,v) - F(v) \tag{3.80}$$

over the space $H^1(\Omega)$. Analogously the problem (3.61) corresponding to mixed boundary conditions can be equivalently formulated as follows: Find $u^0 \in V$ which minimizes the quadratic functional

$$\overline{J}(v) = \frac{1}{2}a(v,v) - F(v) + a(v,\overline{u}) \tag{3.81}$$

over the space V, and then set $u = u^0 + \overline{u}$. In the both cases the function u is called the *variational solution*. This kind of formulation of the problem (3.44)–(3.46) is usually referred as the *primal variational formulation*. The variational solution is also the weak solution. On the other hand, if $\mathcal{A}$ is not symmetric, then the weak solution exists but the variational solution is not defined.

Remark 3.22. In several practical situations, we have to investigate materials with a periodic structure. This may be modelled as follows. Assume that Ω is rectangular, $\Omega = (0, r_1) \times \cdots \times (0, r_d)$ and consider the problem given by (3.44) with the *periodic boundary conditions* (see, e.g., [Temam, 1988, p. 64]):

$$\left.\begin{aligned} u\big|_{S_i} &= u\big|_{S_i'} \\ n^T \mathcal{A} \operatorname{grad} u\big|_{S_i} &= n^T \mathcal{A} \operatorname{grad} u\big|_{S_i'} \end{aligned}\right\} \quad i = 1, \dots, d, \tag{3.82}$$

where S_i and S_i' are opposite faces of $\partial\Omega$, which are perpendicular to the axis x_i, $i = 1, \dots, d$. For the classical formulation we must again assume that $\mathcal{A}$ is smooth enough.

To introduce a nontraditional variational approach of the above problem we set

$$W = \{v \in H^1(\Omega) \mid v|_{S_i} = v|_{S'_i}, \quad i = 1, ..., d\} \tag{3.83}$$

and suppose that some smooth u satisfies (3.44)+(3.82). Multiplying (3.44) by an arbitrary function $v \in W$, integrating this over Ω and then using Green's Theorem 3.10, we find that

$$\int_\Omega (\operatorname{grad} v)^T \mathcal{A} \operatorname{grad} u \, dx - \sum_{i=1}^{d} \int_{S_i \cup S'_i} (n^T \mathcal{A} \operatorname{grad} u) v \, ds = \int_\Omega f v \, dx.$$

Now (3.82) and the relation $n|_{S_i} = -n|_{S'_i}$ imply

$$a(v, u) = F(v) \quad \forall v \in W, \tag{3.84}$$

where $a(\cdot\,,\cdot)$ is defined by (3.55) and

$$F(v) = \int_\Omega f v \, dx, \quad v \in W. \tag{3.85}$$

Since W contains constant functions, we see that $a(1,1) = 0$, i.e., $a(\cdot\,,\cdot)$ is not W-elliptic, and that $F(1) = a(1,u) = 0$. On the contrary the modified form (3.69) is W-elliptic (see the proof of Theorem 3.19) and thus we get that there exists exactly one solution $u \in W$ of the equation

$$\tilde{a}(v, u) = F(v) \quad \forall v \in W. \tag{3.86}$$

This solution satisfies the condition

$$\int_\omega u \, d\omega = 0,$$

since $F(1) = 0$ (see the proof of (3.72)). Similarly to Theorem 3.20, we can prove the connection between the classical problem with periodic boundary conditions and its weak formulation (3.86).

3.5. On the abstract Neumann problem

In this section we generalize the results of Theorem 3.19 into a general abstract Neumann problem, where the bilinear (sesquilinear) form $a(\cdot\,,\cdot)$ may correspond, e.g., to a system of linear elliptic equations, and the space of test functions may contain more than one constraint (cf. (3.67)). We shall use the following assumptions:

(A1) Let V be a real or complex Banach space equipped with the norm $\|\cdot\|_V$ and let $a(\cdot\,,\cdot)$ be a sesquilinear Hermitian (i.e., bilinear symmetric in the real case) continuous form on $V \times V$.

(A2) Let m be a positive integer, let $\varphi_1, ..., \varphi_m$ be linear continuous functionals over V and let

$$W = \{v \in V \mid \varphi_i(v) = 0, \quad i = 1, ..., m\}.$$

(A3) Let $a(\cdot,\cdot)$ be W-elliptic, i.e., there exists a constant $C > 0$ such that

$$a(w,w) \geq C\|w\|_V^2 \quad \forall w \in W.$$

(A4) Let $\dim P = m$, where

$$P = \{p \in V \mid a(p,v) = 0 \quad \forall v \in V\}.$$

(A5) Let $F(\cdot)$ be a linear continuous functional over V such that

$$F(p) = 0 \quad \forall p \in P.$$

We deal with the problem: Find a function $u \in W$ satisfying

$$a(w,u) = F(w) \quad \forall w \in W. \tag{3.87}$$

By the Riesz Theorem 3.2, this problem has a unique solution. Note that a weak formulation of a Neumann problem is typically of the form (3.87). However, due to the constraints occurring in the definition of W, a direct numerical solution of the problem (3.87) can be difficult. We will show that the penalty method or the method of Lagrange's multipliers enables us to solve the problem easily. These methods have very simple forms when applied to the *abstract Neumann problem* (3.87).

Theorem 3.23. *Let (A1)–(A5) hold and let $\lambda_1, ..., \lambda_m$ be arbitrary positive numbers. Then there exists one and only one $u \in V$ such that*

$$a(v,u) + \sum_{i=1}^{m} \lambda_i \varphi_i(v)(\varphi_i(u))^C = F(v) \quad \forall v \in V, \tag{3.88}$$

where $(\cdot)^C$ denotes the conjugate number. Moreover, $\varphi_1, ..., \varphi_m$ are linearly independent and the solution u satisfies the conditions

$$\varphi_i(u) = 0, \quad i = 1, ..., m, \tag{3.89}$$

i.e., $u \in W$ and u is at the same time the unique solution of the original problem (3.87).

Proof. First of all we show that each $v \in V$ can be decomposed so that $v = w + p$ for some $w \in W$ and $p \in P$. So let $v \in V$ be arbitrary. According to (A1) and

(A3), $a(\cdot,\cdot)$ represents a scalar product on W and thus, by the Riesz Theorem 3.2, there exists precisely one $w \in W$ such that

$$a(y,w) = a(y,v) \quad \forall y \in W. \tag{3.90}$$

Let us denote by G the mapping $v \mapsto w$. It is clearly a linear projection operator. Hence, by [Taylor, Chapter 4.8], the space V can be expressed as the direct sum $V = W \oplus \operatorname{Ker}(G)$ and

$$\dim \operatorname{Ker}(G) \leq m, \tag{3.91}$$

(since for the time being we do not know whether $\varphi_1, ..., \varphi_m$ are linearly independent). Putting

$$p = v - w, \tag{3.92}$$

we find, by (3.90), that

$$p \in \operatorname{Ker}(G). \tag{3.93}$$

For any $q \in P$ we have $a(y,q) = 0$ for all $y \in W$ and thus $G(q) = 0$, i.e., $q \in \operatorname{Ker}(G)$ and $P \subset \operatorname{Ker}(G)$. However, from (3.91) and (A4) we see that

$$P = \operatorname{Ker}(G).$$

Consequently, due to (3.93), we obtain $p \in P$.

The left-hand side of (3.88) obviously represents a sesquilinear Hermitian continuous form on $V \times V$. We show now that it is V-elliptic. For any $v \in V$ we have by (3.92), (A4), (A2) and (A3) that

$$\begin{aligned} &a(v,v) + \sum_{i=1}^{m} \lambda_i |\varphi_i(v)|^2 \\ &= a(w,w) + a(p,w) + (a(p,w))^C + a(p,p) + \sum_{i=1}^{m} \lambda_i |\varphi_i(w) + \varphi_i(p)|^2 \\ &= a(w,w) + \sum_{i=1}^{m} \lambda_i |\varphi_i(p)|^2 \geq C\|w\|_V^2 + C_1\|p\|_V^2 \\ &\geq C_2\|w+p\|_V^2 = C_2\|v\|_V^2, \end{aligned}$$

as all norms in the finite dimensional space P are equivalent. Thus the existence of a unique solution $u \in V$ of (3.88) follows again from the Riesz Theorem 3.2.

Finally we check relation (3.89). Since $\dim \operatorname{Ker}(G) = m$, the functionals φ_i are linearly independent. We may thus choose for any $i = 1, ..., m$ an element v_i from the space

$$\{v \in V \mid \varphi_1(v) = ... = \varphi_{i-1}(v) = \varphi_{i+1}(v) = ... = \varphi_m(v) = 0\}$$

for which

$$\varphi_i(v_i) = 1.$$

Hence,

$$\varphi_i(v_j) = \delta_{ij}, \quad i,j = 1,...,m.$$

Putting

$$p_i = v_i - w_i, \quad i = 1,...,m,$$

where $w_i = G(v_i)$, we see that

$$\varphi_i(p_j) = \delta_{ij}, \quad i,j = 1,...,m.$$

From here, (A4), (3.88) and (A5) we get for each $j \in \{1,...,m\}$ that

$$\lambda_j(\varphi_j(u))^C = \sum_{i=1}^{m} \lambda_i \varphi_i(p_j)(\varphi_i(u))^C$$
$$= a(p_j,u) + \sum_{i=1}^{m} \lambda_i \varphi_i(p_j)(\varphi_i(u))^C = F(p_j) = 0.$$

Therefore, $\varphi_j(u) = 0$ for $j = 1,...,m$, i.e., $u \in W$ and by (3.88)

$$a(w,u) = F(w) \quad \forall w \in W.$$

□

Remark 3.24. Define a function $J: V \to \mathbf{R}^1$ by the formula

$$J(v) = a(v,v) - F(v) - (F(v))^C.$$

The problem (3.87) is, by Theorem 3.4, equivalent to the following one: Find $u \in W$ such that

$$J(u) = \min_{w \in W} J(w). \tag{3.94}$$

The *penalty method* for the problem (3.94) consists in finding $u_\lambda \in V$ such that

$$J_\lambda(u_\lambda) = \min_{v \in V} J_\lambda(v), \tag{3.95}$$

where

$$J_\lambda(v) = J(v) + \sum_{i=1}^{m} \lambda_i |\varphi_i(v)|^2,$$

and where $\lambda = (\lambda_1,...,\lambda_m)$, $\lambda_i > 0$. The function u_λ is a solution of the problem (3.95) if and only if u_λ is a solution of the problem: Find $u_\lambda \in V$ such that

$$a(v,u_\lambda) + \sum_{i=1}^{m} \lambda_i \varphi_i(v)(\varphi_i(u_\lambda))^C = F(v) \quad \forall v \in V.$$

It is well-known (see, e.g., [Céa, 1971], [Temam, 1984]) that $u_\lambda \to u$ in V when all $\lambda_i \to \infty$. However, in our case we have, by Theorem 3.23, that $u_\lambda = u$ for arbitrary $\lambda_i > 0$. So the penalty method gives the exact solution of the Neumann problem (3.87) for arbitrary positive penalty parameters.

Remark 3.25. Let V be a real Banach space. The subspace W may also be defined as follows

$$W = \{w \in V \mid (\varphi_i(w))^2 = 0, \quad i = 1, ..., m\}.$$

The functionals $v \mapsto J(v) = a(v,v) - F(v)$ and $v \mapsto (\varphi_i(v))^2$ are continuously Fréchet-differentiable in V (cf. Section 8.1). Hence, from (3.94) and well-known results [Schwartz] it follows that if u is the solution of the problem (3.87) then there exist *Lagrange's multipliers* $\lambda_1, ..., \lambda_m$ so that

$$a(v,u) + \sum_{i=1}^{m} \lambda_i \varphi_i(v) \varphi_i(u) = F(v) \quad \forall v \in V.$$

Comparing this with (3.88), we observe an interesting fact that there is a continuum of Lagrange's multipliers in the case considered in this section. One can choose any set $\lambda_1, ..., \lambda_m$ of positive numbers.

Applications of Theorem 3.23 to several examples can be found in [Křížek, 1993] and [Křížek, Litvinov].

In the next theorem, we show how to satisfy the hypothesis (A4).

Theorem 3.26. *Let q be a positive integer, $\Omega \in \mathcal{L}$ and let*

$$V = \prod_{j=1}^{q} H^{k_j}(\Omega),$$

where $H^{k_j}(\Omega)$ are Sobolev spaces with $k_j \geq 1$. Let $a(\cdot,\cdot)$ be a sesquilinear Hermitian form such that

$$a(v,v) + \|v\|_0^2 \geq C\|v\|_V^2 \quad \forall v \in V, \tag{3.96}$$

where $C > 0$ is independent of v. Then $\dim P < \infty$.

Proof. Assume that P is not a finite-dimensional space. Then by the Riesz lemma (see, e.g., [Taylor, Theorem 3.12-E]) there exists a sequence $\{p_i\}_{i=1}^{\infty} \subset P$ such that $\|p_i\|_V = 1$ and

$$\|p_i - p\|_V > \tfrac{1}{2} \quad \forall p \in \operatorname{span}(p_1, ..., p_{i-1}). \tag{3.97}$$

Using the compactness of the imbedding $V \subset (L^2(\Omega))^q$ – cf. Theorem 3.9, we find that there exists an $(L^2(\Omega))^q$-convergent subsequence of $\{p_i\}$ which will be still denoted by $\{p_i\}$. Hence, there exists an integer i_0 such that

$$\|p_i - p_j\|_0 < \tfrac{1}{2}\sqrt{C}$$

for all $i > j \geq i_0$. However, this contradicts inequality (3.96) for $v = p_i - p_j$, since $a(p_i - p_j, p_i - p_j) = 0$ and $\|p_i - p_j\|_V > \frac{1}{2}$ by (3.97). □

3.6. The space $H(\mathrm{div})$

Let $\Omega \in \mathcal{L}$. By [Girault, Raviart, p. 27], the functional $v \mapsto n \cdot v|_{\partial\Omega}$ defined on $(C^\infty(\overline{\Omega}))^d$ can be extended by continuity to a linear continuous mapping from the space

$$H(\mathrm{div}) = \{v \in (L^2(\Omega))^d \mid \exists f \in L^2(\Omega) : (v, \mathrm{grad}\, z)_0 + (f, z)_0 = 0 \quad \forall z \in C_0^\infty(\Omega)\}$$

into $H^{-1/2}(\partial\Omega)$, the latter being the dual space to $H^{1/2}(\partial\Omega)$, i.e., $H^{-1/2}(\partial\Omega)$ is the space of linear continuous functionals defined on $H^{1/2}(\partial\Omega)$. The function f is called the divergence of v (in the sense of distributions) and we put $\mathrm{div}\, v = f$. Now Green's formula can be rewritten as

$$(\mathrm{div}\, v, z)_0 + (v, \mathrm{grad}\ z)_0 = \langle n \cdot v, z\rangle_{\partial\Omega} \quad \forall v \in H(\mathrm{div}) \quad \forall z \in H^1(\Omega), \tag{3.98}$$

where $n = (n_1, ..., n_d)$ is the outward unit normal and $\langle\cdot,\cdot\rangle_{\partial\Omega}$ denotes the duality pairing between $H^{-1/2}(\partial\Omega)$ and $H^{1/2}(\partial\Omega)$. If v is smooth then

$$\langle n \cdot v, z\rangle_{\partial\Omega} = \int_{\partial\Omega} (n \cdot v) z\, ds.$$

The space $H(\mathrm{div})$ can be formally expressed as follows

$$H(\mathrm{div}) = \{v \in (L^2(\Omega))^d \mid \mathrm{div}\, v \in L^2(\Omega)\}.$$

It is a Hilbert space equipped with the scalar product

$$(v, w)_{H(\mathrm{div})} = (v, w)_0 + (\mathrm{div}\, v, \mathrm{div}\, w)_0$$

and the induced norm

$$\|v\|_{H(\mathrm{div})} = (\|v\|_0^2 + \|\mathrm{div}\, v\|_0^2)^{1/2}.$$

By [Girault, Raviart, p. 27], we have

$$H(\mathrm{div}) = \overline{(C^\infty(\overline{\Omega}))^d},$$

where the closure is taken under the $\|\cdot\|_{H(\mathrm{div})}$-norm.

Next we define several important subspaces of $H(\mathrm{div})$,

$$\begin{aligned} H_0(\mathrm{div}) &= \{v \in H(\mathrm{div}) \mid n \cdot v = 0 \text{ on } \partial\Omega\}, \\ H(\mathrm{div}^0) &= \{v \in H(\mathrm{div}) \mid \mathrm{div}\, v = 0 \text{ in } \Omega\}, \\ H_0(\mathrm{div}^0) &= H_0(\mathrm{div}) \cap H(\mathrm{div}^0). \end{aligned}$$

Note that $H_0(\mathrm{div})$ is the closure of $(C_0^\infty(\Omega))^d$ under the $\|\cdot\|_{H(\mathrm{div})}$-norm.

We show that $H(\mathrm{div}^0)$ contains also functions whose divergence does not exist in the classical sense but exists (and vanishes) merely in the sense of distributions. Let $\Omega = (-1,1) \times (0,1)$ and set

$$q(x_1, x_2) = \begin{cases} (0,0)^T & \text{in } \Omega_1 = (-1,0) \times (0,1), \\ (0,1)^T & \text{in } \Omega_2 = (0,1) \times (0,1). \end{cases}$$

Then by Green's formula (3.29)

$$\begin{aligned}(q, \mathrm{grad}\, z)_{0,\Omega} &= (q, \mathrm{grad}\, z)_{0,\Omega_2} = \left(1, \frac{\partial z}{\partial x_2}\right)_{0,\Omega_2} \\ &= \int_{\partial\Omega_2} z n_2 ds = \int_{\partial\Omega_2 \setminus \partial\Omega} z n_2 ds = 0 \quad \forall z \in H_0^1(\Omega).\end{aligned}$$

Here the last integral is zero, since $n_2 = 0$ on $\partial\Omega_2 \setminus \partial\Omega$. Hence, $q \in H(\mathrm{div}^0)$ but obviously $q \notin (C^1(\overline{\Omega}))^2$.

3.7. Dual variational formulation

In this section we consider problem (3.44)–(3.46) with $\mathcal{A} = (a_{ij})$ symmetric and $\Gamma_1 \neq \emptyset$. Its primal variational formulation reads: Find $u \in H^1(\Omega)$ which minimizes the functional (of potential energy)

$$J(v) = \frac{1}{2} a(v,v) - F(v) \tag{3.99}$$

over the set (affine manifold) $\overline{u} + V = \{\overline{u} + v \in H^1(\Omega) \mid v \in V\}$, where

$$a(v,w) = (\mathcal{A}\, \mathrm{grad}\, v, \mathrm{grad}\, w)_0$$

and

$$F(v) = \int_\Omega f v\, dx + \int_{\Gamma_2} g v\, ds.$$

It is easy to see that this formulation is equivalent to the problem: Find u^0 which minimizes the functional (3.81) over the space V. Recall that

$$u^0 = u - \overline{u} \in V.$$

In many engineering problems the *cogradient* $\mathcal{A}\, \mathrm{grad}\, u$ is often more interesting than the solution u itself, since $\mathcal{A}\, \mathrm{grad}\, u$ represents a flux (e.g., magnetic or heat flux). One way to solve this problem is to apply the primal variational formulation to get u and then differentiate u and evaluate its cogradient. Another way is to use the so-called dual variational formulation of (3.44)–(3.46) which produces the cogradient directly (cf. Theorem 3.27). It consists of finding $p \in Q(f,g)$ which minimizes the functional

$$I(q) = \frac{1}{2}b(q,q) - (q, \operatorname{grad} \overline{u})_0 \tag{3.100}$$

over the set (affine manifold)

$$Q(f,g) = \{q \in (L^2(\Omega))^d \mid (q, \operatorname{grad} v)_0 = F(v) \quad \forall v \in V\},$$

where

$$b(q,q') = (\mathcal{A}^{-1}q, q')_0, \quad q, q' \in (L^2(\Omega))^d, \tag{3.101}$$

is a scalar product. It is easy to show that $\mathcal{A}^{-1}$ is also symmetric and uniformly positive definite.

According to Section 3.6, we see that $Q(f,g) \subset H(\mathrm{div})$. The next theorem (the so-called principle of complementary energy) establishes a relation between the solutions of the primal and dual variational formulation.

Theorem 3.27. *The dual variational formulation has exactly one solution $p \in Q(f,g)$ and we have*

$$p = \mathcal{A} \operatorname{grad} u, \tag{3.102}$$

$$J(u) + I(p) = -F(\overline{u}). \tag{3.103}$$

Proof. The dual variational formulation can be equivalently rewritten as: Find $p^0 \in Q(0,0)$ which minimizes the functional

$$\overline{I}(q) = \frac{1}{2}b(q,q) + b(q,\overline{p}) - (q, \operatorname{grad} \overline{u})_0$$

over the space $Q(0,0)$, where the so-called particular solution $\overline{p} \in H(\mathrm{div})$ satisfies the equation

$$(\overline{p}, \operatorname{grad} v)_0 = F(v) \quad \forall v \in V. \tag{3.104}$$

(This equation represents two equilibrium equations $\operatorname{div} \overline{p} = f$ in Ω and $n \cdot \overline{p} = g$ on Γ_2, and we have $Q(f,g) = \overline{p} + Q(0,0)$.) According to Theorem 3.4 there exists the solution p^0 in the space $Q(0,0)$ and thus we set $p = p^0 + \overline{p}$. The uniqueness of p follows immediately from the strict convexity of I and the convexity of $Q(f,g)$.

Defining $|||q||| = \sqrt{b(q,q)}$, we can write, by (3.101) and (3.100),

$$\begin{aligned}
2I(q) &= b(q,q) - 2(q, \operatorname{grad} \overline{u})_0 = |||q|||^2 - 2(\mathcal{A}^{-1}q, \mathcal{A} \operatorname{grad} \overline{u})_0 \\
&= |||q - \mathcal{A} \operatorname{grad} \overline{u}|||^2 - |||\mathcal{A} \operatorname{grad} \overline{u}|||^2 \\
&= |||q - \mathcal{A} \operatorname{grad} u + \mathcal{A} \operatorname{grad}(u - \overline{u})|||^2 - |||\mathcal{A} \operatorname{grad} \overline{u}|||^2 \\
&= |||q - \mathcal{A} \operatorname{grad} u|||^2 + |||\mathcal{A} \operatorname{grad} u^0|||^2 - |||\mathcal{A} \operatorname{grad} \overline{u}|||^2,
\end{aligned}$$

where the last equality is based on the fact that $q - \mathcal{A}\,\mathrm{grad}\,u$ and $\mathcal{A}\,\mathrm{grad}\,u^0$ are $b(\cdot\,,\cdot)$-orthogonal,

$$b(q - \mathcal{A}\,\mathrm{grad}\,u, \mathcal{A}\,\mathrm{grad}\,u^0) = (q, \mathrm{grad}\,u^0)_0 - (\mathcal{A}\,\mathrm{grad}\,u, \mathrm{grad}\,u^0)_0$$
$$= F(u^0) - F(u^0) = 0 \qquad \forall q \in Q(f,g).$$

Hence, we see that I attains its minimum just at the point $\mathcal{A}\,\mathrm{grad}\,u$, which implies (3.102).

Finally, by (3.99), (3.100), (3.101) and (3.102), we get

$$\begin{aligned} J(u) + I(p) = &J(u) + I(\mathcal{A}\,\mathrm{grad}\,u) = \frac{1}{2}(\mathcal{A}\,\mathrm{grad}\,u, \mathrm{grad}\,u)_0 - F(u) \\ &+ \frac{1}{2}(\mathcal{A}^{-1}\mathcal{A}\,\mathrm{grad}\,u, \mathcal{A}\,\mathrm{grad}\,u)_0 - (\mathcal{A}\,\mathrm{grad}\,u, \mathrm{grad}\,\overline{u})_0 \\ = &(\mathcal{A}\,\mathrm{grad}\,u, \mathrm{grad}(u - \overline{u}))_0 - F(u) = F(u - \overline{u}) - F(u) = -F(\overline{u}). \end{aligned}$$

□

Remark 3.28. Sometimes I is called the functional of complementary energy, since by (3.103) we have

$$J(u) + I(p) = 0 \quad \text{for } \overline{u} = 0.$$

Remark 3.29. If $q \in Q(0,0)$ then $\mathrm{div}\,q = 0$ in Ω (i.e., $Q(0,0) \subset H(\mathrm{div}^0)$), and $n \cdot q = 0$ on Γ_2 for q smooth. Some advantages of the dual variational formulation are outlined in Section 4.3.

3.8. The space $H(\mathrm{rot})$

Throughout this section we shall consider only the case $d = 2$ or $d = 3$. Let again $\Omega \in \mathcal{L}$ and let us introduce the following operators

$$\mathrm{curl}\,z = (\partial_2 z, -\partial_1 z)^T \qquad \text{for } z \in H^1(\Omega),\ d = 2,$$

$$\mathrm{curl}\,z = (\partial_2 z_3 - \partial_3 z_2, \partial_3 z_1 - \partial_1 z_3, \partial_1 z_2 - \partial_2 z_1)^T \qquad \text{for } z = (z_1, z_2, z_3)^T \in (H^1(\Omega))^3,\ d = 3,$$

and the space

$$H(\mathrm{rot}) = \{v \in (L^2(\Omega))^d \mid \exists g \in (L^2(\Omega))^{2d-3} : (v, \mathrm{curl}\,z)_0 = (g, z)_0 \quad \forall z \in (C_0^\infty(\Omega))^{2d-3}\}.$$

Here the function g is called the rotation of v (in the sense of distributions) and we put $\mathrm{rot}\,v = g$. For $v \in (H^1(\Omega))^d$ it is easy check that

$$\mathrm{rot}\,v = \partial_1 v_2 - \partial_2 v_1 \quad \text{for } d = 2,$$
$$\mathrm{rot}\,v = \mathrm{curl}\,v \quad \text{for } d = 3.$$

Let us set $\operatorname{curl} v = \operatorname{rot} v$ for $v \in H(\operatorname{rot})$ and $d = 3$. The operators curl and rot are thus identical in the three-dimensional case. But in the two-dimensional space there are two rotations: curl is a vector operator acting on scalar functions, whereas rot is a scalar operator acting on vector functions.

The space $H(\operatorname{rot})$ can be formally expressed as follows

$$H(\operatorname{rot}) = \{v \in (L^2(\Omega))^d \mid \operatorname{rot} v \in (L^2(\Omega))^{2d-3}\}.$$

It is known that $H(\operatorname{rot})$ is a Hilbert space equipped with the scalar product

$$(v, w)_{H(\operatorname{rot})} = (v, w)_0 + (\operatorname{rot} v, \operatorname{rot} w)_0 \tag{3.105}$$

and the induced norm

$$\|v\|_{H(\operatorname{rot})} = (\|v\|_0^2 + \|\operatorname{rot} v\|_0^2)^{1/2}.$$

Note that (see [Girault, Raviart, p. 34]) the functional $v \mapsto n \wedge v|_{\partial\Omega}$ defined on $(C^\infty(\overline{\Omega}))^d$ can be extended by continuity to a linear continuous mapping from the space $H(\operatorname{rot})$ into $(H^{-1/2}(\partial\Omega))^{2d-3}$, where $n \wedge v = n_1 v_2 - n_2 v_1$ for $d = 2$, and $n \wedge v$ is the usual vector product for $d = 3$, i.e.,

$$n \wedge v = n \times v = (n_2 v_3 - n_3 v_2, n_3 v_1 - n_1 v_3, n_1 v_2 - n_2 v_1)^T.$$

Now Green's formula can be rewritten as

$$(\operatorname{rot} v, z)_0 - (v, \operatorname{curl} z)_0 = \langle n \wedge v, z\rangle_{\partial\Omega} \quad \forall v \in H(\operatorname{rot}) \quad \forall z \in (H^1(\Omega))^{2d-3}, \tag{3.106}$$

where $\langle\cdot,\cdot\rangle_{\partial\Omega}$ denotes the duality pairing between $(H^{-1/2}(\partial\Omega))^p$ and $(H^{1/2}(\partial\Omega))^p$ for $p = 1$ or $p = 3$.

Further, define several important subspaces of $H(\operatorname{rot})$,

$$\begin{aligned} H_0(\operatorname{rot}) &= \{v \in H(\operatorname{rot}) \mid n \wedge v = 0 \text{ on } \partial\Omega\}, \\ H(\operatorname{rot}^0) &= \{v \in H(\operatorname{rot}) \mid \operatorname{rot} v = 0 \text{ in } \Omega\}, \\ H_0(\operatorname{rot}^0) &= H_0(\operatorname{rot}) \cap H(\operatorname{rot}^0). \end{aligned}$$

Note that $H_0(\operatorname{rot})$ is the closure of $(C_0^\infty(\Omega))^d$ in the $\|\cdot\|_{H(\operatorname{rot})}$-norm.

From the density $\overline{C_0^\infty(\Omega)} = H_0^1(\Omega)$ and Green's formulae (3.98) and (3.106) we easily get that

$$\operatorname{grad} z \in H(\operatorname{rot}^0) \quad \text{for } z \in H^1(\Omega), \tag{3.107}$$

$$\operatorname{grad} z \in H_0(\operatorname{rot}^0) \quad \text{for } z \in H_0^1(\Omega), \tag{3.108}$$

$$\operatorname{curl} v \in H(\operatorname{div}^0) \begin{cases} \text{for} \quad v \in H^1(\Omega) \quad \text{and} \quad d = 2, \\ \text{for} \quad v \in H(\operatorname{rot}) \quad \text{and} \quad d = 3, \end{cases} \tag{3.109}$$

and

$$\operatorname{curl} v \in H_0(\operatorname{div}^0) \begin{cases} \text{for} \quad v \in H_0^1(\Omega) \quad \text{and} \quad d = 2, \\ \text{for} \quad v \in H_0(\operatorname{rot}) \quad \text{and} \quad d = 3, \end{cases} \tag{3.110}$$

The relation (3.107) says that the rotation of the gradient is zero. Similarly (3.109) represents the well-known fact that the divergence of the rotation is zero.

Recall that a domain $\Omega \subset \mathbf{R}^2$ is said to be *simply connected* if the set $\mathbf{R}^2 \setminus \overline{\Omega}$ has at most one component. If a domain Ω is not simply connected, it is called *multiply connected*. If there are two components of $\mathbf{R}^2 \setminus \overline{\Omega}$, then Ω is called doubly connected, etc. A simply connected domain in $\mathbf{R}^d$ will be defined in the next section.

Theorem 3.30. *Let $\Omega \in \mathcal{L}$ be a simply connected two-dimensional domain. Then*

(1) *$q \in H(\operatorname{div}^0)$ if and only if there exists a function $z \in H^1(\Omega)$ such that*

$$q = \operatorname{curl} z. \tag{3.111}$$

(2) *$v \in H(\operatorname{rot}^0)$ if and only if there exists a function $z \in H^1(\Omega)$ such that*

$$v = \operatorname{grad} z. \tag{3.112}$$

Proof. The point (1) follows from [Girault, Raviart, p. 37].

(2) Let $\operatorname{rot} v = 0$ for some $v = (v_1, v_2)^T \in H(\operatorname{rot})$. Then $\operatorname{div} q = 0$ for $q = (q_1, q_2)^T = (v_2, -v_1)^T$ and by (1) there exists $z \in H^1(\Omega)$ such that $q = \operatorname{curl} z$. Hence, $v = \operatorname{grad} z$. The converse implication is obvious. □

Remark 3.31. The potential function z in (3.111) is called a *stream function.* It is unique apart from a constant. It need not exist in the case of a multiply connected domain (cf. [Girault, Raviart, p. 37]). Also if $\operatorname{rot} v = 0$ then the associated potential $z \in H^1(\Omega)$ need not exist on multiply connected domains. To see this consider for example the annulus $\Omega = \{x \in \mathbf{R}^2 \mid 1 < \|x\| < 2\}$. By a regularization technique (see [Nečas, 1967, p. 58]), it is easy to construct a function $z \in C^\infty(\overline{\Omega} \setminus \Gamma)$, where $\Gamma = (1,2) \times \{0\}$, such that $z = 1$ for $x_1 > 0$ and $x_2 > 0$, and $z = 0$ for $x_1 > 0$ and $x_2 < 0$. Setting

$$v(x_1, x_2) = \begin{cases} \operatorname{grad} z & \text{in } \Omega \setminus \Gamma, \\ 0 & \text{on } \Gamma, \end{cases}$$

we get that $\operatorname{rot} v = 0$ with $v \in (C^\infty(\overline{\Omega}))^2$, but v is not a potential field globally on Ω, since z has a jump and thus $z \notin H^1(\Omega)$ by Remark 3.13.

Let us point out that the important assumption on the simply connectivity of Ω in Theorem 3.30 is often forgotten in literature.

3.9. Least squares formulation

Suppose that the solution of the system of the linear Maxwell equations from Section 2.2 is independent of time. Then we know that the system disintegrates into two smaller systems. These two systems are independent and formally similar.

In this section we consider the simplest form of the stationary Maxwell equations (cf. (2.4))

$$\operatorname{div} u = f \quad \text{in } \Omega, \tag{3.113}$$

$$\operatorname{rot} u = g \quad \text{in } \Omega, \tag{3.114}$$

$$n \wedge u = 0 \quad \text{on } \partial\Omega, \tag{3.115}$$

where $\Omega \in \mathcal{L}$, $\Omega \subset \mathbf{R}^d$ for $d \in \{2, 3\}$, $f \in L^2(\Omega)$ and $g \in (L^2(\Omega))^{2d-3}$. By (3.106), (3.110), (3.114) and (3.115) we find that

$$(g, 1)_0 = 0 \quad \text{for } d = 2,$$

$$g \in H_0(\operatorname{div}^0) \quad \text{for } d = 3,$$

which are necessary conditions for the solvability of (3.113)–(3.115). The equation (3.115) represents the so-called electric boundary conditions. Later (see (3.131)) we shall also deal with the magnetic boundary conditions $n \cdot u = 0$ (see Figure 3.7).

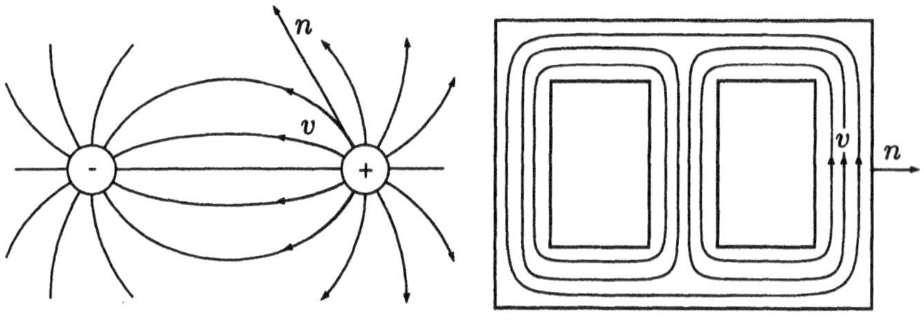

Figure 3.7.

Remark 3.32. If $z \in H_0^1(\Omega)$ is a weak solution of the Poisson equation $-\Delta z = f$ then the function $u = \operatorname{grad} z$ satisfies (3.113)–(3.115).

To introduce a least squares formulation of the problem (3.113)–(3.115) – cf. also Corollary 3.37, we define the space of test functions

$$V = H(\operatorname{div}) \cap H_0(\operatorname{rot})$$

with the norm

$$|||v|||_\Omega = (\|v\|_0^2 + \|\operatorname{div} v\|_0^2 + \|\operatorname{rot} v\|_0^2)^{1/2}, \quad v \in V.$$

Let $v \in V$ be arbitrary. Multiplying (3.113) by $\operatorname{div} v$, (3.114) by $\operatorname{rot} v$, integrating over Ω and then summing up the both equations, we get

$$a(u,v) = F(v) \quad \forall v \in V, \tag{3.116}$$

where

$$a(u,v) = (\operatorname{div} u, \operatorname{div} v)_0 + (\operatorname{rot} u, \operatorname{rot} v)_0$$

is a symmetric bilinear form and

$$F(v) = (f, \operatorname{div} v)_0 + (g, \operatorname{rot} v)_0$$

is a linear form. It is easy to see that the both forms are continuous. In particular,

$$|a(w,v)| \leq \|\operatorname{div} w\|_0 \|\operatorname{div} v\|_0 + \|\operatorname{rot} w\|_0 \|\operatorname{rot} v\|_0 \leq 2 |\!|\!| w |\!|\!|_\Omega |\!|\!| v |\!|\!|_\Omega \quad \forall v, w \in V.$$

To apply the Lax–Milgram Lemma 3.3, we have to prove that $a(\cdot,\cdot)$ is V-elliptic, which will be a consequence of the following theorem. From [Neittaanmäki, Křížek]:

Theorem 3.33. *Let $\Omega \in \mathcal{L}$ and $\Omega \subset \mathbf{R}^d$ for $d \in \{2,3\}$. Then*

$$\|v\|_0 \leq C(\|\operatorname{div} v\|_0 + \|\operatorname{rot} v\|_0) \quad \forall v \in V = H(\operatorname{div}) \cap H_0(\operatorname{rot}) \tag{3.117}$$

if and only if $\partial\Omega$ is connected.

The proof is based on two lemmas and one theorem.

Lemma 3.34. *If $\Omega \in \mathcal{L}$ and $\Omega \subset \mathbf{R}^3$ then*

$$(\operatorname{rot} s, \operatorname{rot} s)_0 = (s, \operatorname{rot} \operatorname{rot} s)_0$$

for all $s \in H(\operatorname{rot})$ such that $\operatorname{rot} s \in H_0(\operatorname{rot})$.

Proof. Let $s \in H(\operatorname{rot})$ with $\operatorname{rot} s \in H_0(\operatorname{rot})$ be given. Since $(C_0^\infty(\Omega))^3$ is dense in $H_0(\operatorname{rot})$ with respect to the $\|\cdot\|_{H(\operatorname{rot})}$-norm (see [Girault, Raviart, p. 32, 42]), there exists a sequence $\phi_j \in (C_0^\infty(\Omega))^3$ such that

$$\|\operatorname{rot} s - \phi_j\|_{H(\operatorname{rot})} \to 0 \quad \text{for } j \to \infty.$$

By the Cauchy–Schwarz inequality

$$|(\operatorname{rot} s - \phi_j, \operatorname{rot} s)_0| \leq \|\operatorname{rot} s - \phi_j\|_0 \|\operatorname{rot} s\|_0,$$
$$|(s, \operatorname{rot}\operatorname{rot} s - \operatorname{rot} \phi_j)_0| \leq \|s\|_0 \|\operatorname{rot}(\operatorname{rot} s - \phi_j)\|_0.$$

Therefore, we conclude that

$$(\operatorname{rot} s, \phi_j)_0 \to (\operatorname{rot} s, \operatorname{rot} s)_0 \tag{3.118}$$

and

$$(s, \operatorname{rot} \phi_j)_0 \to (s, \operatorname{rot}\operatorname{rot} s)_0 \tag{3.119}$$

when $j \to \infty$. By Green's formula (3.106) and the relation $\operatorname{curl} \phi_j = \operatorname{rot} \phi_j$ in $\mathbf{R}^3$, we come to

$$(\operatorname{rot} s, \phi_j)_0 - (s, \operatorname{rot} \phi_j)_0 = \langle n \wedge s, \phi_j \rangle_{\partial\Omega} = 0.$$

This together with (3.118) and (3.119) yields the assertion of the lemma. □

Theorem 3.35. *Let $\Omega \in \mathcal{L}$, $\Omega \subset \mathbf{R}^3$ and let $\partial\Omega$ be connected. Then $q \in H(\mathrm{div}^0)$ if and only if there exists a vector stream function $z \in (H^1(\Omega))^3$ such that*

$$q = \mathrm{rot}\, z. \tag{3.120}$$

Moreover, z can be chosen so that $\mathrm{div}\, z = 0$.

For the proof see [Girault, Raviart, p. 45].

Introduce the spaces

$$\mathcal{H}_D = H_0(\mathrm{div}^0) \cap H(\mathrm{rot}^0),$$
$$\mathcal{H}_R = H(\mathrm{div}^0) \cap H_0(\mathrm{rot}^0),$$

and denote by $X^\perp$ the orthogonal complement of a closed subspace $X \subset (L^2(\Omega))^3$ in $(L^2(\Omega))^3$.

Lemma 3.36. *Let $\Omega \in \mathcal{L}$, $\Omega \subset \mathbf{R}^3$, let $\partial\Omega$ be connected and let $w \in H(\mathrm{div}^0)$ be given. Then there exists a unique $s \in S = Q \cap (\mathcal{H}_D)^\perp$, $Q = H_0(\mathrm{div}^0) \cap H(\mathrm{rot})$, such that*

$$w = \mathrm{rot}\, s$$

and

$$\|s\|_0 \leq C \|\mathrm{rot}\, s\|_0,$$

where $C > 0$ does not depend on s and w.

Proof. By Theorem 3.35 there exists a divergence-free stream function $q' \in H(\mathrm{div}^0) \cap (H^1(\Omega))^3$ (not uniquely determined) such that $w = \mathrm{rot}\, q'$. Since $\langle n \cdot q', 1\rangle_{\partial\Omega} = 0$ by Green's formula (3.98), the weak solution $\psi \in H^1(\Omega)$ of the following Neumann problem

$$\Delta\psi = 0 \quad \text{in } \Omega,$$
$$\frac{\partial\psi}{\partial n} = n \cdot q' \quad \text{on } \partial\Omega$$

exists. Then clearly the function $q = q' - \mathrm{grad}\, \psi$ is from Q and q is also a stream function for w, that is

$$w = \mathrm{rot}\, q.$$

From the orthogonal decomposition in the $(L^2(\Omega))^3$-topology

$$Q = S \oplus \mathcal{H}_D, \tag{3.121}$$

we get $q = s + h$, where $s \in S$ and $h \in \mathcal{H}_D$. Hence, from the definition of $\mathcal{H}_D$ we obtain $w = \mathrm{rot}\, s$. Note that S is the orthogonal complement of $\mathcal{H}_D$ in Q also in the $H(\mathrm{rot})$-topology (see (3.105)), since $\mathrm{rot}\, h = 0$ for all $h \in \mathcal{H}_D$.

Further, let us suppose that $w = \operatorname{rot} s^1 = \operatorname{rot} s^2$ for some $s^1, s^2 \in S$. Then $s^1 - s^2 \in S \cap \mathcal{H}_D$ and (3.121) yields $s^1 - s^2 = 0$, i.e., the stream function $s \in S$ exists and is unique. Therefore, from (3.109) we obtain that the linear operator

$$\operatorname{rot}\colon S \to H(\operatorname{div}^0) \tag{3.122}$$

is bijective. It is obvious that S with the $\|\cdot\|_{H(\mathrm{rot})}$-norm and $H(\operatorname{div}^0)$ with the $\|\cdot\|_0$-norm are Banach spaces. As

$$\|\operatorname{rot} s\|_0 \leq \|s\|_{H(\mathrm{rot})},$$

the linear operator (3.122) is continuous, and we conclude, by Theorem 3.1 on isomorphism, that the inverse operator is continuous, too, and we have

$$\|s\|_0 \leq \|s\|_{H(\mathrm{rot})} \leq C\|\operatorname{rot} s\|_0.$$

□

Proof of Theorem 3.33. Suppose $\partial\Omega$ is not connected and let Γ be one of its components. Let $z \in H^1(\Omega)$ be the weak solution of the problem

$$\begin{aligned} \Delta z &= 0 \quad \text{in } \Omega, \\ z &= 1 \quad \text{on } \Gamma, \\ z &= 0 \quad \text{on } \partial\Omega \setminus \Gamma. \end{aligned}$$

We easily find (cf. (3.108)) that $v = \operatorname{grad} z \in \mathcal{H}_R = H(\operatorname{div}^0) \cap H_0(\operatorname{rot}^0)$ whereas $\|v\|_0 \neq 0$, i.e., (3.117) cannot hold.

Conversely, let $\partial\Omega$ be connected and let $v \in V$ be given. First assume that $d = 3$. Let $z \in H_0^1(\Omega)$ be the weak solution of the problem

$$\begin{aligned} \Delta z &= \operatorname{div} v \quad \text{in } \Omega, \\ z &= 0 \qquad\quad \text{on } \partial\Omega. \end{aligned} \tag{3.123}$$

Thus we have

$$\|z\|_1 \leq C_1\|\operatorname{div} v\|_0. \tag{3.124}$$

From the weak formulation of the problem (3.123) and Green's formula (3.98), and from (3.108) we see that $\operatorname{grad} z \in H(\operatorname{div}) \cap H_0(\operatorname{rot}^0)$, i.e., by (3.123) we get

$$w = v - \operatorname{grad} z \in H(\operatorname{div}^0) \cap H_0(\operatorname{rot}). \tag{3.125}$$

According to Lemma 3.36, there exists a stream function $s \in S \subset H(\operatorname{rot})$ such that

$$w = \operatorname{rot} s \in H_0(\operatorname{rot}),$$

and thus

$$v = \operatorname{grad} z + \operatorname{rot} s. \tag{3.126}$$

Applying now Lemmas 3.34 and 3.36, we come to

$$\begin{aligned} \|\operatorname{rot} s\|_0^2 &= (\operatorname{rot} s, \operatorname{rot} s)_0 = (s, \operatorname{rot}\operatorname{rot} s)_0 \\ &\le \|s\|_0 \|\operatorname{rot}\operatorname{rot} s\|_0 \le C_2 \|\operatorname{rot} s\|_0 \|\operatorname{rot}\operatorname{rot} s\|_0. \end{aligned} \tag{3.127}$$

Thus by (3.125), (3.126), (3.127), (3.124) and (3.107) we observe that (3.117) holds,

$$\begin{aligned} \|v\|_0 &\le \|\operatorname{grad} z\|_0 + \|\operatorname{rot} s\|_0 \le \|z\|_1 + C_2\|\operatorname{rot}\operatorname{rot} s\|_0 \\ &\le C_1\|\operatorname{div} v\|_0 + C_2\|\operatorname{rot} w\|_0 \le C(\|\operatorname{div} v\|_0 + \|\operatorname{rot} v\|_0). \end{aligned}$$

The case $d = 2$ is more trivial. Similarly as in (3.125) and (3.126) we can prove that (cf. also [Girault, Raviart, p. 41]) for any $v \in V$

$$v = \operatorname{grad} z + \operatorname{curl} s, \tag{3.128}$$

where $z \in H_0^1(\Omega)$ is the weak solution of (3.123) and $s \in H^1(\Omega)$ is one of the weak solutions of the Neumann problem

$$\begin{aligned} -\Delta s &= \operatorname{rot} v \quad && \text{in } \Omega, \\ \frac{\partial s}{\partial n} &= 0 && \text{on } \partial\Omega, \end{aligned}$$

where $(\operatorname{rot} v, 1)_0 = 0$ by Green's formula (3.98), since $\langle n \cdot \operatorname{grad} s, 1\rangle_{\partial\Omega} = 0$. Hence, $\|\operatorname{grad} z\|_0 \le C_1\|\operatorname{div} v\|_0$ and $\|\operatorname{curl} s\|_0 \le C_2\|\operatorname{rot} v\|_0$. This together with (3.128) yields (3.117). □

Corollary 3.37. *Let $\Omega \in \mathcal{L}$, $\Omega \subset \mathbf{R}^d$, $d \in \{2,3\}$ and let $\partial\Omega$ be connected. Then the problem (3.116) has exactly one solution $u \in V$. Moreover, the least squares functional*

$$J(v) = \|\operatorname{div} v - f\|_0^2 + \|\operatorname{rot} v - g\|_0^2$$

attains its unique minimum over the space V at u.

Proof. By Theorem 3.33 the bilinear form $a(\cdot,\cdot)$ is a scalar product on V and thus it is V-elliptic. The existence of a unique solution $u \in V$ then follows from the Lax–Milgram Lemma 3.3.

By Theorem 3.4 the problem (3.116) is equivalent to minimization of the functional J over V, since

$$\begin{aligned} J(v) &= (\operatorname{div} v, \operatorname{div} v)_0 - 2(f, \operatorname{div} v)_0 + (f,f)_0 \\ &\quad + (\operatorname{rot} v, \operatorname{rot} v)_0 - 2(g, \operatorname{rot} v)_0 + (g,g)_0 \\ &= a(v,v) - 2F(v) + \text{const}. \end{aligned}$$

□

Remark 3.38. For a bounded convex domain or for a domain with a smooth boundary which is homeomorphic to a ball, we even have (see, e.g., [Křížek, Neittaanmäki, 1984a], [Saranen, 1982])

$$\|v\|_1 \leq C(\|\operatorname{div} v\|_0 + \|\operatorname{rot} v\|_0) \quad \forall v \in V.$$

Remark 3.39. The spaces $\mathcal{H}_D$ and $\mathcal{H}_R$ are finite-dimensional (see, e.g., [Picard, 1983]). From Theorem 3.33 we see that $\mathcal{H}_R$ is trivial if and only if $\partial\Omega$ is connected. The proof of the inequality (3.117) can be modified for $v \in V \cap (\mathcal{H}_R)^\perp$ without any assumptions on the connectivity of $\partial\Omega$.

Next consider the stationary Maxwell equations with the magnetic boundary conditions

$$\operatorname{div} u = f \quad \text{in } \Omega, \tag{3.129}$$

$$\operatorname{rot} u = g \quad \text{in } \Omega, \tag{3.130}$$

$$n \cdot u = 0 \quad \text{on } \partial\Omega, \tag{3.131}$$

where $\Omega \in \mathcal{L}$ and $\Omega \subset \mathbf{R}^d$ for $d \in \{2, 3\}$. By (3.129), Green's formula (3.98) and (3.131) we find that

$$(f, 1)_0 = (\operatorname{div} u, 1)_0 = 0. \tag{3.132}$$

Moreover, from (3.109) we have

$$g \in H(\operatorname{div}^0) \quad \text{for } d = 3. \tag{3.133}$$

The relations (3.132) and (3.133) represent necessary conditions for solvability of (3.129)–(3.131).

Remark 3.40. Let $\partial\Omega$ be not connected and let $\partial\Omega_1, \ldots, \partial\Omega_k$ be the components of $\partial\Omega$. For $v \in H(\operatorname{div})$ define the functional $n \cdot v \in H^{-1/2}(\partial\Omega_i)$, $i \in \{1, \ldots, k\}$, by

$$\langle n \cdot v, z \rangle_{\partial\Omega_i} = (\operatorname{div} v, z)_0 + (v, \operatorname{grad} z)_0, \quad z \in Z_i, \tag{3.134}$$

where

$$Z_i = \{z \in H^1(\Omega) \mid z = 0 \text{ on } \partial\Omega_j \quad \forall j \in \{1, \ldots, k\} \setminus \{i\}\}$$

and $\langle \cdot, \cdot \rangle_{\partial\Omega_i}$ is the duality pairing $H^{-1/2}(\partial\Omega_i)$ and $H^{1/2}(\partial\Omega_i)$. Let us choose now $i \in \{1, \ldots, k\}$ arbitrarily and let $\eta \in C^\infty(\overline{\Omega})$ be such that $\eta = 1$ in a neighbourhood of $\partial\Omega_i$ and $\eta = 0$ in some neighbourhoods of the other components $\partial\Omega_j$, $j \neq i$. That is $\eta \in Z_i$. If $d = 3$ then the relations (3.130), (3.134), (3.109), (3.106) and (3.107) imply

$$\begin{aligned}\langle n \cdot g, 1 \rangle_{\partial\Omega_i} &= \langle n \cdot \operatorname{rot} u, 1 \rangle_{\partial\Omega_i} = (\operatorname{rot} u, \operatorname{grad} \eta)_0 \\ &= \langle n \wedge u, \operatorname{grad} \eta \rangle_{\partial\Omega} = -\langle n \wedge \operatorname{grad} \eta, u \rangle_{\partial\Omega} = 0\end{aligned}$$

for $i = 1, \ldots, k$, which is also a necessary condition for solvability of (3.129)–(1.131).

The least squares formulation of the problem (3.129)–(1.131) reads: Find $u \in V$ such that (cf. (3.116))

$$a(u, v) = F(v) \quad \forall v \in V, \tag{3.135}$$

where

$$V = H_0(\operatorname{div}) \cap H(\operatorname{rot}). \tag{3.136}$$

Definition 3.41. *A domain $\Omega \subset \mathbf{R}^d$, $d \in \{1, 2, ...\}$, is said to be simply connected if it has the following property: Given any simple closed curve $\gamma : x = h(t)$, $t \in [a, b]$, with range in Ω, there exists a continuous function $x = G(r, t)$ defined for $r \in [0, 1]$ and $t \in [a, b]$ such that*

(i) *$G(0, t) = h(t)$ for all $t \in [a, b]$;*

(ii) *$G(1, t) = P$ for all $t \in [a, b]$, where P is some point in Ω;*

(iii) *$G(r, t)$ lies in Ω for all $r \in [0, 1]$ and $t \in [a, b]$;*

(iv) *$G(r, a) = G(r, b)$ for all $r \in [0, 1]$.*

Defining (closed) curves γ_r by $x = G(r, t)$, $t \in [a, b]$, we say that the family $\{\gamma_r\}$ represents a continuous deformation of γ into a point P.

Domains which are not simply connected are called multiply connected.

Remark 3.42. A ball in $\mathbf{R}^d$ is clearly simply connected. The domain

$$\Omega = \{x \in \mathbf{R}^d \mid 1 < \|x\| < 2\}$$

is simply connected for $d = 3$ whereas it is multiply (doubly) connected for $d = 2$. A multiply connected domain in $\mathbf{R}^3$ is for instance a torus (annuloid). Note that any domain with a connected boundary is homeomorphic to a ball with handles, see, e.g., [Moise, p. 161] for details. Domains with no handles are simply connected, domains with one handle are doubly connected, etc. The torus is thus doubly connected.

The solvability of (3.135) is again based on the Lax–Milgram Lemma 3.3 and the following theorem.

Theorem 3.43. *Let $\Omega \in \mathcal{L}$ and $\Omega \subset \mathbf{R}^d$ for $d \in \{2, 3\}$. Then*

$$\|v\|_0 \leq C(\|\operatorname{div} v\|_0 + \|\operatorname{rot} v\|_0) \quad \forall v \in V = H_0(\operatorname{div}) \cap H(\operatorname{rot}) \tag{3.137}$$

if and only if Ω is simply connected.

The proof is similar to that of Theorem 3.33 (the case $d = 3$ is proved in [Křížek, Neittaanmäki, 1985]).

Example 3.44. To show that (3.137) does not hold on multiply connected domains, consider the torus Ω whose boundary is generated by the circle $x_2 = 0$, $(x_1 - 2)^2 + x_3^2 = 1$, rotating about the axis x_3 (see Figure 3.8). Setting

$$v(x_1, x_2, x_3) = \left(\frac{-x_2}{x_1^2 + x_2^2}, \frac{x_1}{x_1^2 + x_2^2}, 0\right)^T \quad \text{for } (x_1, x_2, x_3)^T \in \Omega,$$

we easily find that $\operatorname{div} v = 0$, $\operatorname{rot} v = 0$ in Ω and $n \cdot v = 0$ on $\partial\Omega$, i.e., (3.137) does not hold.

Analogously to Remark 3.31, we can construct another function v on Ω such that $\operatorname{rot} v = 0$ and there is no potential $z \in H^1(\Omega)$ for which $v = \operatorname{grad} z$.

Remark 3.45. From Theorem 3.43 we see that the space $\mathcal{H}_D$ is trivial if and only if the domain Ω is simply connected. When Ω is multiply connected, the bilinear form $a(\cdot, \cdot)$ is a scalar product on $V \cap (\mathcal{H}_D)^\perp$. Thus the solution of (3.135) exists and is unique apart from a function of $\mathcal{H}_D$.

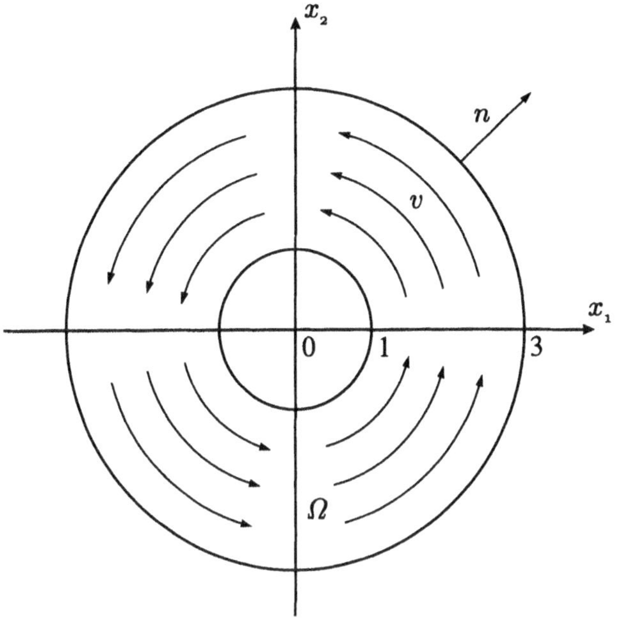

Figure 3.8.

Corollary 3.46. *Let $\Omega \in \mathcal{L}$ and $d \in \{2,3\}$. Then*

$$(L^2(\Omega))^d = H(\mathrm{div}^0) \oplus H_0(\mathrm{rot}^0) \quad \Leftrightarrow \quad \partial\Omega \textit{ is connected},$$
$$(L^2(\Omega))^d = H_0(\mathrm{div}^0) \oplus H(\mathrm{rot}^0) \quad \Leftrightarrow \quad \Omega \textit{ is simply connected},$$

where $\oplus$ denotes the orthogonal decomposition in the $(L^2(\Omega))^d$-topology.

The proof follows from Green's formulae (3.98), (3.106) and Theorems 3.33 and 3.43. For the first equivalence see formulae (3.126) if $d = 3$ and (3.128) if $d = 2$.

Remark 3.47. There are other orthogonal decompositions of $(L^2(\Omega))^d$, $d \in \{1, 2, ...\}$, which are independent of the topology of Ω, for example, (see [Girault, Raviart])

$$(L^2(\Omega))^d = H_0(\mathrm{div}^0) \oplus \{\mathrm{grad}\, z \mid z \in H^1(\Omega)\}.$$

Remark 3.48. For any $\Omega \in \mathcal{L}$ we have (cf., e.g., [Křížek, Neittaanmäki, 1984a, p. 22])

$$(H_0^1(\Omega))^2 = H_0(\mathrm{div}) \cap H_0(\mathrm{rot}) \quad \text{for } d = 2.$$

4. Finite elements

In this chapter we shall briefly describe some basic principles of finite element approximations, and then, in Sections 4.3–4.6, we introduce several special but important themes. For a thorough presentation of the standard finite element method we refer to [Akin], [Axelsson, Barker], [Ciarlet], [Ciarlet, Lions], [Johnson], [Mitchell, Wait], [Silvester, Ferrari], [Strang, Fix], [White], [Zienkiewicz], [Zienkiewicz, Morgan].

4.1. The main idea of the finite element method

For simplicity let us restrict ourselves to solution of the Poisson equation

$$-\Delta u = f \quad \text{in } \Omega \tag{4.1}$$

with the homogeneous Dirichlet boundary condition

$$u = 0 \quad \text{on } \partial\Omega, \tag{4.2}$$

where Ω is a polyhedral domain in $\mathbf{R}^d$ ($d \geq 1$) with the boundary $\partial\Omega$, $\Delta u = \partial_{11}u + ... + \partial_{dd}u$ and $f \in L^2(\Omega)$ is a given function. The analytical solution of this problem is not known, in general, and thus we have to employ an approximate method. The finite element method (FEM) for the solution of the problem (4.1)-(4.2) is based on its weak (variational) formulation. To this end we introduce the space (cf. Section 3.3)

$$V = H_0^1(\Omega) = \{v \in H^1(\Omega) \mid v = 0 \text{ on } \partial\Omega\}.$$

Multiplying (4.1) by an arbitrary function $v \in V$ and then integrating over Ω, we get, by (4.2) and Green's Theorem 3.10,

$$(\operatorname{grad} u, \operatorname{grad} v)_0 = (f, v)_0 \quad \forall v \in V, \tag{4.3}$$

where $(\cdot\,,\cdot)_0$ is the standard scalar product on $(L^2(\Omega))^r$, $r = 1, 2, \dots$. By Theorem 3.17 we already know that there exists precisely one weak solution $u \in V$ satisfying (4.3).

The infinite dimensional problem (4.3) will be now replaced by a finite dimensional problem with the help of the Galerkin method. Let $V_h \subset V$ be a non-empty finite dimensional subspace and let us look for $u_h \in V_h$ so that

$$(\operatorname{grad} u_h, \operatorname{grad} v_h)_0 = (f, v_h)_0 \quad \forall v_h \in V_h. \tag{4.4}$$

Since V_h is evidently a Hilbert space, we may again use Theorem 3.3 to see that there is precisely one solution $u_h \in V_h$.

Seeking $u_h \in V_h$ as a linear combination of basis functions $v^1, ..., v^N$ of the space V_h ($N = \dim V_h$),

$$u_h = \sum_{j=1}^{N} c_j v^j, \tag{4.5}$$

then by (4.4) and (4.5), we get a system of linear algebraic equations for unknowns $c_1, ..., c_N \in \mathbf{R}^1$,

$$\sum_{j=1}^{N} (\operatorname{grad} v^i, \operatorname{grad} v^j)_0 \, c_j = (f, v^i)_0, \quad i = 1, ..., N, \tag{4.6}$$

whose matrix is symmetric and positive definite. However, this matrix is full, in general.

The finite element method can be characterized as a special case of the Galerkin method, in which the space V_h and its basis functions are chosen so that the matrix

$$A = \left((\operatorname{grad} v^i, \operatorname{grad} v^j)_0 \right)_{i,j=1}^{N} \tag{4.7}$$

of the system (4.6) is sparse, i.e., only $\mathcal{O}(N)$ entries of A are different from zero. This enables us to solve the system (4.6) much more quickly and to store the matrix (4.7) with less computer memory than in the case of a full matrix which has $\mathcal{O}(N^2)$ of nonzero entries, in general.

Let us note that the matrix (4.7) in the FEM is traditionally called the *stiffness matrix*. Problems of linear statics, which were solved by the FEM in the first place, lead to matrices that express "stiffness" – the relation between a load and the corresponding deformation of the body.

The sparsity of the stiffness matrix is achieved by the following three steps:

(i) A decomposition (triangulation) $\mathcal{T}_h$ of $\overline{\Omega}$ to simpler closed subdomains is established. These subdomains are called elements and they are usually line segments, triangles, quadrangles, tetrahedra, pentahedra or hexahedra. (In Figure 4.1 we see a decomposition of a 1,2,3-dimensional domain into elements.)

(ii) The space V_h is chosen so that each function $v_h \in V_h$ is of a very simple form (usually polynomial) on each element. Such a space is called the *space of finite elements*.

(iii) A basis $v^1, ..., v^N$ in V_h is established so that all v^i have "small" supports (i.e., the measure of the set, where v^i is not zero, must be small when compared with the measure of Ω).

To illustrate steps (i)-(iii) we introduce the most common space of continuous and piecewise linear functions for $d = 2$. In this case the triangulation $\mathcal{T}_h$ will be a finite set of triangles such that

$$\overline{\Omega} = \bigcup_{K \in \mathcal{T}_h} K$$

Figure 4.1.

and that any two different triangles from $\mathcal{T}_h$ have just one side or just one vertex in common or they are disjoint. The symbol h (the so-called discretization parameter) characterizes the maximum size of triangles in the triangulation $\mathcal{T}_h$.

To fulfill requirement (ii), we set

$$V_h = \{v_h \in V \mid v_h|_K \in P_1(K) \quad \forall K \in \mathcal{T}_h\},$$

where $P_1(K)$ is the space of linear polynomials over the triangle K. Note that functions from V_h are continuous (cf. Theorem 4.1). Figure 4.2 shows an example of such a function $v_h \in V_h$ over a triangulated domain.

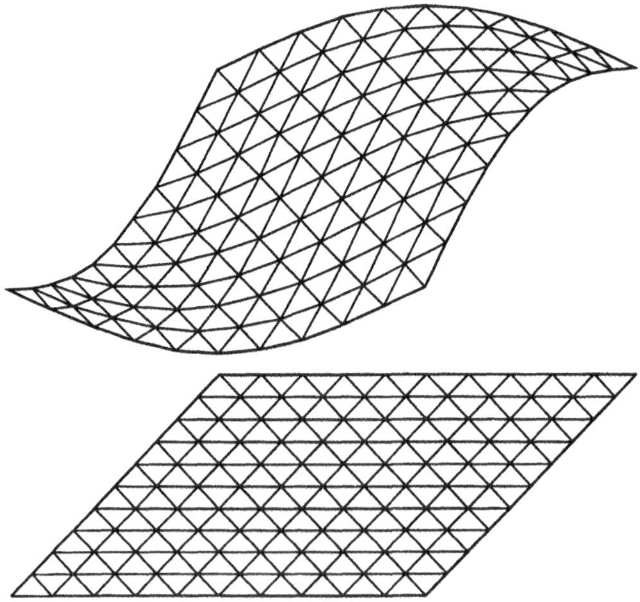

Figure 4.2.

To satisfy the most important assumption (iii) we denote by $B_1, ..., B_N \in \Omega$ the internal nodes (i.e., vertices of triangles) in the triangulation. Then there exists a basis $v^1, ..., v^N$ of the space V_h such that

$$v^i(B_j) = \delta_{ij} = \begin{cases} 1 & \text{if } i = j, \\ 0 & \text{if } i \neq j, \end{cases} \quad \text{(Kronecker's symbol).} \tag{4.8}$$

These functions satisfy (iii) and are called the Courant (or Lagrange) basis functions. A typical shape of v^i corresponding to the node B_i for $d = 2$ is sketched in Figure 4.3. In this case the support of v^i is a hexagon composed of six triangles. We see that if B_i and B_j $(i \neq j)$ are not vertices of the same triangle then

$$(\operatorname{grad} v^i, \operatorname{grad} v^j)_0 = 0, \tag{4.9}$$

which guarantees the sparsity of the stiffness matrix (4.7). Let us number the nodes B_i in Figure 4.3 row-wise. Then from (4.7) and (4.8) we can easily derive (see, e.g., [Křížek, Neittaanmäki, 1990, p. 64] for details) that the stiffness matrix associated to the triangulation of Figure 4.3 is of the form

$$\begin{pmatrix}
4 &&&&&&&&&&&&&&& \\
-1 & 4 &&&&&&&&&&&&& \mathrm{s} & \\
0 & -1 & 4 &&&&&&&&&&&& \mathrm{y} & \\
0 & 0 & -1 & 4 &&&&&&&&&&& \mathrm{m} & \\
-1 & 0 & 0 & 0 & 4 &&&&&&&&&&& \\
0 & -1 & 0 & 0 & -1 & 4 &&&&&&&&&& \\
0 & 0 & -1 & 0 & 0 & -1 & 4 &&&&&&&&& \\
0 & 0 & 0 & -1 & 0 & 0 & -1 & 4 &&&&&&&& \\
0 & 0 & 0 & 0 & -1 & 0 & 0 & 0 & 4 &&&&&&& \\
0 & 0 & 0 & 0 & 0 & -1 & 0 & 0 & -1 & 4 &&&&&& \\
0 & 0 & 0 & 0 & 0 & 0 & -1 & 0 & 0 & -1 & 4 &&&&& \\
0 & 0 & 0 & 0 & 0 & 0 & 0 & -1 & 0 & 0 & -1 & 4 &&&& \\
0 & 0 & 0 & 0 & 0 & 0 & 0 & 0 & -1 & 0 & 0 & 0 & 4 &&& \\
0 & 0 & 0 & 0 & 0 & 0 & 0 & 0 & 0 & -1 & 0 & 0 & -1 & 4 && \\
0 & 0 & 0 & 0 & 0 & 0 & 0 & 0 & 0 & 0 & -1 & 0 & 0 & -1 & 4 & \\
0 & 0 & 0 & 0 & 0 & 0 & 0 & 0 & 0 & 0 & 0 & -1 & 0 & 0 & -1 & 4
\end{pmatrix}$$

Moreover, from relations (4.5) and (4.8) we have

$$u_h(B_i) = \sum_{j=1}^{N} c_j v^j(B_i) = c_i, \quad i = 1, ..., N, \tag{4.10}$$

i.e., solving the system (4.6), we directly get the values of the approximate solution u_h at the nodal points.

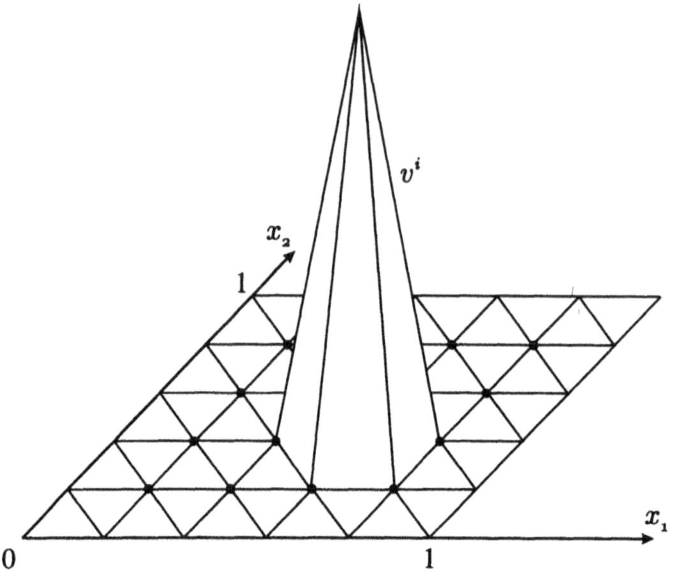

Figure 4.3.

4.2. Finite element spaces

A finite element approximation of the problem (3.58)–(3.59) consists of finding u_h such that

$$u_h - \overline{u} \in V_h$$

and

$$a(v_h, u_h) = F(v_h) \quad \forall v_h \in V_h, \tag{4.11}$$

where

$$V_h = \{v_h \in V \mid v_h|_K \in P_K \quad \forall K \in \mathcal{T}_h\}$$

is a finite element space and where P_K is some finite dimensional space of functions defined over the element K. Note that u_h need not be in V_h. Anyway, analogously to (3.61), the above problem can be equivalently rewritten as: Find $u_h^0 \in V_h$ such that

$$a(v_h, u_h^0) = F(v_h) - a(v_h, \overline{u}) \quad \forall v_h \in V_h.$$

By Theorem 3.3, we obtain a unique solution u_h^0 and then we set $u_h = u_h^0 + \overline{u}$.

In this section, we briefly describe how to construct subspaces V_h of V. Then (4.11) is called the conforming finite element method.

First we establish a *triangulation* $\mathcal{T}_h$ over the set $\overline{\Omega}$. We subdivide $\overline{\Omega}$ into a finite number of subsets (called *elements*) so that the following assumptions hold:

(1) $\overline{\Omega} = \bigcup_{K \in \mathcal{T}_h} K$,

(2) for each $K \in \mathcal{T}_h$, the set K is closed and its interior K^0 is non-empty,

(3) for each distinct $K_1, K_2 \in \mathcal{T}_h$, we have $K_1^0 \cap K_2^0 = \emptyset$,

(4) for each $K \in \mathcal{T}_h$, the boundary ∂K is Lipschitz continuous.

Sometimes the tringulation is also called a decomposition, division, grid, lattice, mesh, net, network, partition, space discretization, subdivision, etc.

Later on we shall usually deal with elements which are convex polygons or polyhedra (then, of course, $\overline{\Omega}$ is a polygon or polyhedron). In this case we add two more assumptions:

(5) any face of any $K \in \mathcal{T}_h$ is either a subset of the boundary $\partial\Omega$, or a face of another element $K' \in \mathcal{T}_h$,

(6) the interior of any face of any $K \in \mathcal{T}_h$ is disjoint with $\overline{\Gamma}_1 \cap \overline{\Gamma}_2$, where Γ_1 and Γ_2 satisfy (3.48).

From assumption (5) we observe that the situation of Figure 4.4 is not allowed. The assumption (6) is called the *consistency condition* for mixed boundary conditions (see Figure 4.5). Note that triangulations which do not satisfy (5) are called nonmatching. They are commonly used in domain decomposition finite element methods which will be not treated in this book.

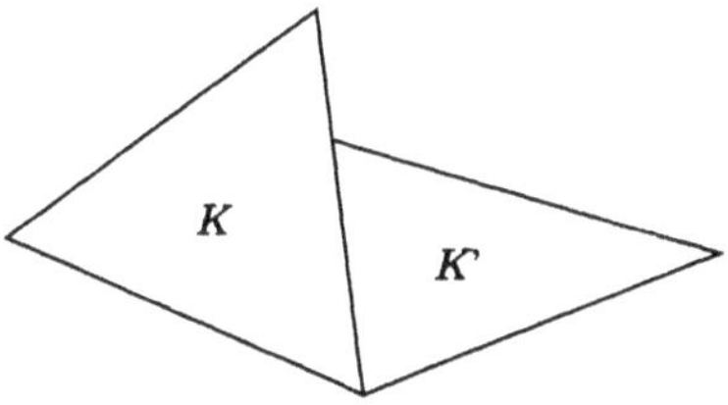

Figure 4.4.

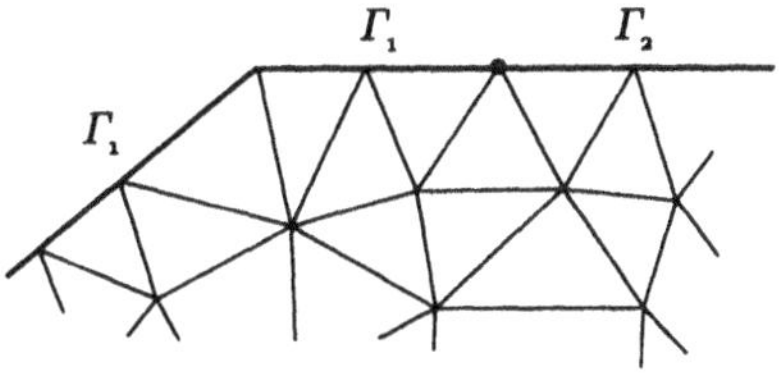

Figure 4.5.

Although elements are closed, we shall write for simplicity $H^k(K)$ instead of $H^k(K^0)$ throughout the whole book. By [Schulte, p. 902], convex compact sets which tile the space $\mathbf{R}^d$ are convex polyhedra. Hence, convex elements $K \in \mathcal{T}_h$ which have at most one common point with the boundary $\partial\Omega$ are convex polyhedra in $\mathbf{R}^d$. Elements having at least two common points with $\partial\Omega$ can have curved boundary, in general. Next we introduce an important theorem.

Theorem 4.1. *Let $\mathcal{T}_h$ be a triangulation of $\Omega \subset \mathbf{R}^d$ formed by convex elements. Let V_h be a subspace of $L^2(\Omega)$ such that the space*

$$P_K = \{v_h|_K \mid v_h \in V_h\}$$

consists of polynomial functions for any $K \in \mathcal{T}_h$. Then

$$V_h \subset H^1(\Omega) \Longleftrightarrow V_h \subset C(\overline{\Omega}),$$

i.e., a piecewise polynomial function is from $H^1(\Omega)$ if and only if it is continuous.

For the proof see [Křížek, Neittaanmäki, 1990, p. 27]. The fact that $v_h \in V_h \subset H^1(\Omega)$ is continuous means, more precisely, that the equivalence class of v_h contains a continuous function among all elements.

To achieve the continuity of functions from V_h, the concept of finite elements is employed. The formal definition of a *finite element* is a triple (K, P_K, Σ_K), where K is a closed subset of $\mathbf{R}^d$ with a nonempty interior and a Lipschitz continuous boundary, P_K is an m-dimensional space of functions defined over K and Σ_K is a set of m linearly independent linear forms defined over a space which contains P_K. These forms are called *degrees of freedom* (or parameters) of the finite element.

In this book we use especially the *Lagrange finite elements* for which the associated space P_K is formed by polynomials over a convex set $K \subset \mathbf{R}^d$. Their degrees of freedom are of the form $p \mapsto p(B)$ for $p \in P_K$ and $B \in K$. The point B is called a *node*. Suppose now that $B_1, ..., B_m$ are given nodes in K. Then the Lagrange finite element (K, P_K, Σ_K) is said to be P_K-*unisolvent*, if for any given $(\alpha_1, ..., \alpha_m) \in \mathbf{R}^m$, $m = \dim P_K$, there exists just one polynomial $p \in P_K$ such that

$$p(B_i) = \alpha_i, \quad i = 1, ..., m.$$

In Figure 4.6, we see an example of a quadratic tetrahedral finite element with 10 degrees of freedom.

If we want to joint two adjacent polygonal or polyhedral elements $(K_1, P_{K_1}, \Sigma_{K_1})$ and $(K_2, P_{K_2}, \Sigma_{K_2})$, we moreover need that

$$\{p^1|_S \mid p^1 \in P_{K_1}\} = \{p^2|_S \mid p^2 \in P_{K_2}\}, \tag{4.12}$$

where $S = K_1 \cap K_2$ is their common face. The nodes must be chosen so that if the nodes of both the elements on S coincide (see Figure 4.7), the equality (4.12) is achieved (cf. (4.13)).

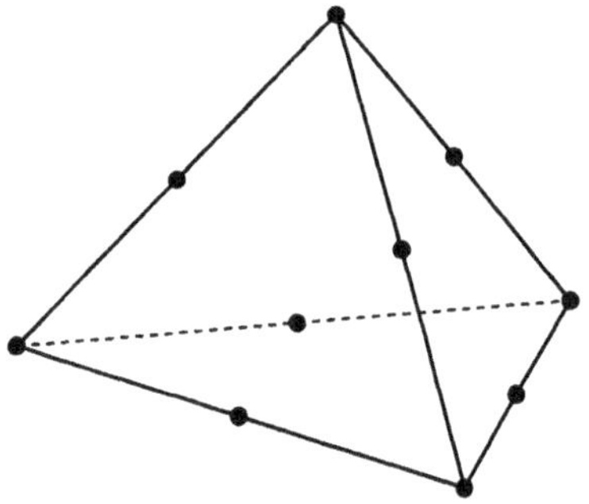

Figure 4.6.

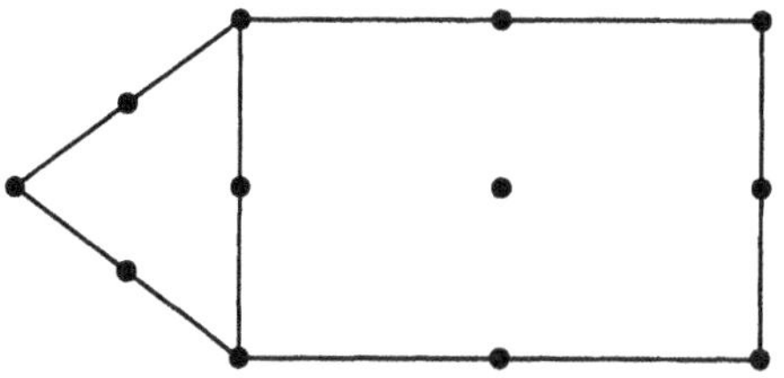

Figure 4.7.

Denote by P_S the space of polynomial functions from P_K restricted to the face S of K, and let $\mathcal{N}_S$ be the set of nodes on S. We, moreover, assume that

$$p \in P_S, \quad p(B) = 0 \quad \forall B \in \mathcal{N}_S \implies p = 0 \quad \text{on } S. \tag{4.13}$$

Set

$$\mathcal{N}_h = \bigcup_{K \in \mathcal{T}_h} \mathcal{N}_K,$$

where, for each finite element (K, P_K, Σ_K), the symbol $\mathcal{N}_K$ denotes the set of nodes. The associated finite element space is given by

$$X_h = \{v_h \in C(\overline{\Omega}) \mid v_h|_K \in P_K \quad \forall K \in \mathcal{T}_h\}.$$

Therefore, a function v_h from the space X_h is uniquely determined by the set

$$\Sigma_h = \{v_h(B) \mid B \in \mathcal{N}_h\},$$

which is called the *set of degrees of freedom of* X_h.

According to Theorem 4.1 and the consistency condition (6), we define a finite element subspace of the space $V = \{v \in H^1(\Omega) \mid v = 0 \text{ on } \Gamma_1\}$ by

$$V_h = \{v_h \in X_h \mid v_h(B) = 0 \quad \text{for } B \in \mathcal{N}_h \cap \overline{\Gamma}_1\}.$$

This is really a subspace of V, since $v_h = 0$ on Γ_1 due to (4.13).

To guarantee the requirement (iii) from Section 4.1 we choose the basis $\{v^i\}_{i=1}^N \subset V_h$ like in (4.8) so that $v^i(B_j) = \delta_{ij}$, $i, j = 1, \ldots, N$, where $N = \dim V_h$, $B_i \in \mathcal{N}_h$ and $B_i \notin \overline{\Gamma}_1$. Two procedures for assembling the associated stiffness matrix

$$A = (a(v^i, v^j))_{i,j=1}^N$$

are described in [Křížek, Neittaanmäki, 1990]. Numerical quadrature formulae for evaluation of the integrals $a(v^i, v^j)$ are also presented there.

4.3. A posteriori error estimates

In many theoretical books on the finite element method we meet the so-called *a priori error estimates* which are usually of the form (see Remark 4.35, or [Ciarlet], [Ciarlet, Lions], [Johnson], [Strang, Fix], etc.),

$$\|u - u_h\|_1 \leq Ch^k \|u\|_{k+1}, \quad k \in \{1, 2, \ldots\}.$$

They show the asymptotic behaviour of the error for $h \to 0$. Anyway, we often do not know a realistic bound for the constant C, i.e., we are not able to establish "how far" is the finite element solution u_h from the true solution u for a given triangulation.

The aim of this section is to introduce the so-called *a posteriori error estimates*, which give us some information about the error after computation. To this end we will apply the primal and dual finite element methods for solving the problem (3.44)–(3.46),

$$\begin{aligned} -\operatorname{div}(\mathcal{A} \operatorname{grad} u) &= f \quad \text{in } \Omega, \\ u &= \overline{u} \quad \text{on } \Gamma_1, \\ n^T \mathcal{A} \operatorname{grad} u &= g \quad \text{on } \Gamma_2, \end{aligned} \tag{4.14}$$

with $\mathcal{A} = (a_{ij})$ symmetric, $\Gamma_1 \neq \emptyset$ and $\Gamma_2 \neq \emptyset$.

The primal and dual variational formulation of (4.14) is given in Section 3.7 and we keep the notation introduced there; in particular,

$$a(v, w) = (\mathcal{A} \operatorname{grad} v, \operatorname{grad} w)_0$$

and

$$F(v) = \int_\Omega fv\, dx + \int_{\Gamma_2} gv\, ds.$$

As we shall see later the simultaneous use of primal and dual (equilibrium) finite element methods enables us to obtain a posteriori error estimates, two-sided bounds of energy and, moreover, we can apply the hypercircle method [Synge].

First let us equip the space $(L^2(\Omega))^d$ with the scalar product

$$b(q, q') = (\mathcal{A}^{-1} q, q')_0, \quad q, q' \in (L^2(\Omega))^d. \tag{4.15}$$

According to [Haslinger, Hlaváček, p. 46], the spaces

$$Q = \{q \in (L^2(\Omega))^d \mid (q, \operatorname{grad} v)_0 = 0 \quad \forall v \in V\}, \tag{4.16}$$
$$Q' = \{q' \in (L^2(\Omega))^d \mid \exists v \in V : q' = \mathcal{A} \operatorname{grad} v\} \tag{4.17}$$

are closed subspaces of $(L^2(\Omega))^d$. By Sections 3.6 and 3.7, we have $Q \subset H(\operatorname{div}^0)$. Now we will prove that Q' is the orthogonal complement of Q in $(L^2(\Omega))^d$ with respect to $b(\cdot\,,\cdot)$.

Lemma 4.2. $(L^2(\Omega))^d = Q \oplus Q'$.

Proof. Let $q \in Q$ and $q' \in Q'$ be given. Then there exists $v \in V$ such that $q' = \mathcal{A} \operatorname{grad} v$ and we have

$$b(q, q') = (\mathcal{A}^{-1} q, \mathcal{A} \operatorname{grad} v)_0 = (q, \operatorname{grad} v)_0 = 0. \tag{4.18}$$

Denote by $(Q')^\perp$ the orthogonal complement of Q' in $(L^2(\Omega))^d$ with respect to $b(\cdot\,,\cdot)$. Referring to (4.18), we see that $Q \perp Q'$, and thus $Q \subset (Q')^\perp$.

Conversely, let $q \in (Q')^\perp$. Then

$$b(q, \mathcal{A} \operatorname{grad} v) = 0 \quad \forall v \in V, \tag{4.19}$$

which implies that

$$(q, \operatorname{grad} v)_0 = 0 \quad \forall v \in V,$$

i.e., $q \in Q$. □

The weak solution of a boundary elliptic problem is sometimes not so interesting as its cogradient, e.g., the knowledge of magnetic flux is more important than the corresponding magnetic potential. We can compute the cogradient of u_h by differentiation of u_h and then multiplying by $\mathcal{A}$. Another way is to use directly the dual finite element method.

Let $V_h \subset V$ and $Q_h \subset Q$ be some finite element spaces. A *primal (and dual) finite element method* for the problem (4.14) consists of finding u_h (and p_h) which minimizes the functional (3.99) (and (3.100)) over the set $\overline{u} + V_h$ (and $\overline{p} + Q_h$), respectively. The functions u_h and p_h are called finite element solutions.

Remark 4.3. Let $\Omega \subset \mathbf{R}^2$ be simply connected, $\Omega \in \mathcal{L}$, and let

$$W = \{v \in H^1(\Omega) \mid v = 0 \text{ on } \Gamma_2\}.$$

If $\{w^i\}_{i=1}^m$ is a basis of some finite element space $W_h \subset W$, then the functions $q^i = \operatorname{curl} w^i = (\partial_2 w^i, -\partial_1 w^i)^T$, $i = 1, ..., m$, form a basis in the space

$$Q_h = \operatorname{curl} W_h \subset Q.$$

This inclusion is a consequence of the equality

$$Q = \operatorname{curl} W = \{q \in (L^2(\Omega))^2 \mid \exists w \in W : q = \operatorname{curl} w\},$$

which is derived in [Křížek, 1983]. Its proof is based on the fact that curl: $W \to Q$ is a one-to-one mapping for $\Gamma_2 \neq \emptyset$.

Suppose that $\mathcal{A}$ is constant, for instance. Then, since

$$\mathcal{A}^{-1} = \frac{1}{\det \mathcal{A}} \begin{pmatrix} a_{22} & -a_{12} \\ -a_{21} & a_{11} \end{pmatrix},$$

we have

$$\begin{aligned} b(q^i, q^j) &= (\mathcal{A}^{-1} q^i, q^j)_0 = (\mathcal{A}^{-1} \operatorname{curl} w^i, \operatorname{curl} w^j)_0 \\ &= (\mathcal{A} \operatorname{grad} w^i, \operatorname{grad} w^j)_0 / \det \mathcal{A} = a(w^i, w^j) / \det \mathcal{A}. \end{aligned} \tag{4.20}$$

This relation simplifies assembling the stiffness (and flexibility) matrix of the primal (and dual) finite element method, respectively. If the family $\{W_h\}$ is dense in W in the $\|\cdot\|_1$-norm, then $\{Q_h\}$ is dense in Q in the $\|\cdot\|_0$-norm. For more details about this technique, see [Křížek, 1983]. (The case $d = 3$ is investigated in [Hlaváček, Křížek, 1993b] and [Křížek, Neittaanmäki, 1986].)

Let us now put

$$|||q||| = \sqrt{b(q,q)} \quad \forall q \in (L^2(\Omega))^d. \tag{4.21}$$

This norm is equivalent with the $\|\cdot\|_0$-norm, since $\mathcal{A}^{-1}$ is uniformly positive definite matrix and its entries belong to $L^\infty(\Omega)$ (compare (3.47)), i.e.,

$$c_1 \|q\|_0 \leq |||q||| \leq c_2 \|q\|_0 \quad \forall q \in (L^2(\Omega))^d. \tag{4.22}$$

Theorem 4.4. *(A posteriori error estimates) Let u and p be the solutions of the primal and dual variational formulation of the problem (4.14), respectively (cf. (3.99) and (3.100)). Let u_h and p_h be the corresponding finite element solutions. Then there exist constants $C_1, C_2 > 0$ independent of V_h and Q_h such that*

$$C_1\|u - u_h\|_1 \leq \|\mathcal{A}\,\mathrm{grad}\, u_h - p_h\|_0, \tag{4.23}$$
$$C_2\|p - p_h\|_0 \leq \|\mathcal{A}\,\mathrm{grad}\, u_h - p_h\|_0. \tag{4.24}$$

Proof. The dual variational formulation of (4.14) consists of minimizing the functional (3.100) over the affine manifold $Q(f,g)$, see Section 3.7. So let $q \in Q(f,g)$ be an arbitrary element. Then by (3.102), $\mathcal{A}\,\mathrm{grad}\, u - q \in Q = Q(0,0)$ and clearly $\mathcal{A}\,\mathrm{grad}(u_h - u) \in Q'$.

According to Lemma 4.2, the spaces Q and Q' are $b(\cdot,\cdot)$-orthogonal and thus

$$\begin{aligned} |\!|\!|\mathcal{A}\,\mathrm{grad}\, u_h - q|\!|\!|^2 &= |\!|\!|\mathcal{A}\,\mathrm{grad}\, u_h - \mathcal{A}\,\mathrm{grad}\, u + \mathcal{A}\,\mathrm{grad}\, u - q|\!|\!|^2 \\ &= |\!|\!|\mathcal{A}\,\mathrm{grad}(u_h - u)|\!|\!|^2 + |\!|\!|\mathcal{A}\,\mathrm{grad}\, u - q|\!|\!|^2. \end{aligned} \tag{4.25}$$

Setting $q = p_h$, we see by (3.102) and (4.22) that (4.24) holds.

From (4.25), (4.21), (4.15) and the V-ellipticity of $a(\cdot,\cdot)$, we obtain

$$|\!|\!|\mathcal{A}\,\mathrm{grad}\, u_h - p_h|\!|\!|^2 \geq |\!|\!|\mathcal{A}\,\mathrm{grad}(u - u_h)|\!|\!|^2 = a(u - u_h, u - u_h) \geq C'\|u - u_h\|_1^2.$$

Hence, (4.23) is valid due to (4.22).

Remark 4.5. If we apply both the primal and dual finite element method to solve the problem (4.14), then the right-hand side of inequalities (4.23) and (4.24) is known, and it can be used to estimate the errors. Notice that $C_2 = 1$ in the case of the Laplace operator.

Theorem 4.6. *(The hypercircle method) Under the assumptions of Theorem 4.4 we have*

$$|\!|\!|\mathcal{A}\,\mathrm{grad}\, u - \tfrac{1}{2}(\mathcal{A}\,\mathrm{grad}\, u_h + p_h)|\!|\!| = \tfrac{1}{2}|\!|\!|\mathcal{A}\,\mathrm{grad}\, u_h - p_h|\!|\!|. \tag{4.26}$$

Proof. By Lemma 4.2, the vectors $p - p_h \in Q$ and $\mathcal{A}\,\mathrm{grad}(u - u_h) \in Q'$ are $b(\cdot,\cdot)$-orthogonal. Therefore, from (3.102) we get

$$\begin{aligned} &|\!|\!|2\mathcal{A}\,\mathrm{grad}\, u - \mathcal{A}\,\mathrm{grad}\, u_h - p_h|\!|\!|^2 = |\!|\!|\mathcal{A}\,\mathrm{grad}\, u - \mathcal{A}\,\mathrm{grad}\, u_h + \mathcal{A}\,\mathrm{grad}\, u - p_h|\!|\!|^2 \\ &= |\!|\!|\mathcal{A}\,\mathrm{grad}(u - u_h)|\!|\!|^2 + |\!|\!|\mathcal{A}\,\mathrm{grad}\, u - p_h|\!|\!|^2 = |\!|\!|\mathcal{A}\,\mathrm{grad}\, u_h - p_h|\!|\!|^2. \end{aligned}$$

□

Remark 4.7. The cogradient $\mathcal{A}\,\mathrm{grad}\, u$ of the true solution u lies on the hypercircle (see Figure 4.8) with center $s_h = \frac{1}{2}(\mathcal{A}\,\mathrm{grad}\, u_h + p_h)$ and radius $\frac{1}{2}|\!|\!|\mathcal{A}\,\mathrm{grad}\, u_h - p_h|\!|\!|$. If $\frac{1}{2}(\mathcal{A}\,\mathrm{grad}\, u_h + p_h)$ is considered as an approximate solution, then the error is exactly equal to the radius of the hypercircle (see (4.26)).

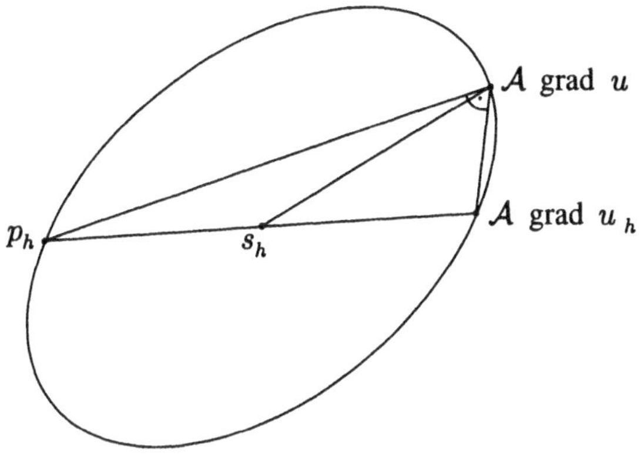

Figure 4.8.

Theorem 4.8. *(Two-sided energy estimates) Under the assumptions of Theorem 4.4 we have*

$$-I(p_h) - F(\overline{u}) \leq J(u) \leq J(u_h), \tag{4.27}$$

and

$$a(u_h - \overline{u}, u_h - \overline{u}) \leq a(u - \overline{u}, u - \overline{u}) \leq \|p_h - \mathcal{A}\,\mathrm{grad}\,\overline{u}\|. \tag{4.28}$$

Proof. Since $V_h \subset V$ and $Q_h \subset Q$, we see that $\overline{u}+V_h \subset \overline{u}+V$ and $\overline{p}+Q_h \subset \overline{p}+Q$, and thus, by (3.103),

$$-I(p_h) - F(\overline{u}) \leq -I(p) - F(\overline{u}) = J(u) \leq J(u_h).$$

Hence, (4.27) holds.

Set

$$w = u - \overline{u} \in V \quad \text{and} \quad w_h = u_h - \overline{u} \in V.$$

From the symmetry of $a(\cdot\,,\cdot)$, (3.59) and (4.11) we get the well-known *orthogonality condition*

$$a(u - u_h, v_h) = 0 \quad \forall v_h \in V_h,$$

which implies that $a(w - w_h, w_h) = 0$. Consequently,

$$\begin{aligned} 0 &\leq a(w - w_h, w - w_h) = a(w,w) - 2a(w, w_h) + 2a(w_h, w_h) - a(w_h, w_h) \\ &= a(w,w) - a(w_h, w_h), \end{aligned}$$

and the first inequality in (4.28) is proved.

Referring to (3.99) and (3.59), we get

$$\begin{aligned} J(u) &= \frac{1}{2}a(u,u) - F(\overline{u}) - F(w) = \frac{1}{2}a(u,u) - F(\overline{u}) - a(u,w) \\ &= \frac{1}{2}a(u, u - 2w) - F(\overline{u}) = \frac{1}{2}a(\overline{u} + w, \overline{u} - w) - F(\overline{u}) \\ &= \frac{1}{2}(a(\overline{u},\overline{u}) - a(w,w)) - F(\overline{u}). \end{aligned} \tag{4.29}$$

On the other hand, by (4.27), we find that

$$-J(u) - F(\overline{u}) \leq I(p_h).$$

From here and (4.29) it follows that

$$\frac{1}{2}(a(w,w) - a(\overline{u},\overline{u})) \leq I(p_h),$$

where I is defined by (3.100). Now taking into account that

$$a(\overline{u},\overline{u}) = |\!|\!|\mathcal{A}\operatorname{grad}\overline{u}|\!|\!|^2, \tag{4.30}$$

we obtain the second inequality of (4.28),

$$\begin{aligned} a(w,w) &\leq 2I(p_h) + a(\overline{u},\overline{u}) = |\!|\!|p_h|\!|\!|^2 - 2b(p_h, \mathcal{A}\operatorname{grad}\overline{u}) + |\!|\!|\mathcal{A}\operatorname{grad}\overline{u}|\!|\!|^2 \\ &= |\!|\!|p_h - \mathcal{A}\operatorname{grad}\overline{u}|\!|\!|^2. \end{aligned}$$

□

Example 4.9. Let $\Omega = (0,1) \times (0,1)$ and let us employ the standard linear triangular elements over uniform triangulations of $\overline{\Omega}$. Let $\mathcal{A}$ in (4.14) be the identity matrix, $f = 0$, $\overline{u}(x_1, x_2) = x_1 x_2$ and $\Gamma_1 = \partial\Omega$. According to (4.20), the entries of the corresponding flexibility matrix of the dual method are given by

$$b(q^i, q^j) = a(w^i, w^j),$$

i.e., they equal to entries associated to internal nodes of the stiffness matrix of the primal method. Since $\overline{u}$ is a harmonic function ($\Delta\overline{u} = 0$), it is also the true solution of (4.14).

Hence,

$$J(u) = J(\overline{u}) = \frac{1}{2}a(\overline{u},\overline{u}) = \frac{1}{2}\int_\Omega (x_1^2 + x_2^2)\,dx_1\,dx_2 = \frac{1}{3}.$$

In Figure 4.9, we see the lower and upper bound of the potential energy $J(u)$ given by the inequalities (4.27) for $h = \frac{1}{2}, \frac{1}{4}, \frac{1}{8}$ and $\frac{1}{16}$. Note that two-sided bounds are useful tools for testing the validity of a program. For other a posteriori techniques see, e.g., [Babuška, Miller], [Bank, Weiser], [Johnson], and for a nonlinear problem [Weisz].

4.4. Convergence of the finite element method

To show the main idea of convergence proofs for the finite element method, we restrict ourselves in this section only to linear second order elliptic problems in the

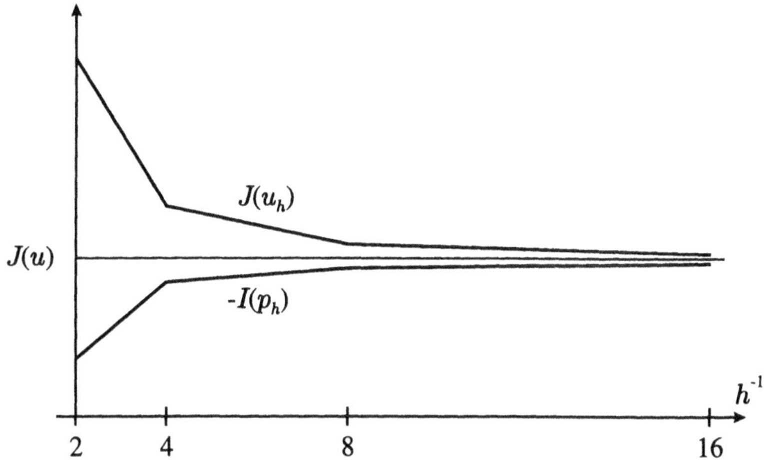

Figure 4.9.

plane solved by the linear triangular elements. We prove the convergence in the $H^1(\Omega)$-norm under the so-called *maximum angle condition* (see (4.33)). Convergence results in another norms and for other finite elements are only summarized in Remarks 4.19, 4.34 and 4.35.

Consider a polygonal domain $\Omega \subset \mathbf{R}^2$ and define the space

$$V = \{v \in H^1(\Omega) \mid v = 0 \text{ on } \Gamma_1\}, \tag{4.31}$$

where $\Gamma_1 \neq \emptyset$ satisfies (3.48). Let the forms $a(\cdot\,,\cdot)$ and $F(\cdot)$ fulfill the hypotheses of the Lax–Milgram Lemma 3.3 and let $u \in V$ be the (unique) solution of the problem

$$a(v, u) = F(v) \quad \forall v \in V. \tag{4.32}$$

Denote by $\mathcal{T}_h$ a triangulation of $\overline{\Omega}$ formed by triangles which satisfy the requirements (1)–(6) from Section 4.2. A set of triangulations $\mathcal{F}$ is called a *family of triangulations* if for every $\varepsilon > 0$ there exists $\mathcal{T}_h \in \mathcal{F}$ with $h < \varepsilon$.

Definition 4.10. *A family $\mathcal{F}$ of triangulations is said to satisfy the maximum angle condition if there exists a constant γ_0 such that for any $\mathcal{T}_h \in \mathcal{F}$ and any $K \in \mathcal{T}_h$ we have*

$$\gamma_K \leq \gamma_0 < \pi, \tag{4.33}$$

where γ_K is the maximum angle of the triangle K.

To any $\mathcal{T}_h \in \mathcal{F}$ we assign the problem: Find $u_h \in V_h$ such that

$$a(v_h, u_h) = F(v_h) \quad \forall v_h \in V_h, \tag{4.34}$$

where

$$V_h = \{v_h \in V \mid v|_K \in P_1(K) \quad \forall K \in \mathcal{T}_h\}. \tag{4.35}$$

Under the condition (4.33) we shall prove that u_h converge to u when $h \to 0$, i.e.,

$$\lim_{h\to 0} \|u - u_h\|_1 \to 0. \tag{4.36}$$

The proof will be based on the following five lemmas. The first lemma is a special case of Theorem 7.6, which we prove later.

Lemma 4.11. *(Céa's lemma) There exists a constant $C > 0$ such that for any space $V_h \neq \emptyset$ we have*

$$\|u - u_h\|_1 \leq C \inf_{v_h \in V_h} \|u - v_h\|_1. \tag{4.37}$$

For the proof see also [Ciarlet, p. 104].

This lemma enables us to transform the question of convergence $\|u - u_h\|_1 \to 0$ for $h \to 0$ to a study of the approximation properties of the spaces V_h. To estimate the infimum in (4.37), we shall construct, in each V_h, an element which is close to u in the $H^1(\Omega)$-norm. To this end let us introduce a linear interpolation operator $\pi_h \colon C(\overline{\Omega}) \to V_h$ which is uniquely determined by the relation

$$\pi_h v(B_i) = v(B_i) \tag{4.38}$$

for all vertices B_i of all $K \in \mathcal{T}_h$. Notice that $\pi_h v$ (cf. Figure 4.5) is well-defined also for $v \in H^2(\Omega)$ due to the Sobolev imbedding (cf. (3.30))

$$H^2(\Omega) \hookrightarrow C(\overline{\Omega}). \tag{4.39}$$

Let us deal first with one (closed) reference triangle $\widehat{K}$ with the vertices $\widehat{A}_0 = (0,0)^T$, $\widehat{A}_1 = (1,0)^T$, $\widehat{A}_2 = (0,1)^T$. Analogously to (4.38) we define $\hat{\pi}\hat{v} \in P_1(\widehat{K})$ by

$$(\hat{\pi}\hat{v})(\widehat{A}_j) = \hat{v}(\widehat{A}_j), \quad j = 0,1,2, \ \hat{v} \in C(\widehat{K}). \tag{4.40}$$

Lemma 4.12. *There exists a constant $\widehat{C} > 0$ such that*

$$\|\hat{\partial}_j(\hat{v} - \hat{\pi}\hat{v})\|_{0,\widehat{K}} \leq \widehat{C}|\hat{\partial}_j\hat{v}|_{1,\widehat{K}} \quad \forall \hat{v} \in H^2(\widehat{K}), \quad j = 1,2, \tag{4.41}$$

where $\hat{\partial}_j\hat{v} = \dfrac{\hat{\partial}\hat{v}}{\hat{\partial}\hat{x}_j}$.

Proof. For simplicity we omit the symbol ^ throughout the whole proof. Define the operator

$$Q \colon H^1(K) \to P_0(K) \tag{4.42}$$

by

$$Qz = \int_0^1 z(s,0)\,ds. \tag{4.43}$$

According to the Trace Theorem 3.8, we have $z|_{\partial K} \in L^2(\partial K)$ for $z \in H^1(K)$, i.e., Q is well-defined. Since Qz is constant, we get by Cauchy–Schwarz inequality (3.10) and (3.26) that

$$\begin{aligned} \|Qz\|_{0,K}^2 &= \operatorname{meas} K \left(\int_0^1 z(s,0)\, ds \right)^2 \\ &\leq \frac{1}{2} \int_0^1 z^2(s,0)\, ds \leq C_1 \|z\|_{1,K}^2 \quad \forall z \in H^1(K). \end{aligned} \tag{4.44}$$

Therefore, Q is continuous.

By (4.43), we immediately see that

$$Qz = z \quad \forall z \in P_0(K). \tag{4.45}$$

Thus due to the Bramble–Hilbert lemma from [Ciarlet, p. 121, 192] there exists $C > 0$ such that

$$\|z - Qz\|_{0,K} \leq C|z|_{1,K} \quad \forall z \in H^1(K). \tag{4.46}$$

Let $v \in H^2(K)$ be arbitrary now. Setting $z = \partial_1 v \in H^1(K)$, we find that

$$Qz = \int_0^1 \partial_1 v(s,0)\, ds = v(1,0) - v(0,0) = \partial_1(\pi v), \tag{4.47}$$

and from (4.46) we get (4.41) for $j = 1$. The case $j = 2$ can be proved similarly. □

Consider now an arbitrary triangle K with the vertices A_j, $j = 0,1,2$ (A_j-column vectors). Assume that the maximum angle γ_K is at the vertex A_0 (see Figure 4.10).

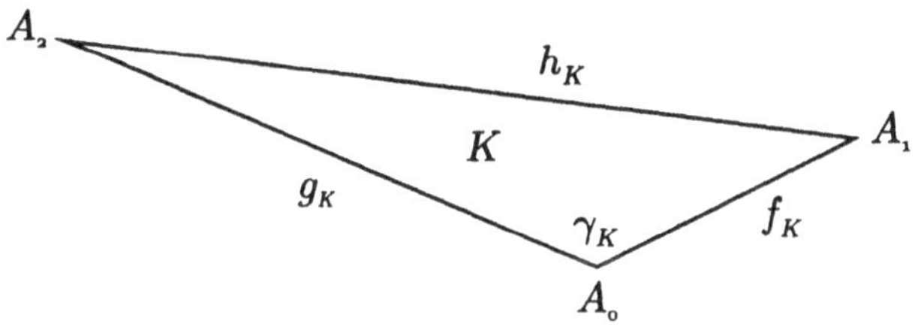

Figure 4.10.

As usual let $h_K = \operatorname{diam} K$ and let f_K and g_K denote the lengths of the sides A_0A_1 and A_0A_2, respectively. Define an affine one-to-one mapping $F_K \colon \widehat{K} \to K$ (see Figure 4.11) by

$$F_K(\hat{x}) = B_K \hat{x} + A_0, \quad \hat{x} = (\hat{x}_1, \hat{x}_2)^T \in \widehat{K}, \tag{4.48}$$

where

$$B_K = (B_{kl})_{k,l=1}^2 = (A_1 - A_0, A_2 - A_0) \tag{4.49}$$

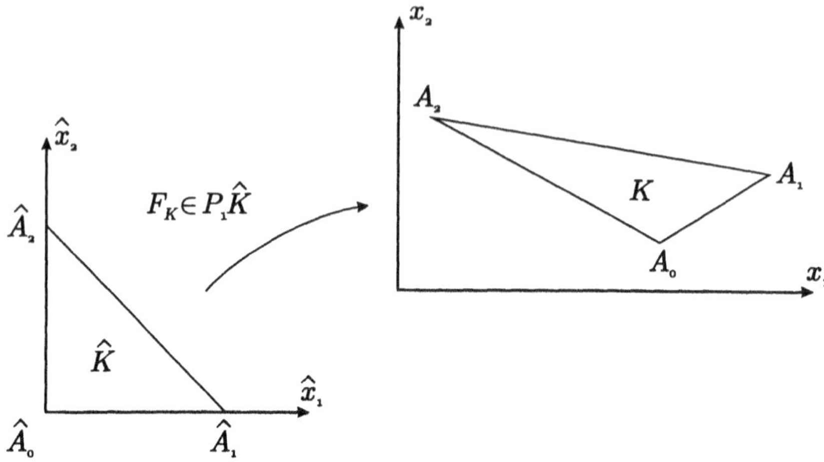

Figure 4.11.

is a nonsingular 2×2 matrix as

$$F_K(\widehat{A}_j) = A_j, \quad j = 0, 1, 2. \tag{4.50}$$

From (4.49) we see that

$$\begin{aligned} B_{11}^2 + B_{21}^2 = f_K^2, \quad B_{12}^2 + B_{22}^2 = g_K^2, \\ |B_{11}| \le f_K, \qquad |B_{21}| \le f_K, \\ |B_{12}| \le g_K, \qquad |B_{22}| \le g_K, \\ \max\big(|B_{11}|, |B_{12}|, |B_{21}|, |B_{22}|\big) \le h_K. \end{aligned} \tag{4.51}$$

Note that f_K and g_K may considerably differ (see, e.g., Figure 4.12). For every $v \in L^2(K)$ and almost every $\hat{x} \in \widehat{K}$ let us set

$$\hat{v}(\hat{x}) = v(x), \tag{4.52}$$

where $x = F_K(\hat{x})$. Thus we have a one-to-one correspondence between $\hat{v}$ and v. From (4.48) and (4.52) we can directly derive the next two relations. If $\hat{v} \in H^1(\widehat{K})$ and $v \in H^1(K)$ then

$$(\hat{\partial}_1 \hat{v}(\hat{x}), \hat{\partial}_2 \hat{v}(\hat{x}))^T = B_K^T(\partial_1 v(x), \partial_2 v(x))^T \tag{4.53}$$

for almost every $\hat{x} \in \widehat{K}$ and corresponding $x \in K$. If $\hat{v} \in H^2(\widehat{K})$ and $v \in H^2(K)$ then similarly for the second derivatives we have

$$\begin{bmatrix} \hat{\partial}_{11}\hat{v} & \hat{\partial}_{12}\hat{v} \\ \hat{\partial}_{12}\hat{v} & \hat{\partial}_{22}\hat{v} \end{bmatrix} = B_K^T \begin{bmatrix} \partial_{11}v & \partial_{12}v \\ \partial_{12}v & \partial_{22}v \end{bmatrix} B_K. \tag{4.54}$$

Lemma 4.13. *For any $v \in H^1(K)$ and $\hat{v} \in H^1(\widehat{K})$ satisfying (4.52) we have*

$$|v|_{1,K}^2 \le 2|\det B_K^{-1}| \left(g_K^2 \|\hat{\partial}_1 \hat{v}\|_{0,\widehat{K}}^2 + f_K^2 \|\hat{\partial}_2 \hat{v}\|_{0,\widehat{K}}^2 \right). \tag{4.55}$$

Proof. Let us denote the entries of B_K^{-1} by C_{kl}, that is

$$B_K^{-1} = \frac{1}{\det B_K} \begin{bmatrix} B_{22} & -B_{12} \\ -B_{21} & B_{11} \end{bmatrix} = \begin{bmatrix} C_{11} & C_{12} \\ C_{21} & C_{22} \end{bmatrix}. \tag{4.56}$$

Using now the substitution $x = F_K(\hat{x})$, we get using (4.53), (4.56) and the inequality $|2ab| \le a^2 + b^2$ that

$$\begin{aligned} |v|_{1,K}^2 &= \int_K \left((\partial_1 v)^2 + (\partial_2 v)^2 \right) dx \qquad (4.57) \\ &= |\det B_K| \int_{\widehat{K}} \left((C_{11}\hat{\partial}_1 \hat{v} + C_{21}\hat{\partial}_2 \hat{v})^2 + (C_{12}\hat{\partial}_1 \hat{v} + C_{22}\hat{\partial}_2 \hat{v})^2 \right) d\hat{x} \\ &\le 2|\det B_K| \left((C_{11}^2 + C_{12}^2)\|\hat{\partial}_1 \hat{v}\|_{0,\widehat{K}}^2 + (C_{21}^2 + C_{22}^2)\|\hat{\partial}_2 \hat{v}\|_{0,\widehat{K}}^2 \right) \\ &= \frac{2}{|\det B_K|} \left((B_{22}^2 + B_{12}^2)\|\hat{\partial}_1 \hat{v}\|_{0,\widehat{K}}^2 + (B_{21}^2 + B_{11}^2)\|\hat{\partial}_2 \hat{v}\|_{0,\widehat{K}}^2 \right). \end{aligned}$$

Thus (4.55) follows from (4.51). □

Lemma 4.14. *For any $\hat{v} \in H^2(\widehat{K})$ and $v \in H^2(K)$ satisfying (4.52) we have*

$$\begin{aligned} |\hat{\partial}_1 \hat{v}|_{1,\widehat{K}}^2 &\le 16|\det B_K^{-1}|\, f_K^2 h_K^2\, |v|_{2,K}^2, \\ |\hat{\partial}_2 \hat{v}|_{1,\widehat{K}}^2 &\le 16|\det B_K^{-1}|\, g_K^2 h_K^2\, |v|_{2,K}^2. \end{aligned} \tag{4.58}$$

Proof. Using the substitution $\hat{x} = F_K^{-1}(x) = B_K^{-1}x - B_K^{-1}A_0$, (4.54) and (4.51), we find that

$$\begin{aligned} |\hat{\partial}_1 \hat{v}|_{1,\widehat{K}}^2 &= \sum_{j=1}^{2} \int_{\widehat{K}} (\hat{\partial}_{1j}\hat{v})^2 \, d\hat{x} \\ &= |\det B_K^{-1}| \sum_{j=1}^{2} \int_K \left(\sum_{k,l=1}^{2} B_{k1} B_{lj} \partial_{kl} v \right)^2 dx \\ &\le 4|\det B_K^{-1}| \sum_{j=1}^{2} \sum_{k,l=1}^{2} B_{k1}^2 B_{lj}^2 \int_K (\partial_{kl} v)^2 \, dx \\ &\le 8|\det B_K^{-1}| f_K^2 h_K^2 \sum_{k,l=1}^{2} \|\partial_{kl} v\|_{0,K}^2. \end{aligned} \tag{4.59}$$

From here we obtain the first relation of (4.58). The second one can be proved analogously. □

For any $K \in \mathcal{T}_h$ and any $v \in C(\overline{\Omega})$ we set

$$\pi_K v = (\pi_h v)\big|_K. \tag{4.60}$$

Using the transformation (4.48), we easily deduce (see, e.g., [Ciarlet, p. 124]) that

$$\widehat{\pi_K v} = \hat{\pi}(\widehat{v\big|_K}). \tag{4.61}$$

Theorem 4.15. *Let $\mathcal{F}$ be a family of triangulations satisfying the maximum angle condition (4.33). Then there exists a constant $C > 0$ such that for any $\mathcal{T}_h \in \mathcal{F}$ with $h \le 1$ we have*

$$\|v - \pi_h v\|_1 \le Ch|v|_2 \quad \forall v \in H^2(\Omega). \tag{4.62}$$

Proof. Let $\mathcal{T}_h \in \mathcal{F}$ and $K \in \mathcal{T}_h$ be given. According to well-known relations from planar geometry and (4.33), we have (see Figure 4.10)

$$|\det B_K| = 2\,\mathrm{meas}\, K = f_K g_K \sin\gamma_K \ge f_K g_K \min\left(\sin\gamma_0, \frac{\sqrt{3}}{2}\right), \tag{4.63}$$

since if $\gamma_K < \frac{\pi}{2}$ then $\sin\gamma_K \ge \sin\frac{\pi}{3} = \frac{\sqrt{3}}{2}$ else $\sin\gamma_K \ge \sin\gamma_0$.

Now let $v \in H^2(\Omega)$ be arbitrary. Writing $v - \pi_K v$ in (4.55) instead of v, we get by (4.61), (4.41), (4.58) and (4.63) that

$$\begin{aligned}
|v - \pi_K v|_{1,K}^2 &\le 2|\det B_K^{-1}|\left(g_K^2\|\hat{\partial}_1(\hat{v} - \widehat{\pi_K v})\|_{0,\widehat{K}}^2 + f_K^2\|\hat{\partial}_2(\hat{v} - \widehat{\pi_K v})\|_{0,\widehat{K}}^2\right)\\
&\le 2\widehat{C}^2|\det B_K^{-1}|\left(g_K^2|\hat{\partial}_1\hat{v}|_{1,\widehat{K}}^2 + f_K^2|\hat{\partial}_2\hat{v}|_{1,\widehat{K}}^2\right)\\
&\le 64\widehat{C}^2|\det B_K^{-1}|^2 f_K^2 g_K^2 h_K^2 |v|_{2,K}^2 \le C_1^2 h_K^2 |v|_{2,K}^2.
\end{aligned} \tag{4.64}$$

Summing this over all triangles $K \in \mathcal{T}_h$, we find that

$$|v - \pi_h v|_1 \le C_1 h|v|_2 \quad \forall v \in H^2(\Omega), \tag{4.65}$$

where C_1 is independent of h. For the $L^2(\Omega)$-norm, we have by [Ciarlet, p. 118, 120] and (4.51) that

$$\begin{aligned}
\|v - \pi_K v\|_{0,K}^2 &= |\det B_K|\|\hat{v} - \widehat{\pi_K v}\|_{0,\widehat{K}}^2\\
&\le C_2|\det B_K||\hat{v}|_{2,K}^2 = C_3|\det B_K|\|B_K\|^4|\det B_K^{-1}||v|_{2,K}^2\\
&\le C_0^2 h_K^4|v|_{2,K}^2,
\end{aligned} \tag{4.66}$$

where $\|\cdot\|$ is the Euclidean norm. This yields

$$\|v - \pi_h v\|_0 \le C_0 h^2|v|_2 \quad v \in H^2(\Omega) \tag{4.67}$$

(without the assumption (4.33)).

Finally from (4.67) and (4.65) we have

$$\|v - \pi_h v\|_1 \le \|v - \pi_h v\|_0 + |v - \pi_h v|_1 \le (C_0 h^2 + C_1 h)|v|_2, \tag{4.68}$$

and the desired estimate (4.62) follows for $h \le 1$. □

Theorem 4.16. *Let the maximum angle condition (4.33) hold. Then*

$$\lim_{h\to 0} \|u - u_h\|_1 = 0, \tag{4.69}$$

where u and u_h are the solutions of (4.32) and (4.34), respectively.

Proof. Let $\varepsilon > 0$ be arbitrary. According to [Doktor, p. 618], there exists $v \in C^\infty(\overline{\Omega}) \cap V$ such that

$$\|u - v\|_1 \le \frac{\varepsilon}{2}, \tag{4.70}$$

and by (4.62)

$$\|v - \pi_h v\|_1 \le \frac{\varepsilon}{2} \tag{4.71}$$

for sufficiently small h. Now from (4.37), (4.70) and (4.71) we easily get the required convergence

$$\|u - u_h\|_1 \le C\|u - \pi_h v\|_1 \le C(\|u - v\|_1 + \|v - \pi_h v\|_1) \le C\varepsilon.$$

□

If the true solution $u \in V \subset H^1(\Omega)$ is piecewise smooth we may even achieve the linear rate of convergence.

Theorem 4.17. *Let there exist mutually disjoint polygonal domains Ω_j, $j = 1, ..., q$, such that*

$$u\big|_{\Omega_j} \in H^2(\Omega_j), \quad j = 1, ..., q, \tag{4.72}$$

and

$$\overline{\Omega} = \bigcup_{j=1}^{q} \overline{\Omega}_j. \tag{4.73}$$

Let $\mathcal{F}$ be a family of triangulations satisfying the maximum angle condition (4.33) and let

$$\partial\Omega_j \cap K^0 = \emptyset \quad \forall K \in \mathcal{T}_h \quad \forall \mathcal{T}_h \in \mathcal{F}, \tag{4.74}$$

where K^0 is the interior of K. Then there exists a constant $C > 0$ such that for any $\mathcal{T}_h \in \mathcal{F}$ with $h \le 1$ we have

$$\|u - u_h\|_1 \le Ch \sum_{j=1}^{q} |u|_{2,\Omega_j}. \tag{4.75}$$

Proof. The solution $u \in H^1(\Omega)$ is by the imbedding (4.39) and (4.72) continuous on each $\overline{\Omega}_j$. First we show that it is continuous on the whole $\overline{\Omega}$.

So let u be not continuous. Then there exists two domains Ω_k, Ω_l and an open ball $B \subset \overline{\Omega}_k \cup \overline{\Omega}_l$ such that

$$B \cap S \ne \emptyset \quad \text{and} \quad u\big|_{\overline{\Omega}_k} > u\big|_{\overline{\Omega}_l} \text{ on } B \cap S, \tag{4.76}$$

where $S = \overline{\Omega}_k \cap \overline{\Omega}_l$ is a line segment. Renumbering $\overline{\Omega}_j$ appropriately, we may assume that $k = 1$ and $l = 2$. Let $w \in C_0^\infty(\Omega)$ be a "hill" function such that

$$w(x) > 0 \quad \forall x \in B \quad \text{and} \quad w(x) = 0 \quad \forall x \in \Omega \setminus B. \tag{4.77}$$

Let $n^j = (n_1^j, n_2^j)^T$ be the outward unit normal to $\partial\Omega_j$. Obviously there exists $p \in \{1,2\}$ such that

$$n_p^1 \neq 0 \quad \text{on } S. \tag{4.78}$$

Thus by Green's formula (3.29), we arrive at

$$\begin{aligned} 0 &= \int_\Omega w\partial_p u\,dx + \int_\Omega u\partial_p w\,dx = \sum_{j=1}^2 \left(\int_{\Omega_j} w\partial_p u\,dx + \int_{\Omega_j} u\partial_p w\,dx \right) \\ &= \sum_{j=1}^2 \left(-\int_{\Omega_j} u\partial_p w\,dx + \int_{\partial\Omega_j} u|_{\overline{\Omega}_j} w n_p^j\,ds + \int_{\Omega_j} u\partial_p w\,dx \right) \\ &= \int_S \left(u|_{\overline{\Omega}_1} - u|_{\overline{\Omega}_2} \right) w n_p^1\,ds = n_p^1 \int_{B\cap S} \left(u|_{\overline{\Omega}_1} - u|_{\overline{\Omega}_2} \right) w\,ds. \end{aligned}$$

However, this is a contradiction, since $n_p^1 \neq 0$ and the last integral is positive by (4.76) and (4.77). Hence, u is continuous and by (4.38) its interpolation $\pi_h u$ is well-defined.

Due to (4.72) and (4.74), we may apply the estimate (4.62) to each Ω_j, i.e.,

$$\|u - \pi_h u\|_{1,\Omega_j} \leq Ch|u|_{2,\Omega_j}, \quad j = 1, \ldots, q. \tag{4.79}$$

Finally from (4.37) and (4.79) we have

$$\begin{aligned} \|u - u_h\|_1 \leq C_1\|u - \pi_h u\|_1 &\leq C_1 \sum_{j=1}^q \|u - \pi_h u\|_{1,\Omega_j} \\ &\leq Ch \sum_{j=1}^q |u|_{2,\Omega_j}. \end{aligned}$$

□

Example 4.18. Assume that $\Omega = (0,1) \times (0,1)$,

$$v(x_1, x_2) = 2x_1^2 - x_1 x_2 - 3x_2^2$$

and consider the sequence $\mathcal{F} = \{\mathcal{T}^j\}_{j=0}^\infty$ of triangulations in Figure 4.12. Sides parallel to the axes x_1 and x_2 have the lengths 2^{-j} and 2^{-2j}, respectively. We immediately see that the maximum angle condition (4.33) is fulfilled with $\gamma_0 = \dfrac{\pi}{2}$.

Table 4.1 confirms the theoretical approximation order $\mathcal{O}(h)$ of linear elements as stated in Lemma 4.15. To evaluate the norms $\|v - \pi_h v\|_1$, we have used the

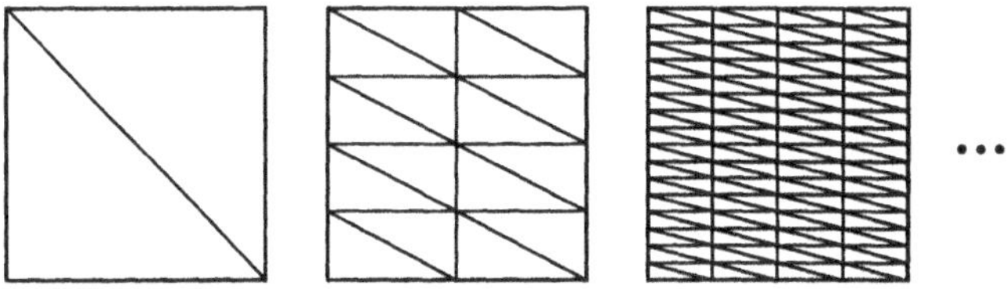

Figure 4.12.

numerical integration formula from [Křížek, Neittaanmäki, 1990, p. 58] which is exact for all quintic polynomials on each triangle. Thus these norms were calculated exactly (except for rounding errors).

j	h	$\|v - \pi_h v\|_1$	$h^{-1}\|v - \pi_h v\|_1 / \|v\|_2$
0	1.414214	2.100264	0.203996
1	0.559017	0.732818	0.180067
2	0.257694	0.318461	0.169751
3	0.125973	0.153600	0.167485
4	0.062622	0.076344	0.167462

Table 4.1.

Remark 4.19. In [Babuška, Aziz, 1976, p. 223], [Strang, Fix, p. 138] or [Ženíšek, 1969, p. 365] there are examples for which the maximum angle condition (4.33) does not hold and the linear triangular elements loose the approximation order in (4.62). The fact that the greatest angle should be bounded away from π when $h \to 0$, has been known since 1957 (see [Synge, p. 211]). Related papers on this topic include [Apel, Dobrowolski], [Babuška, Aziz, 1976], [Barnhill, Gregory], [Gregory], [Jamet], [Křížek, 1991], where the convergence $\pi_h v \to v$ (under the maximum angle condition) is obtained in the norm $\|\cdot\|_{1,p}$ for $p \in [1, \infty]$. Note that the question of convergence of the finite element method for some nonlinear problems can be transformed (see Theorem 7.6) to the investigation of approximation properties of π_h (cf. (4.62)).

Let us recall the following definition due to Zlámal (see [Zlámal, 1968, p. 397]). A family $\mathcal{F}$ of triangulations is said to satisfy the *minimum angle condition* if there exists a constant α_0 such that for any $\mathcal{T}_h \in \mathcal{F}$ and any $K \in \mathcal{T}_h$ it holds that

$$0 < \alpha_0 \leq \alpha_K, \tag{4.80}$$

where α_K is the minimum angle of K. It is clear that the minimum angle condition implies the maximum angle condition. The sequence of triangulations from Figure 4.12 shows that the converse implication is not true.

In standard convergence proofs, the determinant (4.63) is usually estimated in this way

$$|\det B_K| = 2\,\mathrm{meas}\,K \geq Ch_K^2. \tag{4.81}$$

Here the last inequality follows from the well-known inscribed ball condition (see, e.g., [Ciarlet, p. 124] or Definition 4.23) which is equivalent to (4.80) for triangular elements. However, estimate (4.81) cannot be used when we need to employ the maximum angle condition (4.33). Consider for instance the right-angled triangle of Figure 4.13. Estimating the area of K by the area of the inscribed ball with radius r, we have

$$\mathrm{meas}\,K > \pi r^2 > \pi h^4/8,$$

since $r \in (h^2/4, h^2/2)$ for $h \in (0,1)$. This estimate is not "optimal" since, by (4.63), $\mathrm{meas}\,K = h^3/2$.

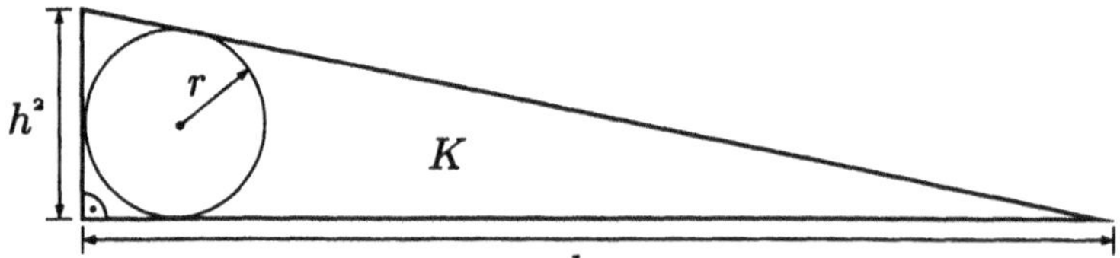

Figure 4.13.

Remark 4.20. Due to Lemma 4.15 we need not be afraid of using triangles with small angles which can be useful for developing FE-software for adaptive mesh refinement since we need not prescribe any lower positive bound upon the minimum angle. Thus we may employ triangular element which are almost flat (degenerate). This can be useful for covering thin slots, gaps or strips of different materials to satisfy assumption (4.74). In Figure 4.14 a) we see for instance a cross section of two copper wires covered by a thin insulation in an electric engine (cf. Chapter 11). A possible triangulation with a small number of nodes is sketched in Figure 4.14 b). "Flat" triangles are also commonly used to approximate functions that change much more rapidly in one direction than in another direction.

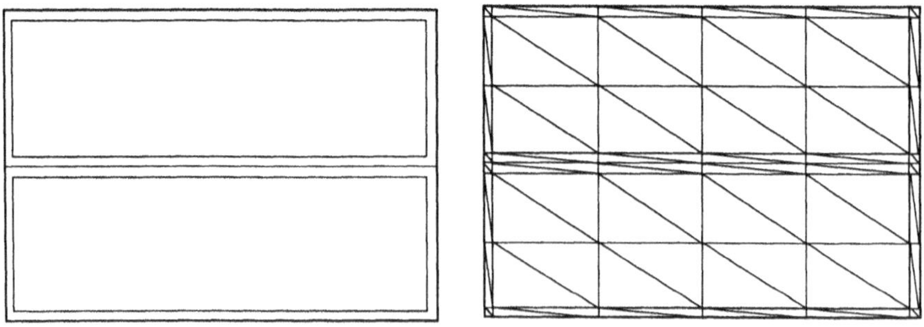

a) b)

Figure 4.14.

When the maximum angle condition holds then a big triangle may have a common vertex with a very small triangle (see Figure 4.15). This is not possible assuming only the minimum angle condition (4.80).

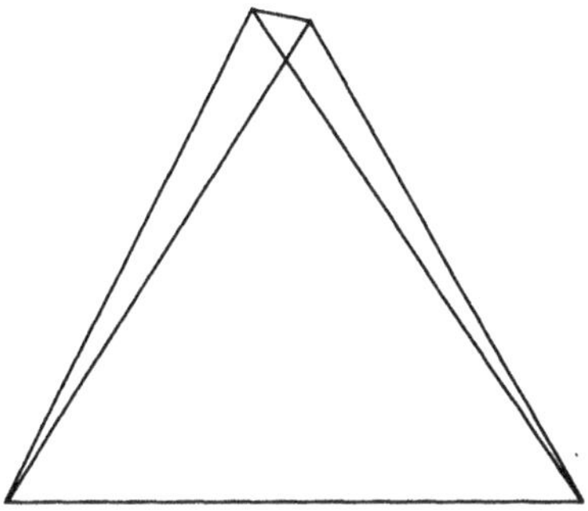

Figure 4.15.

Sometimes it is necessary to employ flat triangles. For instance, in Figure 4.16, we are not able to cover domains with "cusp-points" (see [Ženíšek, 1993]) by triangles satisfying (4.80) for $h \to 0$.

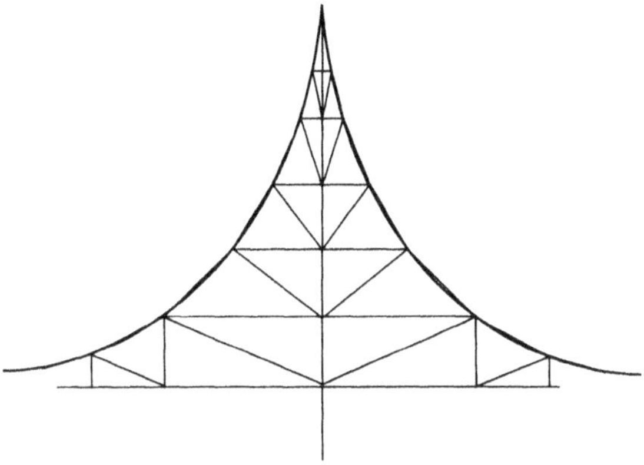

Figure 4.16.

Figure 4.17 shows a refinement of a triangle which fulfills the maximum angle condition (4.33) but not the minimum angle condition (4.80).

4.5. Linear interpolation on polyhedra

In this section we generalize some results of Section 4.4 to the three-dimensional case. For instance we shall show how to modify the maximum angle condition (4.33) for tetrahedra to preserve the optimal approximation order $\mathcal{O}(h)$ of the linear interpolation in the $\|\cdot\|_1$-norm. Recall that a *polyhedron* in $\mathbb{R}^3$ is a bounded domain whose boundary can be expressed as a finite union of polygons, and a

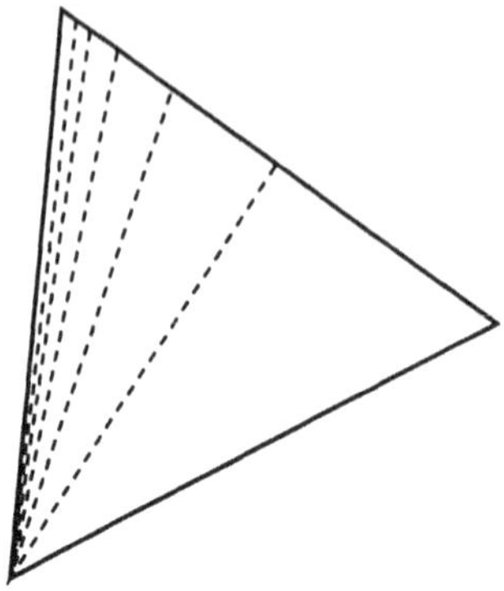

Figure 4.17.

polygon is a bounded plane domain whose boundary can be expressed as a finite union of straight line segments. Recall that a convex polyhedron is usually defined as a convex hull of a finite set of points in $\mathbf{R}^3$.

Theorem 4.21. *For any polyhedron there exists a decomposition into tetrahedra.*

The proof is constructive, so it may help in practical situations. It is based on an auxiliary Lemma 4.22. Note that computer generators of tetrahedral divisions are usually based on the Delaunay triangulation, which is uniquely determined by prescribing nodes only (cf., e.g., [George], [Weatherill, Hassan]). For a local refinement of tetrahedral meshes, see [Křížek, Strouboulis].

Lemma 4.22. *For any polyhedron $\overline{\Omega}$ there exist convex polyhedra $\overline{\Omega}_1, ..., \overline{\Omega}_r$ such that*

(y) the union of all $\overline{\Omega}_p$, $p = 1, ..., r$, is $\overline{\Omega}$,

(yy) the interiors of these convex polyhedra are mutually disjoint,

(yyy) any face of any polyhedron $\overline{\Omega}_p$, $p \in \{1, ..., r\}$, is either a face of another polyhedron $\overline{\Omega}_q$, $q \neq p$, or a subset of the boundary $\partial\Omega$.

Proof of Lemma 4.22. Let $\overline{\Omega}$ be an arbitrary polyhedron and let $P^1, ..., P^k$ be polygons the union of which is $\partial\Omega$. Let $R^1, ..., R^k$ be planes such that $P^i \subset R^i$, $i = 1, ..., k$. Finally, let $\Omega_1, ..., \Omega_r \subset \mathbf{R}^3$ be all components of the set $\overline{\Omega} \setminus \cup_{i=1}^k R^i$ (i.e., the components which arise by "cutting" $\overline{\Omega}$ by the planes R^i). We show that $\overline{\Omega}_p$, $p = 1, ..., r$, are the convex polyhedra sought (their number is finite, because k planes divide the space $\mathbf{R}^3$ into 2^k parts at most).

Since $\partial\Omega \subset \cup_{i=1}^k R^i$, it follows that

$$\overline{\Omega} \setminus \bigcup_{i=1}^{k} R^i = \Omega \setminus \bigcup_{i=1}^{k} R^i.$$

This set is open since Ω is open and $\cup_{i=1}^k R^i$ is closed, i.e., Ω_p are open connected sets.

Let $p \in \{1, ..., r\}$ be fixed. Any plane R^i, $i = 1, ..., k$, splits the space $\mathbf{R}^3$ into two halfspaces. Denoting by Q^i that closed halfspace with the boundary plane R^i which contains $\overline{\Omega}_p$, it is easy to show

$$\overline{\Omega}_p = \bigcap_{i=1}^k Q^i.$$

Hence, $\overline{\Omega}_p$ is a convex polyhedron, since the set $\overline{\Omega}_p$ is bounded and contains at least one interior point.

Using the definition formula $\cup_{p=1}^r \Omega_p = \overline{\Omega} \setminus \cup_{i=1}^k R^i$, we find that the condition (y) holds. Since any two components Ω_p, Ω_q, $p \neq q$, are separated by at least one plane R^i, (yy) holds. It remains to verify (yyy).

Let x be an interior point of a face S of the convex polyhedron $\overline{\Omega}_p$ and let $x \in \partial\Omega_q$, $q \neq p$. Suppose, for a moment, that x lies on an edge of the convex polyhedron $\overline{\Omega}_q$. Then x must lie on at least two different planes R^s, R^t, $s, t \in \{1, ..., k\}$. But this is a contradiction, since x is an interior point of S. Hence, x is also an interior point of a face S' of the polyhedron $\overline{\Omega}_q$ and we deduce that S and S' have common interior points, i.e., $S = S'$. If the face S of $\overline{\Omega}_p$ does not coincide with a face of any other polyhedron $\overline{\Omega}_q$, $q \neq p$, then it is easy to see that $S \subset \partial\Omega$.

□

Proof of Theorem 4.21. Let $\overline{\Omega} \subset \mathbf{R}^3$ be an arbitrary polyhedron and let $\overline{\Omega}_1, ..., \overline{\Omega}_r$ be the convex polyhedra from Lemma 4.22. We shall divide all these convex polyhedra into tetrahedra in the following way. Let $p \in \{1, ..., r\}$ be arbitrary. As has been said, all faces of the convex polyhedron $\overline{\Omega}_p$ are convex polygons. Denoting by $B_1, ..., B_j$ (for instance, counter-clockwise) the vertices of any face, we can divide this face into the triangles $B_1B_2B_3$, $B_1B_3B_4$,...,$B_1B_{j-1}B_j$. Let $\{S_p^l\}_{l=1}^{m_p}$ be the set of all triangles which are obtained in this way on the surface of the polyhedron $\overline{\Omega}_p$ (see Figure 4.18).

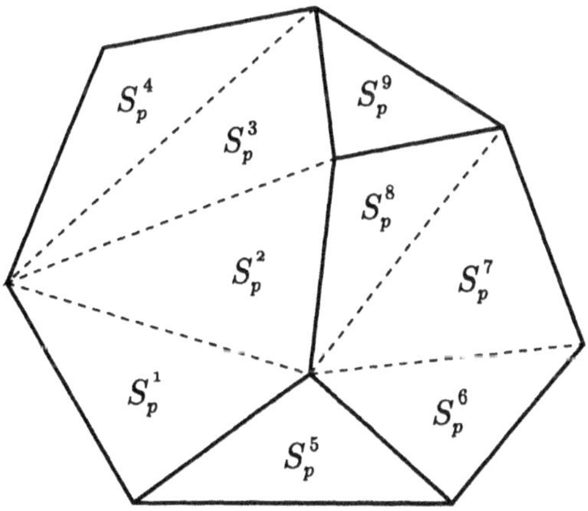

Figure 4.18.

In addition, we require that common faces of two convex polyhedra (i.e., faces $S \subset \overline{\Omega}_p$, $S' \subset \overline{\Omega}_q$, $p \neq q$, such that $S = S'$) be divided into triangles "in the same manner". Let A_p be an arbitrary interior point of the polyhedron $\overline{\Omega}_p$. The convex hull K_p^l of the triangle S_p^l and the point A_p is a tetrahedron. Introduce the set

$$\mathcal{T} = \{K_p^l \mid p = 1, \ldots, r, \quad l = 1, \ldots, m_p\}.$$

Utilizing the conditions (y), (yy), (yyy) from Lemma 4.22 as well as the fact that a possible common face of two convex polyhedra $\overline{\Omega}_p$, $\overline{\Omega}_q$, $p \neq q$, is divided into triangles in the same manner, it is easy to see that the finite set $\mathcal{T}$ represents a decomposition of $\overline{\Omega}$. □

Definition 4.23. *A family of decompositions $\mathcal{F}$ of a polyhedron $\overline{\Omega}$ into tetrahedra is said to be regular (strongly regular) if there exists a constant $\varkappa > 0$ such that for any decomposition $\mathcal{T}_h \in \mathcal{F}$ and for any tetrahedron $K \in \mathcal{T}_h$ there exists a sphere S_K of radius ϱ_K such that $S_K \subset K$ and*

$$\varkappa h_K \leq \varrho_K \quad (\varkappa h \leq \varrho_K).$$

Remark 4.24. Obviously, any strongly regular family is regular. Recall that a strongly regular family of triangulations of a polygon (the definition is analogous) is easy to obtain due to the fact that any triangle in the triangulation is divided by midlines into four congruent triangles, which are similar to the original one. In three-dimensional space the situation is considerably more complicated, since it may not be possible to divide any tetrahedron into coinciding similar (congruent) tetrahedra (cf. [Goldberg]).

Theorem 4.25. *For any polyhedron there exists a strongly regular family of decompositions into tetrahedra.*

Proof. The proof will be composed of three parts – a), b), c) – see [Křížek, 1982].

a) First, we prove the theorem for the simplest polyhedron – a tetrahedron, which will be chosen in a particular way. So let $\overline{\Omega} = \widetilde{K}$, where $\widetilde{K}$ is the tetrahedron with the vertices $\widetilde{A}$, $\widetilde{B}$, $\widetilde{C}$, $\widetilde{D}$ having the coordinates $(\frac{1}{2}, 0, 0)^T$, $(-\frac{1}{2}, 0, 0)^T$, $(0, \frac{1}{2}, \frac{1}{2})^T$, $(0, -\frac{1}{2}, \frac{1}{2})^T$, respectively (see Figure 4.19). The length of the opposite edges $\widetilde{A}\widetilde{B}$ and $\widetilde{C}\widetilde{D}$ is equal to 1 and the length of all other edges is $\sqrt{3}/2$. Denote by $\widetilde{M}_1$, $\widetilde{M}_2$, $\widetilde{M}_3$, $\widetilde{M}_4$, $\widetilde{M}_5$, $\widetilde{M}_6$ the midpoints of $\widetilde{A}\widetilde{B}$, $\widetilde{A}\widetilde{C}$, $\widetilde{A}\widetilde{D}$, $\widetilde{B}\widetilde{C}$, $\widetilde{B}\widetilde{D}$, $\widetilde{C}\widetilde{D}$, respectively and divide the tetrahedron $\widetilde{K}$ into eight tetrahedra (see Figure 4.19 for exploded tetrahedra in an angular displacement of the coordinate system):

$$\widetilde{A}\widetilde{M}_1\widetilde{M}_2\widetilde{M}_3, \quad \widetilde{B}\widetilde{M}_1\widetilde{M}_4\widetilde{M}_5, \quad \widetilde{C}\widetilde{M}_6\widetilde{M}_2\widetilde{M}_4, \quad \widetilde{D}\widetilde{M}_6\widetilde{M}_3\widetilde{M}_5$$
$$\widetilde{M}_1\widetilde{M}_6\widetilde{M}_2\widetilde{M}_3, \quad \widetilde{M}_1\widetilde{M}_6\widetilde{M}_2\widetilde{M}_4, \quad \widetilde{M}_1\widetilde{M}_6\widetilde{M}_3\widetilde{M}_5, \quad \widetilde{M}_1\widetilde{M}_6\widetilde{M}_4\widetilde{M}_5$$

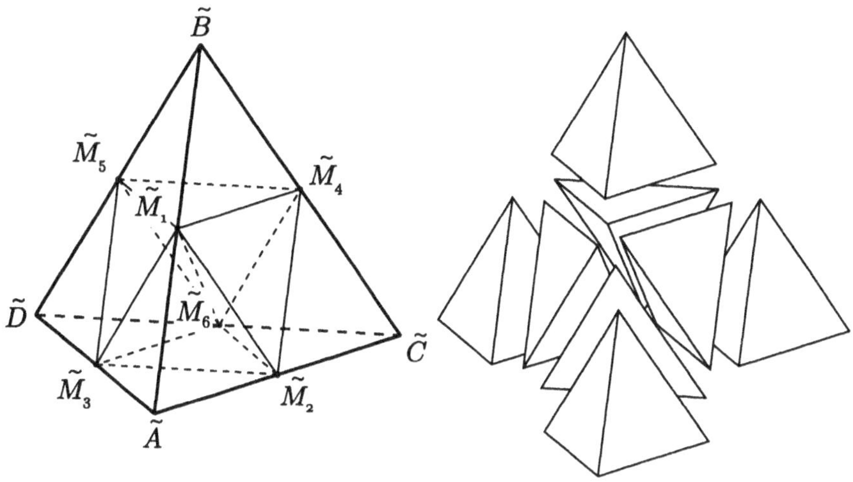

Figure 4.19.

It is easily seen that these tetrahedra form a decomposition of $\widetilde{K}$. We denote by $\widetilde{\mathcal{T}}_1$ this decomposition and put $\widetilde{\mathcal{T}}_0 = \{\widetilde{K}\}$. The length of the edge $\widetilde{M}_1\widetilde{M}_6$ is $\frac{1}{2}$, since the coordinates of the end points are $(0,0,0)^T$, $(0,0,\frac{1}{2})^T$. The length of all the edges which are the midlines of the external faces of $\widetilde{K}$ and which are parallel with $\widetilde{A}\widetilde{B}$ or $\widetilde{C}\widetilde{D}$ is $\frac{1}{2}$ as well. Therefore, the length of the edges

$$\widetilde{A}\widetilde{M}_1,\ \widetilde{B}\widetilde{M}_1,\ \widetilde{M}_2\widetilde{M}_4,\ \widetilde{M}_3\widetilde{M}_5,\ \widetilde{C}\widetilde{M}_6,\ \widetilde{D}\widetilde{M}_6,\ \widetilde{M}_2\widetilde{M}_3,\ \widetilde{M}_4\widetilde{M}_5,\ \widetilde{M}_1\widetilde{M}_6$$

is equal to $\frac{1}{2}$ and for each tetrahedron of $\widetilde{\mathcal{T}}_1$ precisely two of these edges are opposite (e.g., $\widetilde{A}\widetilde{M}_1$ is opposite to $\widetilde{M}_2\widetilde{M}_3$ is the tetrahedron $\widetilde{A}\widetilde{M}_1\widetilde{M}_2\widetilde{M}_3$). The length of all remaining edges of the tetrahedra of $\widetilde{\mathcal{T}}_1$ is $\sqrt{3}/4$. Hence, all tetrahedra of $\widetilde{\mathcal{T}}_1$ are congruent and similar to the original tetrahedron $\widetilde{K}$. If $\tilde{\varrho}$ is the radius of the inscribed sphere of $\widetilde{K}$ then the radius of the inscribed sphere of any tetrahedron of $\widetilde{\mathcal{T}}_1$ is evidently $\frac{1}{2}\tilde{\varrho}$. Any tetrahedron of $\widetilde{\mathcal{T}}_1$ can be divided in the same way into eight coincident tetrahedra similar to $\widetilde{K}$ again and we obtain the next decomposition $\widetilde{\mathcal{T}}_2$. Repeating this process to infinity, we get the family of decompositions $\{\widetilde{\mathcal{T}}_m\}_{m=0}^{\infty}$. The diameter of each tetrahedron from $\widetilde{\mathcal{T}}_m$ is 2^{-m}. Moreover, this family is strongly regular since the corresponding coefficient $\tilde{\varkappa}$ can be chosen as $\tilde{\varkappa} = 2^{-m}\tilde{\varrho}/2^{-m} = \tilde{\varrho}$.

b) Consider $\overline{\Omega} = K$, where K is an arbitrary tetrahedron with vertices A, B, C, D, the coordinates of which are

$$a = (a_1, a_2, a_3)^T, \quad b = (b_1, b_2, b_3)^T, \quad c = (c_1, c_2, c_3)^T, \quad d = (d_1, d_2, d_3)^T.$$

We introduce an affine one-to-one mapping $F\colon \widetilde{K} \mapsto K$ given by

$$F(\tilde{x}) = Q\tilde{x} + q, \quad \tilde{x} \in \widetilde{K}, \tag{4.82}$$

where $q = (a+b)/2$ and $Q = (a-b, c-d, c+d-a-b)$ is a nonsingular matrix as $F(\widetilde{A}) = A, \ldots, F(\widetilde{D}) = D$. This mapping transforms the edges of $\widetilde{K}$ onto the corresponding edges of K, the midlines of the faces of $\widetilde{K}$ onto the midlines of the corresponding faces of K and so on. Therefore, for $m = 0, 1, 2, \dots$ the set

$$\mathcal{T}_m = \{F(\widetilde{K}') \mid \widetilde{K}' \in \widetilde{\mathcal{T}}_m\}$$

is also a decomposition of K. Let e_m be the largest edge of all tetrahedra from $\mathcal{T}_m$, its length being h_m, and let $\tilde{h}_m$ be the length of the corresponding edge $\tilde{e}_m = F_{\mathcal{T}}^{-1}(e_m)$. According to (4.82) and part a), we have

$$h_m \le \|Q\| \tilde{h}_m \le \|Q\| 2^{-m}, \quad m = 0, 1, 2, \dots \tag{4.83}$$

Thus, $\{\mathcal{T}_m\}_{m=0}^{\infty}$ is a family of decompositions of K.

Denote by $\widetilde{\mathcal{S}}$ the inscribed sphere of $\widetilde{K}$, i.e.,

$$\widetilde{\mathcal{S}} = \{\tilde{x} \mid \|\tilde{x} - \tilde{x}_0\| \le \tilde{\varrho}\} \subset \widetilde{K},$$

where $\tilde{x}_0 = (0, 0, \frac{1}{4})^T$ is the centre. Then

$$F(\widetilde{\mathcal{S}}) = \mathcal{E} = \{x \mid \|Q^{-1}(x - q) - \tilde{x}_0\| \le \tilde{\varrho}\} \subset K$$

is the ellipsoid with the centre $x_0 = Q\tilde{x}_0 + q$ which is inscribed in K. Denote by ϱ the length of the shortest semi-axis of $\mathcal{E}$ and let $\mathcal{S}$ be the sphere with the centre x_0 and the radius ϱ. Then $\mathcal{S} \subseteq \mathcal{E} \subset K$. According to a), the radius of all the inscribed spheres of the tetrahedra from $\widetilde{\mathcal{T}}_m$ is $2^{-m}\tilde{\varrho}$. Hence, the mapping F transforms all these spheres onto ellipsoids that are congruent to $\mathcal{E}$. These ellipsoids will be inscribed in the corresponding tetrahedra from $\mathcal{T}_m$ and, obviously, the length of their shortest semi-axes will be $2^{-m}\varrho$. One immediately sees that for any $K' \in \mathcal{T}_m$ there exists a sphere $\mathcal{S}'$ with radius $2^{-m}\varrho$ such that $\mathcal{S}' \subset K'$. Consequently, the family $\{\mathcal{T}_m\}$ is strongly regular and since (4.83) implies

$$\frac{2^{-m}\varrho}{h_m} \ge \frac{2^{-m}\varrho}{2^{-m}\|Q\|} = \frac{\varrho}{\|Q\|},$$

the corresponding coefficient $\varkappa$ can be chosen as $\varkappa = \varrho / \|Q\|$.

c) Let $\overline{\Omega}$ be an arbitrary polyhedron and let $\mathcal{T}$ be an arbitrary decomposition of $\overline{\Omega}$ into tetrahedra — see Theorem 4.21. According to b), a strongly regular family $\{\mathcal{T}_m(K)\}_{m=0}^{\infty}$ with a coefficient $\varkappa_K$ corresponds to any $K \in \mathcal{T}$. Setting

$$\overline{\mathcal{T}}_m = \bigcup_{K \in \mathcal{T}} \mathcal{T}_m(K), \quad m = 0, 1, 2, \dots,$$

we can easily verify that $\overline{\mathcal{T}}_m$ is a decomposition of $\overline{\Omega}$. Let $h_m(K)$ be the maximum diameter of tetrahedra from $\mathcal{T}_m(K)$. Then $\max_{K \in \mathcal{T}} h_m(K)$ represents the discretization parameter of $\overline{\mathcal{T}}_m$. Thus, $\{\overline{\mathcal{T}}_m\}_{m=0}^{\infty}$ is a family of decompositions of $\overline{\Omega}$, since by (4.83)

$$h_m(K) \to 0 \quad \text{as } m \to \infty$$

for any K from the finite set $\mathcal{T}$. This family is strongly regular, since the corresponding coefficient $\overline{\varkappa}$ can be chosen as $\overline{\varkappa} = \min_{K \in \mathcal{T}} \varkappa_K$. □

Definition 4.26. *A family $\mathcal{F} = \{\mathcal{T}_h\}$ of decompositions of a polyhedron into tetrahedra is said to be semiregular if there exists a constant $\bar{\gamma} < \pi$ such that for any $\mathcal{T}_h \in \mathcal{F}$ and any $K \in \mathcal{T}_h$ we have*

$$\gamma_K \leq \bar{\gamma}, \tag{4.84}$$
$$\varphi_K \leq \bar{\gamma}, \tag{4.85}$$

where γ_K is the maximum angle of all triangular faces of the tetrahedron K, and φ_K is the maximum angle between faces of K.

In Examples 4.32 and 4.33, we show that the maximum angle conditions (4.84) and (4.85) are independent.

Theorem 4.27. *Any regular family of decompositions of a polyhedron into tetrahedra is semiregular.*

For the proof see [Křížek, 1992, p. 514].

Remark 4.28. The converse theorem is not true. One may construct a sequence of decompositions, e.g., of a cube analogously to Figure 4.12. When $\mathcal{F}$ is semiregular, tetrahedra may degenerate to "triangles" or "needles" (see Figure 4.20). Such tetrahedra do not fulfill the requirements of Definition 4.23 but fulfill Definition 4.26.

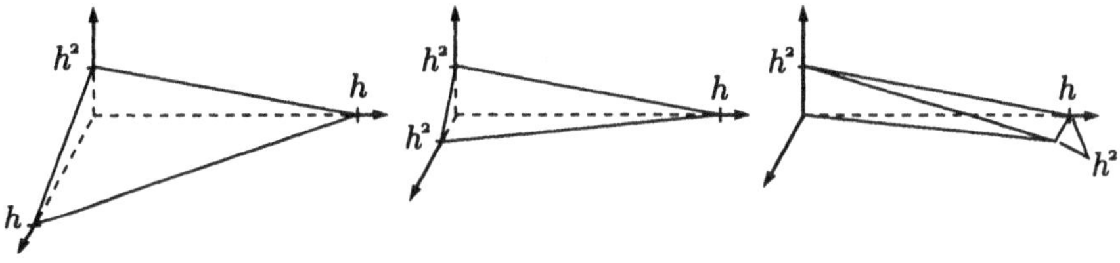

Figure 4.20.

Lemma 4.29. *Let $\alpha \leq \beta \leq \gamma$ be angles of an arbitrary face of an arbitrary tetrahedron for which (4.84) holds. Then $\gamma \geq \pi/3$ and*

$$\beta, \gamma \in \left[\frac{\pi - \bar{\gamma}}{2}, \bar{\gamma}\right]. \tag{4.86}$$

Proof. As $\alpha + \beta + \gamma = \pi$ and $\gamma \leq \bar{\gamma}$, we have $\alpha + \beta \geq \pi - \bar{\gamma}$. From here and the inequalities $\alpha \leq \beta \leq \gamma$ and (4.84), we come to

$$\frac{\pi - \bar{\gamma}}{2} \leq \beta \leq \gamma \leq \bar{\gamma},$$

i.e., (4.86) holds. □

Lemma 4.30. *Let A be an arbitrary vertex of an arbitrary tetrahedron K for which (4.85) holds and let $\chi \le \psi \le \varphi$ be angles between faces passing through A. Then $\varphi > \pi/3$ and*

$$\psi, \varphi \in \left(\frac{\pi - \bar{\gamma}}{2}, \bar{\gamma} \right]. \tag{4.87}$$

Proof. The intersection of K and a sufficiently small sphere with the centre A is a spatially curved triangle with the angles χ, ψ, φ – see Figure 4.21. From the Riemannian geometry we know that $\chi + \psi + \varphi > \pi$. Therefore, as $\varphi \le \bar{\gamma}$, we have $\chi + \psi > \pi - \bar{\gamma}$, i.e., (4.87) holds, and clearly $\varphi > \pi/3$. □

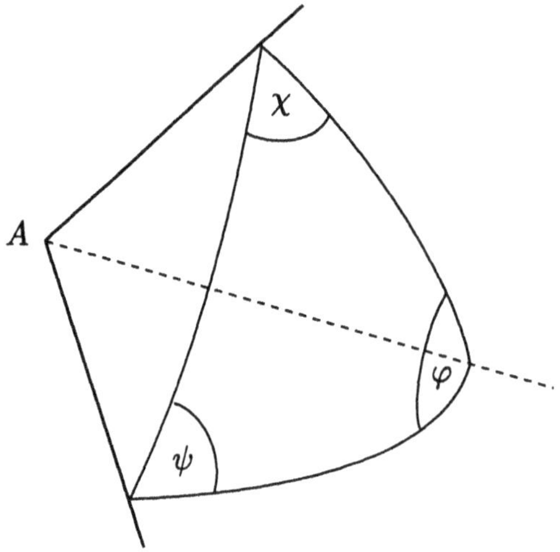

Figure 4.21.

From Lemma 4.30 and Definition 4.26 we see that $\bar{\gamma} \in (\frac{\pi}{3}, \pi)$.

As usual, for any tetrahedron K and any $v \in C(K)$ we denote by $\pi_K v$ the standard linear interpolant of v on K, i.e., $\pi_K v$ is a linear polynomial and $\pi_K v = v$ at all vertices of K. We derive its approximation properties under the maximum angle conditions (4.84) and (4.85). Recall that $W^k_\infty(K)$, $k \in \{0, 1, ...\}$, is the Sobolev space of functions whose generalized derivatives up to order k belong to $L^\infty(K)$.

To derive approximation properties of the linear interpolant we use a different technique to that of Section 4.4, since it is difficult to modify Lemma 4.12 for the three-dimensional case.

Theorem 4.31. *Let $\mathcal{F}$ be a semiregular family of decompositions of a polyhedron into tetrahedra. Then there exists a constant $C > 0$ such that for any $\mathcal{T}_h \in \mathcal{F}$ and any $K \in \mathcal{T}_h$ we have*

$$\|v - \pi_K v\|_{1,\infty} \le C h_K |v|_{2,\infty} \quad \forall v \in C^2(K). \tag{4.88}$$

Proof. Let $K \in \mathcal{T}_h \in \mathcal{F}$, $v \in C^2(K)$ be arbitrary and let e be an arbitrary edge of K. Set

$$w = v - \pi_K v \quad \text{on } K. \tag{4.89}$$

Since $w = 0$ at all vertices of K, there exists by Rolle's theorem a point $Q \in e$ such that

$$t^T \operatorname{grad} w(Q) = 0, \tag{4.90}$$

where t is a unit vector along e. Set

$$z = t^T \operatorname{grad} w \quad \text{in } K. \tag{4.91}$$

Let P be an arbitrary fixed point in the interior of K and let s be a unit vector along QP. Then by (4.90) and (4.91), we have $z(Q) = 0$, and since z is a Lipschitz continuous function on QP, we further have

$$z(P) = \int_Q^P s^T \operatorname{grad} z \, ds = \int_Q^P s^T (\operatorname{hes} w) t \, ds,$$

where $\operatorname{hes} w$ is the matrix of the second derivatives of w. From here, (4.91) and (4.89) we obtain

$$|t^T \operatorname{grad} w(P)| \leq h_K |w|_{2,\infty} = h_K |v|_{2,\infty}, \tag{4.92}$$

i.e., the orthogonal projection of the vector $\operatorname{grad} w$ on any one of the six edges of K cannot exceed $h_K |v|_{2,\infty}$. Consider now three such edges and the corresponding unit vectors t_1, t_2, t_3 parallel with them (they will be specified later). Writing

$$\operatorname{grad} w(P) = \sum_{j=1}^{3} c_j t_j, \quad c_j \in \mathbf{R}^1, \tag{4.93}$$

we see that the coefficients c_j fulfill the simultaneous equations

$$t_i^T \operatorname{grad} w(P) = \sum_{j=1}^{3} c_j t_i^T t_j, \quad i = 1, 2, 3, \tag{4.94}$$

with the symmetric matrix $M = (t_i^T t_j)_{i,j=1}^3$. Denote by M^* the matrix of cofactors of the entries of M, and by $\|\cdot\|$ the Euclidean norm. Then

$$M^{-1} = (M^{-1})^T = \frac{1}{\det M} M^*, \quad \|M^*\| \leq 6, \tag{4.95}$$

since $\|t_i\| = 1$ for $i = 1, 2, 3$. Consequently, by (4.92), (4.93) (4.94) and (4.95),

$$\|\operatorname{grad} w(P)\| \leq \sum_{j=1}^{3} |c_j| \leq 9 \|M^{-1}\| \max_i |t_i^T \operatorname{grad} w(P)| \leq 9 \left| \frac{6}{\det M} \right| h_K |v|_{2,\infty}. \tag{4.96}$$

Let $A_0A_1A_2$ be an arbitrary face of K and A_3 the opposite vertex. Without lost of generality we may assume that the maximum angle, say γ, of the face $A_0A_1A_2$ is at the vertex A_0. From (4.86) we find that

$$\overline{C} := \min\left(\sin\frac{\pi-\bar{\gamma}}{2}, \sin\bar{\gamma}\right) \leq \sin\gamma. \tag{4.97}$$

Let t_1 and t_2 be the unit vectors along A_0A_1 and A_0A_2. Let the angle, say ω, between the faces $A_0A_1A_2$ and $A_0A_1A_3$ be not less than the angle between $A_0A_1A_2$ and $A_0A_2A_3$ (otherwise we exchange the notation of the vertices A_1 and A_2, see Figure 4.22 a)). Then, by (4.87) and (4.97), we come to

$$\sin\omega \geq \overline{C}. \tag{4.98}$$

Furthermore, consider the triangle $A_0A_1A_3$. Choose $k \in \{0,1\}$ so that the angle, say ξ, at the vertex A_k is not less than that at the vertex A_{1-k}. Then by (4.86) and (4.97)

$$\sin\xi \geq \overline{C}. \tag{4.99}$$

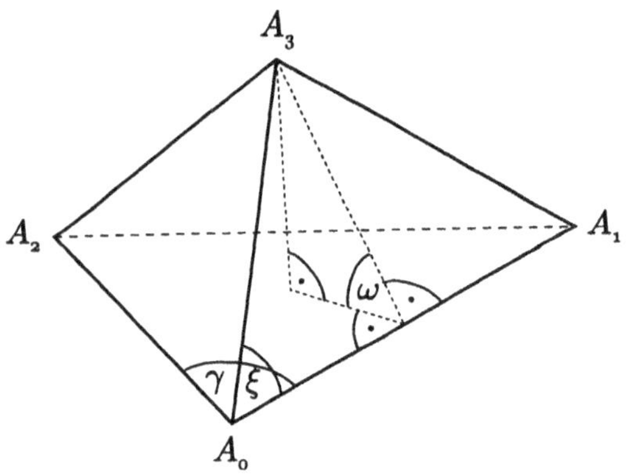

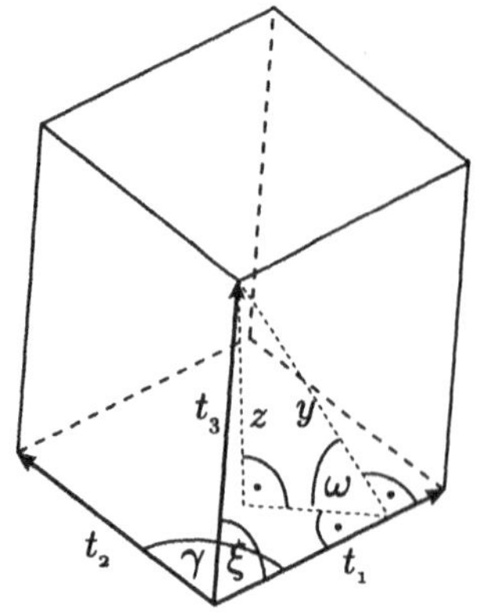

Figure 4.22.

Finally, let t_3 be the unit vector along A_kA_3. It is known that $|\det M|$ is equal to the volume of the parallelepiped generated by the unit vectors t_1, t_2, t_3 (see Figure 4.22 b) for the case $k = 0$). Denote by z the length of its spatial altitude perpendicular to t_1 and t_2. Denote by y the length of the altitude perpendicular to t_1, which lies in the plane given by t_1, t_3. Since $\|t_i\| = 1$ for $i = 1,2,3$, we get by (4.97), (4.98) and (4.99),

$$|\det M| = z\sin\gamma = y\sin\omega\sin\gamma = \sin\xi\sin\omega\sin\gamma \geq \overline{C}^3 > 0.$$

From here and (4.96), we obtain

$$|w|_{1,\infty} \leq Ch_K|v|_{2,\infty} \quad \forall v \in C^2(K),$$

since P was an arbitrary point inside K. By [Ciarlet, p. 118, p. 120],

$$|w|_{0,\infty} \leq Ch_K^2|v|_{2,\infty} \quad \forall v \in W_\infty^2(K)$$

without any regularity assumptions upon the family $\mathcal{F}$. □

In the next two examples we will illustrate that the maximum angle conditions (4.84) and (4.85) are independent. Moreover, we show that if (4.84) or (4.85) is violated, then (4.88) need not be valid.

Example 4.32. Let $A = (0,0,0)$, $B = (h,0,0)$, $C = (h,0,h^3)$ and $D = (-h,h^3,0)$ be the vertices of K (see Figure 4.23 a)) for some $h \in (0,1)$. We observe that the angle DAB tends to π as $h \to 0$, i.e., (4.84) is violated and we show that (4.85) remains true. Consider four outward unit normals of ∂K:

$$n_1 = (0,-1,0)^T, \quad n_2 = (0,0,-1)^T, \quad n_3 = \frac{1}{\sqrt{h^4+4}}(h^2,2,0)^T,$$
$$n_4 = \frac{1}{\sqrt{1+2h^{-4}}}(-1,-h^{-2},h^{-2})^T.$$

Since their mutual scalar products are always less than $\sqrt{2}/2$, we have

$$-\cos\varphi_{ij} = \cos(\pi-\varphi_{ij}) = \|n_i\|\|n_j\|\cos(\pi-\varphi_{ij}) = n_i^Tn_j < \sqrt{2}/2, \quad (i<j)$$

where φ_{ij} is the angle between those faces which are perpendicular to n_i and n_j, i.e., $\cos\varphi_{ij} > -\sqrt{2}/2$ and thus $\varphi_K < \frac{3}{4}\pi$. Putting

$$v(x_1,x_2,x_3) = x_1^2,$$

we find that $v(A) = 0$, $v(B) = v(C) = v(D) = h^2$. Hence, as $\pi_K v$ is linear,

$$\frac{\partial}{\partial x_2}(v-\pi_K v) = -\frac{\partial(\pi_k v)}{\partial x_2} = \frac{(\pi_K v)(A) - (\pi_K v)((B+D)/2)}{\frac{h^3}{2}}$$
$$= -\frac{v(B)+v(D)}{h^3} = -\frac{2h^2}{h^3} \to -\infty$$

for $h \to 0$. Hence, (4.88) does not hold.

Example 4.33. Let $A = (-h,0,0)$, $B = (0,h^3,-h)$, $C = (h,0,0)$, $D = (0,h^3,h)$ be the vertices of K (see Figure 4.23 b)) for some $h \in (0,1)$. We observe that (4.84) holds as $\gamma_K < \pi/2$. However, $\varphi_K \to \pi$ when $h \to 0$, where φ_K is the angle between faces ABD and BCD. For the same v as in Example 4.32 we have

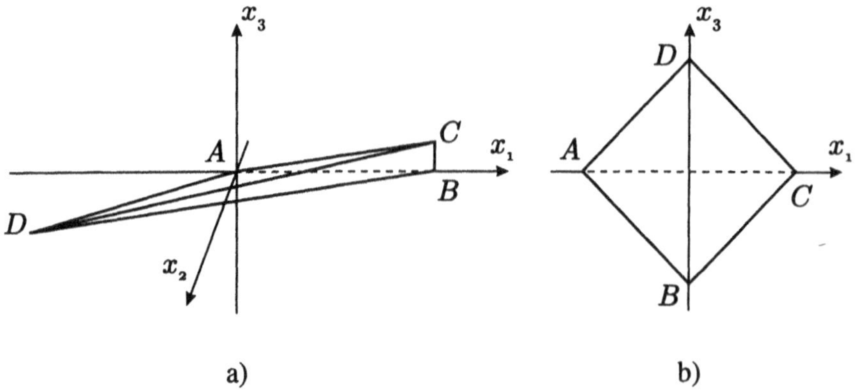

Figure 4.23.

$v(A) = v(C) = h^2$, $v(B) = v(D) = 0$. Thus again from the linearity of $\pi_K v$, we get

$$\frac{\partial}{\partial x_2}(v - \pi_K v) = -\frac{\partial(\pi_K v)}{\partial x_2} = \frac{(\pi_K v)((A+C)/2) - (\pi_K v)((B+D)/2)}{h^3}$$
$$= \frac{v(A) + v(C) - v(B) - v(D)}{2h^3} = h^{-1} \to \infty \quad \text{as } h \to 0.$$

Remark 4.34. Let $\mathcal{F}$ be a semiregular family of decompositions of a polyhedron $\overline{\Omega}$ whose boundary is Lipschitz continuous. Then by (4.88) for any $p \geq 1$,

$$\|v - \pi_h v\|_{1,p} \leq C_p \|v - \pi_h v\|_{1,\infty} \leq C_p C h |v|_{2,\infty} \quad \forall v \in C^2(\overline{\Omega}),$$

where $\pi_h v$ is the standard linear interpolant of v over $\mathcal{T}_h \in \mathcal{F}$ and $C_p > 0$ is a constant characterizing the topological imbedding $W^1_\infty(\Omega) \subset W^1_p(\Omega)$. In particular, for $p = 2$ we get

$$\|v - \pi_h v\|_1 = \mathcal{O}(h)$$

for v sufficiently smooth. This estimate can be used to modify Theorem 4.16 for linear tetrahedral elements. Note that C_p for any $p \in [1, \infty]$ can be bounded by a constant independent of p, since

$$\| \cdot \|_{0,p} \leq (\operatorname{meas} \Omega)^{1/p} \| \cdot \|_{0,\infty} \leq \max(1, \operatorname{meas} \Omega) \| \cdot \|_{0,\infty}.$$

Remark 4.35. Analogously to Remark 4.19, almost degenerate tetrahedra can be used to cover thin slots or strips, which saves computer memory. They should also be used along polyhedron edges, when the true solution of the boundary value problem is "smooth" along edges and "nonsmooth" across edges. An elegant generalization of the maximum angle condition for other elements in $\mathbf{R}^d$ is presented in [Jamet, p. 50]. Roughly speaking, we can still achieve the optimal approximation order if elements degenerate, except the case when all element edges are almost parallel to some hyperplane in $\mathbf{R}^d$.

If we look for u_h in the space

$$V_h = \{v_h \in V \mid v|_K \in P_K \quad \forall K \in \mathcal{T}_h\},$$

where $P_K \supseteq P_k(K)$, $k \in \{1, 2, ...\}$, then under certain assumptions, see, e.g., [Ciarlet], [Jamet], [Johnson], [Koukal, 1973a], [Strang, Fix], we can derive that

$$\|u - u_h\|_{1,p} = \mathcal{O}(h^k), \quad \|u - u_h\|_{0,p} = \mathcal{O}(h^{k+1}),$$

provided u is smooth enough, $p \in [1, \infty)$ and $d \in \{1, 2, ...\}$. For estimates in $L^\infty(\Omega)$-norm see also [Rannacher, Scott], [Whiteman].

The accuracy of FE-approximations can be increased by various superconvergence and postprocessing techniques, see, e.g., [Chen], [Chen, Huang], [Hlaváček, Křížek, 1987], [Hlaváček, Křížek, Pištora], [Křížek, Neittaanmaki, 1987], [Lin, Zhu], [Wahlbin].

4.6. Affine curved elements

In many applications we have to deal with domains whose boundaries are piecewise smooth. First of all we specify what we mean by "piecewise smooth", because this notion is often understood only heuristically – especially in the three-dimensional space. The following definition is similar to [Bernardi, p. 1231].

Definition 4.36. *We say that a domain $\Omega \subset \mathbb{R}^d$, $d \geq 2$, has a piecewise C^k boundary, $k \in \{1, 2, ...\}$ or $k = \infty$, if there exists a finite number of mappings $\psi^l = (\psi_1^l, ..., \psi_d^l)\colon \bar{\omega}_l \to \mathbb{R}^d$, $l = 1, ..., q$, where ω_l is a bounded domain in $\mathbb{R}^{d-1}$, such that each ψ^l belongs to $(C^k(\bar{\omega}_l))^d$,*

$$\partial\Omega = \bigcup_{l=1}^{q} \{\psi^l(x) \mid x \in \bar{\omega}_l\}$$

and the $(d-1) \times d$ matrices

$$D\psi^l(x) = \left(\frac{\partial \psi_j^l(x)}{\partial x_i}\right)_{\substack{i=1,...,d-1 \\ j=1,...,d}}$$

have the maximum rank $d - 1$ for any $l = 1, ..., q$ and any $x \in \bar{\omega}_l$.

Remark 4.37. The last requirement on the maximum rank is essential. Set for instance $d = 2$, $\psi_1(x) = x^3$ and $\psi_2(x) = |x^3|$ for $x \in \bar{\omega} = [-1, 1]$. Then $\psi = (\psi_1, \psi_2) \in (C^2(\bar{\omega}))^2$ but the rank of the matrix $D\psi(0) = (0, 0)$ is not equal to 1. In this case we find that the graph of $\psi\colon \bar{\omega} \to \mathbb{R}^2$ consists of two straight line segments which are perpendicular at the origin (see Figure 4.24), i.e., the boundary has a vertex at $(0, 0)$ and is thus not C^2 there, even though it is described by a

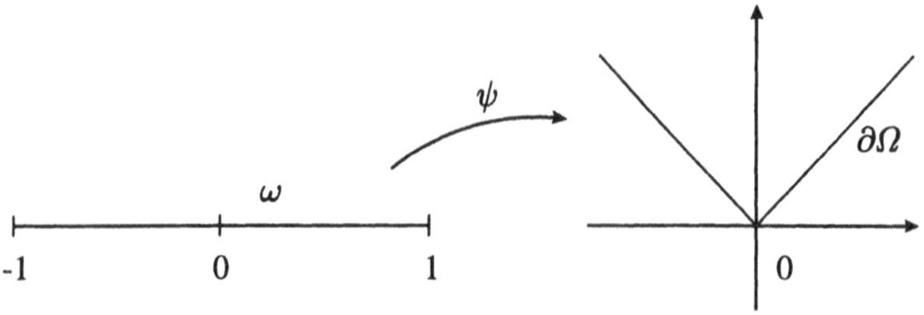

Figure 4.24.

smooth function ψ. Similarly we could construct another smooth ψ_1, ψ_2 which yield, e.g., infinitely many vertices for which the rank of the corresponding matrix $D\psi^l(x)$ is zero.

There are several ways of applying the finite element method to domains with piecewise smooth boundaries. One way is to approximate them by polygonal (polyhedral) domains and then use uncurved elements from Section 4.1. However, if for instance we employ the common straight triangular quadratic elements (see Figure 4.25 a)) to the problem (3.44)–(3.46), we are not able to achieve the optimal order of convergence $\mathcal{O}(h^2)$ in $H^1(\Omega)$-norm (cf. Remark 4.35). We can obtain, in general, at most $\mathcal{O}(h^{3/2})$ due to the rough approximation of the curved boundary $\partial\Omega$ (see [Thomée, 1973, p. 33]). Using a piecewise parabolic approximation of the boundary, the optimal order $\mathcal{O}(h^2)$ may be preserved if we employ the isoparametric quadratic elements [Ciarlet], [Johnson] – see Figure 4.25 b) or the ideal Zlámal element [Zlámal, 1973]. The function spaces of these curved elements consist of nonpolynomial functions which cause several computational difficulties. For instance, we always have to integrate nonpolynomial functions over the reference element to compute the stiffness matrix (see [Křížek, Neittaanmäki, 1990, p. 76]). We shall present an almost unknown approach to this problem (see [Koukal, 1973b]), where it suffices to integrate only polynomials over a special triangulation (see Figure 4.25 c)) to obtain the stiffness matrix corresponding, e.g., to the Poisson equation with Dirichlet boundary conditions.

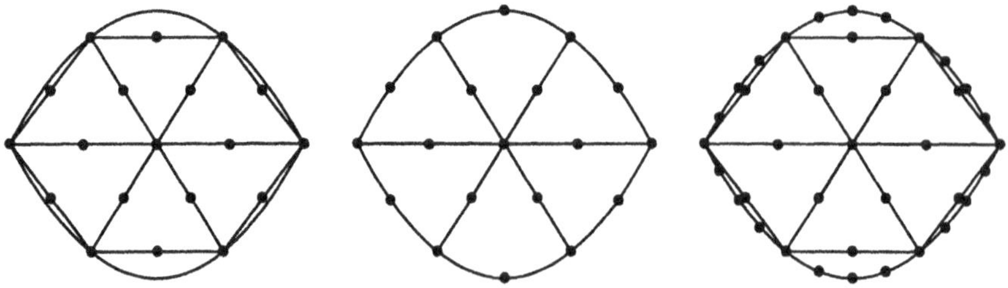

a) b) c)

Figure 4.25.

We shall examine curved quadratic finite elements which provide (under some assumptions) the optimal interpolation order $\mathcal{O}(h^2)$ in the H^1-norm globally. They can be transformed to the reference elements by means of affine mappings, i.e., no nonlinear transformation is necessary (as it is for isoparametric quadratic elements). This property simplifies assembling the stiffness matrices.

Koukal's biangular element (K, P_K, Σ_K). The set K is a curved closed biangle which is the bounded domain formed from a parabola and a straight line that intersects the parabola twice, $P_K = P_2(K)$ is the space of polynomials of degree 2 at most and the set of the degrees of freedom reads

$$\Sigma_K = \{p(A_j),\ j = 1, ..., 6 \mid p \in P_K\},$$

where A_1, A_2 are the vertices of K, A_4 is the midpoint of the straight line segment A_1A_2 and the other A_j (see Figure 4.26) will be defined later. By A_j we shall also mean the column vector of coordinates of the point A_j.

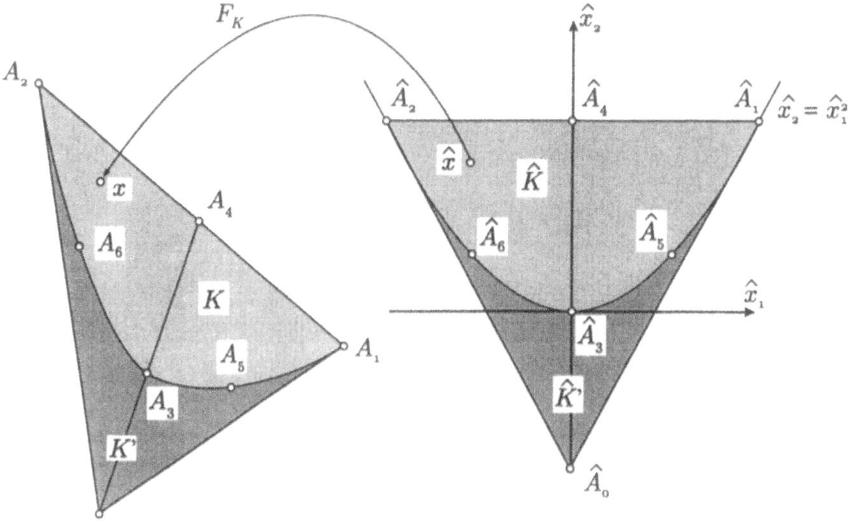

Figure 4.26.

Now let us introduce the reference biangle

$$\widehat{K} = \{(\hat{x}_1, \hat{x}_2)^T \in \mathbf{R}^2 \mid \hat{x}_1^2 \le \hat{x}_2 \le 1\},$$

the points

$$\widehat{A}_0 = (0, -1)^T, \quad \widehat{A}_1 = (1, 1)^T, \quad \widehat{A}_2 = (-1, 1)^T, \quad \widehat{A}_3 = (0, 0)^T,$$
$$\widehat{A}_4 = (0, 1)^T, \quad \widehat{A}_5 = (\tfrac{1}{2}, \tfrac{1}{4})^T, \quad \widehat{A}_6 = (-\tfrac{1}{2}, \tfrac{1}{4})^T,$$

and such an affine mapping F_K that

$$A_j = F_K(\widehat{A}_j), \quad j = 0, 1, 2,$$

where A_0 is the intersection of the tangents to the curved side of K at the vertices A_1 and A_2 as sketched in Figure 4.26. Analogously to (4.48)–(4.49),

$$F_K(\hat{x}) = B_K\hat{x} + \tfrac{1}{4}(A_1 + A_2 + 2A_0), \tag{4.100}$$

where the matrix

$$B_K = \begin{pmatrix} b_{11} & b_{12} \\ b_{21} & b_{22} \end{pmatrix} = \tfrac{1}{2}(A_1 - A_2, \tfrac{1}{2}(A_1 + A_2) - A_0) \tag{4.101}$$

is nonsingular.

Since F_K is an affine mapping, we have $A_4 = F_K(\widehat{A}_4)$ and

$$K = F_K(\widehat{K}) \tag{4.102}$$

in such a way that the curved side of $\widehat{K}$ is mapped onto the curved side of K, $A_3 = F(\widehat{A}_3)$ is simultaneously the midpoint of the straight line segment A_0A_4 and the intersection of this segment and the curved side of K. Setting $A_5 = F_K(\widehat{A}_5)$ and $A_6 = F_K(\widehat{A}_6)$, we see that

$$A_j = F_K(\widehat{A}_j), \quad j = 0, 1, ..., 6. \tag{4.103}$$

Due to (4.102), computation of the stiffness matrix of the element K may be transformed to the reference element $\widehat{K}$. To this end we set for $(\hat{x}_1, \hat{x}_2)^T \in \widehat{K}$

$$\begin{aligned}
\hat{p}_1(\hat{x}_1, \hat{x}_2) &= \tfrac{1}{2}\hat{x}_1^2 + \tfrac{2}{3}\hat{x}_1\hat{x}_2 + \tfrac{2}{3}\hat{x}_2^2 - \tfrac{1}{6}\hat{x}_1 - \tfrac{2}{3}\hat{x}_2, \\
\hat{p}_2(\hat{x}_1, \hat{x}_2) &= \hat{p}_1(-\hat{x}_1, \hat{x}_2), \\
\hat{p}_3(\hat{x}_1, \hat{x}_2) &= 4(\hat{x}_2 - 1)(\hat{x}_2 - \tfrac{1}{4}), \\
\hat{p}_4(\hat{x}_1, \hat{x}_2) &= \hat{x}_2 - \hat{x}_1^2, \\
\hat{p}_5(\hat{x}_1, \hat{x}_2) &= \tfrac{4}{3}(1 - \hat{x}_2)(\hat{x}_1 + 2\hat{x}_2), \\
\hat{p}_6(\hat{x}_1, \hat{x}_2) &= \hat{p}_5(-\hat{x}_1, \hat{x}_2).
\end{aligned} \tag{4.104}$$

A direct calculation leads now to

$$\hat{p}_i(\widehat{A}_j) = \delta_{ij}, \quad i,\ j = 1, ..., 6. \tag{4.105}$$

Hence, $\hat{p}_1, ..., \hat{p}_6$ are linearly independent. Since $\hat{p}_i \in P_2(\widehat{K})$ and $\dim P_2(\widehat{K}) = 6$, the functions $\{\hat{p}_i\}_{i=1}^6$ represent a basis of the reference element $(\widehat{K}, P_2(\widehat{K}), \Sigma_{\widehat{K}})$. Their contours are sketched in Figure 4.27.

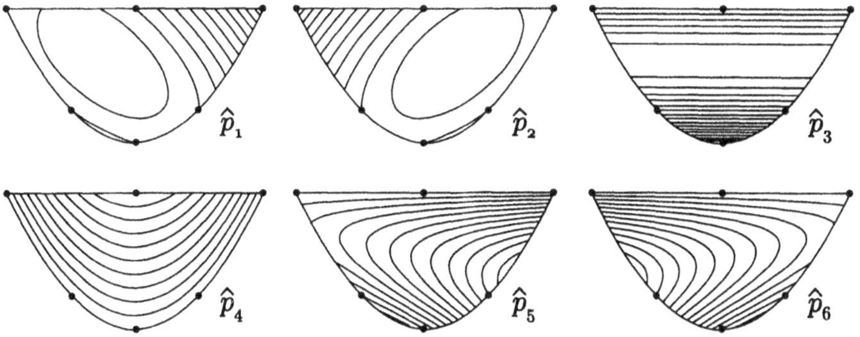

Figure 4.27.

For an arbitrary biangle element K we define the basis functions $p_1, ..., p_6$ by

$$p_i(x) = \hat{p}_i\left(F_K^{-1}(x)\right), \quad x \in K, \quad i = 1, ..., 6. \tag{4.106}$$

Due to (4.102) and (4.106), the computation of the stiffness matrix of the element K can be replaced by the computation of the stiffness matrix of the reference element $\widehat{K}$. For example, in view of (4.106) and (4.100) we have

$$\operatorname{grad} p_i(x) = B_K^{-T} \operatorname{grad} \hat{p}_i\left(F_K^{-1}(x)\right), \quad x \in K, \tag{4.107}$$

where

$$B_K^{-T} = (B_K^{-1})^T = \frac{1}{\det B_K}\begin{pmatrix} b_{22} & -b_{21} \\ -b_{12} & b_{11} \end{pmatrix} = \frac{1}{\det B_K} B_K^* \tag{4.108}$$

and B_K^* is the matrix of cofactors of the entries of B_K. By virtue of (4.102) and (4.107) the entries of the stiffness matrix of the element K corresponding to the Dirichlet problem for the Poisson equation can be expressed as

$$\begin{aligned}(\operatorname{grad} p_i, \operatorname{grad} p_j)_{0,K} &= (B_K^{-T} \operatorname{grad} \hat{p}_i, B_K^{-T} \operatorname{grad} \hat{p}_j |\det B_K|)_{0,\widehat{K}} \\ &= \frac{1}{|\det B_K|}(B_K^* \operatorname{grad} \hat{p}_i, B_K^* \operatorname{grad} \hat{p}_j)_{0,\widehat{K}}, \quad i,j = 1, ..., 6,\end{aligned} \tag{4.109}$$

i.e., the mapping F_K^{-1} enables us to transform all computations to the reference biangle $\widehat{K}$. The last scalar product in (4.109) is, in fact, the integral of a quadratic function over $\widehat{K}$. For its evaluation the following calculation may be useful (cf. Figure 4.28)

$$\begin{aligned}\int_{\widehat{K}} \hat{x}_1^r \hat{x}_2^s \, d\hat{x} &= \int_{-1}^{1} \int_{\hat{x}_1^2}^{1} \hat{x}_1^r \hat{x}_2^s \, d\hat{x}_2 \, d\hat{x}_1 = \int_{-1}^{1} \hat{x}_1^r \left[\frac{\hat{x}_2^{s+1}}{s+1}\right]_{\hat{x}_1^2}^{1} d\hat{x}_1 \\ &= \frac{1}{s+1}\int_{-1}^{1} \hat{x}_1^r (1 - \hat{x}_1^{2s+2}) \, d\hat{x}_1 \\ &= \frac{1}{s+1}\left[\frac{\hat{x}_1^{r+1}}{r+1} - \frac{\hat{x}_1^{2s+r+3}}{2s+r+3}\right]_{-1}^{1} \\ &= \frac{1-(-1)^{r+1}}{s+1}\left(\frac{1}{r+1} - \frac{1}{2s+r+3}\right),\end{aligned} \tag{4.110}$$

where $r, s \geq 0$ are integers. Thus no numerical integration is needed. This also applies to the computation of the stiffness matrix of the element K corresponding to the Dirichlet problem for an elliptic equation with polynomial coefficients on K.

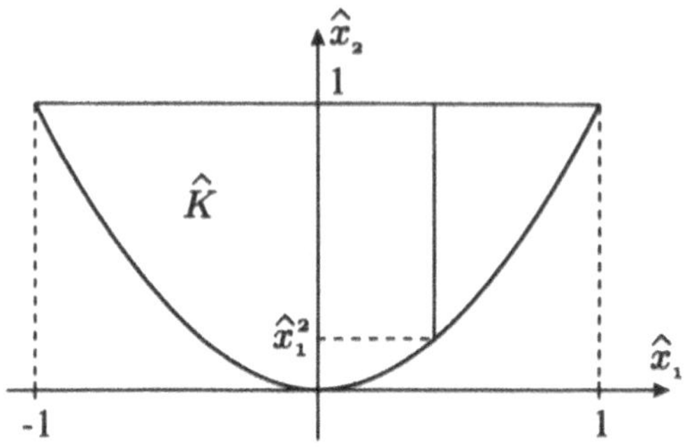

Figure 4.28.

Koukal's triangular element $(K', P_{K'}, \Sigma_{K'})$. The set K' is a triangle with one curved side A_1A_2, which is a part of a parabola and two straight sides A_0A_1 and A_0A_2 which are tangents to the parabola at vertices A_1 and A_2 (see Figure 4.29), $P_{K'} = P_2(K')$, and the set $\Sigma_{K'} = \{\phi_i\}_{i=0}^5$ contains, unlike the biangle element, two directional derivative degrees of freedom:

$$\begin{aligned} \Phi_i(p) &= p(A_i), \qquad i = 0, 1, 2, 3, \\ \Phi_{i+3}(p) &= \frac{\partial p(A_i)}{\partial \tau_i}, \quad i = 1, 2, \end{aligned} \tag{4.111}$$

where $p \in P_{K'}$ and $\tau_i = A_0 - A_i$, $i = 1, 2$.

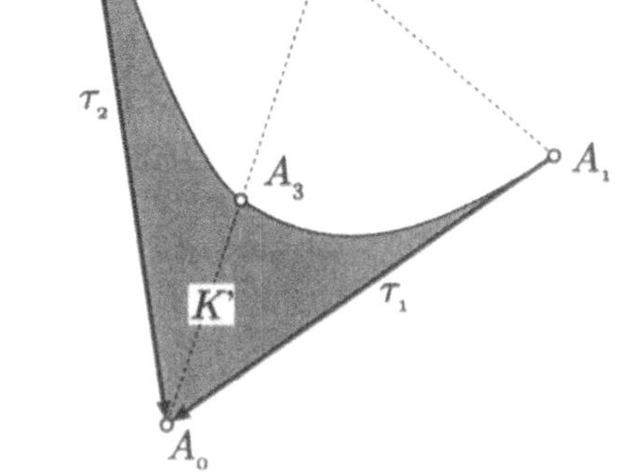

Figure 4.29.

The affine mapping (4.100) again represents a one-to-one correspondence between the reference curved triangle

$$\widehat{K}' = \{(\hat{x}_1, \hat{x}_2)^T \in \mathbf{R}^2 \mid 2|\hat{x}_1| - 1 \leq \hat{x}_2 \leq \hat{x}_1^2\}$$

and K'. The basis functions over K' satisfying

$$\widehat{\Phi}_i(\hat{p}^j) = \delta_{ij}, \quad i,\ j = 0, ..., 5,$$

are defined as follows

$$\begin{aligned}
\hat{p}^0(\hat{x}_1, \hat{x}_2) &= \hat{x}_1^2 - \hat{x}_2, \\
\hat{p}^1(\hat{x}_1, \hat{x}_2) &= \frac{1}{4}(2\hat{x}_1 + \hat{x}_2 + 1)(3\hat{x}_1 - 2\hat{x}_2), \\
\hat{p}^2(\hat{x}_1, \hat{x}_2) &= \hat{p}^1(-\hat{x}_1, \hat{x}_2), \\
\hat{p}^3(\hat{x}_1, \hat{x}_2) &= (2\hat{x}_1 + \hat{x}_2 + 1)(-2\hat{x}_1 + \hat{x}_2 + 1), \\
\hat{p}^4(\hat{x}_1, \hat{x}_2) &= \frac{\sqrt{5}}{4}(2\hat{x}_1 + \hat{x}_2 + 1)(\hat{x}_1 - \hat{x}_2), \\
\hat{p}^5(\hat{x}_1, \hat{x}_2) &= \hat{p}^4(-\hat{x}_1, \hat{x}_2).
\end{aligned} \tag{4.112}$$

Their contours are sketched in Figure 4.30.

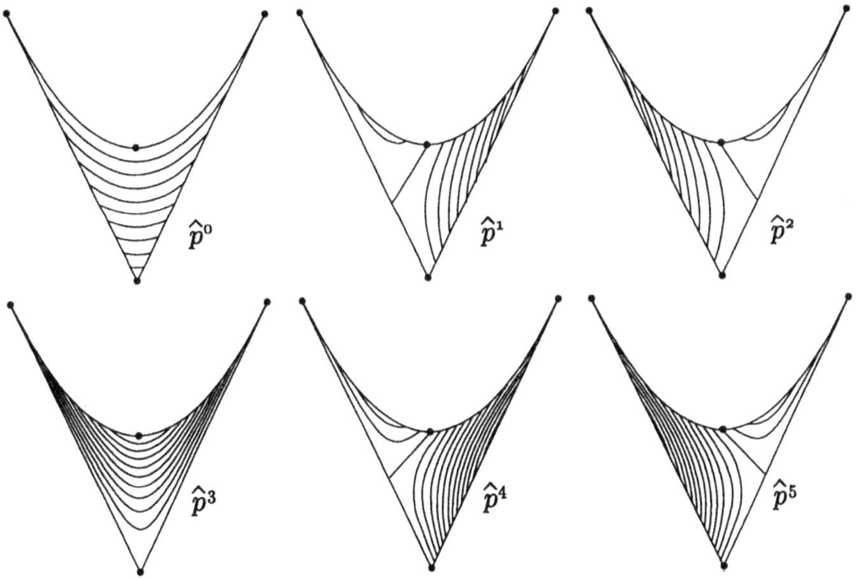

Figure 4.30.

Computation of the stiffness matrix of the element K' can be now transformed to the reference triangle $\widehat{K}'$ as in (4.109).

Consider now a bounded plane domain $\Omega \neq \emptyset$ with a Lipschitz continuous boundary which consists of a finite number of parabolic arcs and straight line segments. By a triangulation of $\overline{\Omega}$ we shall mean a set $\mathcal{T}_h$ of elements satisfying the following properties:

(1) each element $K \in \mathcal{T}_h$ is biangle or curved or straight triangle,

(2) the interiors of these elements are mutually disjoint,

(3) $\overline{\Omega} = \bigcup_{K \in \mathcal{T}_h} K$,

(4) any side of any $K \in \mathcal{T}_h$ is a side of another element from $\mathcal{T}_h$ or a subset of the boundary $\partial\Omega$,

(5) curved sides of elements belong only to curved parts of $\partial\Omega$ (see Figure 4.31).

It is clear that the points of inflexion of $\partial\Omega$ coincide with some vertices of elements from $\mathcal{T}_h$. The finite element space X_h associated to $\mathcal{T}_h$ is given by

$$X_h = \left\{ v_h \in C(\overline{\Omega}) \mid v_h\big|_K \in P_2(K) \quad \forall K \in \mathcal{T}_h \right\}.$$

Its degrees of freedom are symbolically sketched in Figure 4.31.

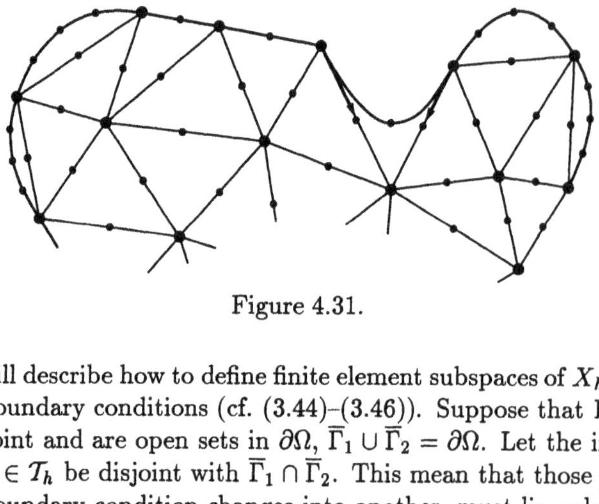

Figure 4.31.

Next we will describe how to define finite element subspaces of X_h for problems with mixed boundary conditions (cf. (3.44)–(3.46)). Suppose that Γ_1 and Γ_2 are mutually disjoint and are open sets in $\partial\Omega$, $\overline{\Gamma}_1 \cup \overline{\Gamma}_2 = \partial\Omega$. Let the interior of any side of any $K \in \mathcal{T}_h$ be disjoint with $\overline{\Gamma}_1 \cap \overline{\Gamma}_2$. This mean that those points, where one type of boundary condition changes into another, must lie only in the set of vertices of all $K \in \mathcal{T}_h$.

So let now $S = \partial K \cap \overline{\Gamma}_1$ be the curved side of some $K \in \mathcal{T}_h$. Notice that the basis functions $\hat{p}_4$ and $\hat{p}^0$ vanish on the curved side of the reference elements (see (4.104), (4.112) and also Figures 4.27 and 4.30). Hence, if all degrees of freedom of some $p \in P_2(K)$ associated to the curved side S vanish, then $p = 0$ on the whole S. This fact enables us to construct easily the space

$$V_h = \left\{ v_h \in X_h \mid v_h = 0 \text{ on } \overline{\Gamma}_1 \right\}$$

which appears in the variational formulation of the discrete problem (4.34).

Remark 4.38. In a standard way we may define the V_h-interpolant $\pi_h v$ for a sufficiently smooth function v so that $\|v - \pi_h v\|_1 = \mathcal{O}(h^2)$ (cf. [Koukal, Křížek]). This estimate can be generalized (see [Koukal, 1973b]) to the case when $\partial\Omega$ consists of a finite number of parts of conic sections. However, in this case the function space of each element contains rational functions, which are, in general, nonpolynomial.

For further study of curved elements we refer to [Barrett, Elliott], [Bernardi], [Ciarlet], [Johnson], [Křížek, Neittaanmäki, 1990], [Lenoir], [Ženíšek, 1978] and [Zlámal, 1973].

5. Conjugate gradients

5.1. Preliminary remarks

Our concern in this chapter is with the solution of a system of real or complex linear algebraic equations

$$Ax = b, \tag{5.1}$$

where A is a nonsingular $N \times N$ matrix, $x = (x_1, ..., x_N)^T$ is the vector of unknowns and $b = (b_1, ..., b_N)^T$.

The problem (5.1) was first completely solved by Carl Friedrich Gauss (1777–1855). Recall that his well-known elimination method requires, in general, $\mathcal{O}(N^3)$ of arithmetic operations. One of its first computer implementations is described in [Verzuh, 1949], where 4 hours were necessary to solve (5.1) for $N = 10$. Nowadays modern parallel architecture computers enable us to solve millions of simultaneous linear algebraic equations, see [Ortega, Voigt], [Rice].

Let us point out that there exist direct solvers for (5.1) demanding less that $\mathcal{O}(N^{2.81})$ operations (see [Strassen, 1969]), or even $\mathcal{O}(N^{2.38})$ (see [Coppersmith, Winograd], [Strassen, 1990]). However, these solvers are not used in practical computations, since they are recursive, complicated and unstable.

From Chapter 4 we know that the use of the finite element method yields a system like (5.1), in which the matrix A is sparse. This means that only $\mathcal{O}(N)$ entries of A are different from zero. This fact makes iterative methods for (5.1) more applicable than direct methods based on the Gaussian elimination when N is very large. After elimination, the band of a sparse matrix is filled in by nonzero entries in general. In solving actual practical problems the dimension N can be so large that we are not able to solve (5.1) by direct methods due to limitations of the internal fast computer memory (see, e.g., Table 5.1 for $d = 3$). Note that the width of the band of A for $d = 1, 2, 3$ is typically $\mathcal{O}(1)$, $\mathcal{O}(N^{1/2})$, $\mathcal{O}(N^{2/3})$, respectively. On the other hand, iterative methods enable us to store only nonzero entries of A and, of course, some information about their positions in A. At the present time the most efficient iterative methods seem to be the conjugate gradient method and its various variants (cf. Example 5.12). In Table 5.1 we see the computer storage for one, two and three-dimensional problems when employing the Gaussian elimination (Gauss), the conjugate gradient method (CG) and the preconditioned conjugate gradient method (PCG) – cf. Section 5.3.

COMPUTER STORAGE			
Method	$d = 1$	$d = 2$	$d = 3$
Gauss	$\mathcal{O}(N)$	$\mathcal{O}(N^{3/2})$	$\mathcal{O}(N^{5/3})$
CG	$\mathcal{O}(N)$	$\mathcal{O}(N)$	$\mathcal{O}(N)$
PCG	$\mathcal{O}(N)$	$\mathcal{O}(N)$	$\mathcal{O}(N)$

Table 5.1.

COMPUTATIONAL COST			
Method	$d=1$	$d=2$	$d=3$
Gauss	$\mathcal{O}(N)$	$\mathcal{O}(N^2)$	$\mathcal{O}(N^{7/3})$
CG	$\mathcal{O}(N^2)$	$\mathcal{O}(N^{3/2})$	$\mathcal{O}(N^{4/3})$
PCG	$\mathcal{O}(N^{3/2})$	$\mathcal{O}(N^{5/4})$	$\mathcal{O}(N^{7/6})$

Table 5.2.

Table 5.2 shows the computational cost of the methods considered here when A is positive definite. Note that assembling the stiffness matrix requires only $\mathcal{O}(N)$ operations. The number of arithmetic operations for Gaussian elimination is proportional to the square of the width of the band multiplied by N. We again see that the use of the Gaussian elimination is not too efficient when $d > 1$, especially when $d = 3$ which corresponds to great values of N.

Anyhow, if we have enough computer time and memory, it is usually better to use direct methods than iterative methods (see Remark 5.25). Moreover, for any iterative method we have

$$\|u - \tilde{u}_h\|_{L^\infty(\Omega)} \leq \|u - u_h\|_{L^\infty(\Omega)} + \|u_h - \tilde{u}_h\|_{L^\infty(\Omega)},$$

where $\tilde{u}_h \in V_h$ is the V_h-interpolant corresponding to the last iteration. The right-hand side of this inequality is the sum of the discretization and iteration error. The iteration process should be stopped when these two errors have approximately the same size. Thus the last two rows of Table 5.2 illustrate how many arithmetic operations are necessary to reduce the iteration error down to the level of discretization error. However, the discretization error may converge very slowly if the weak solution u has, e.g., singularities. Hence, the last two rows of Table 5.2 hold if the weak solution is sufficiently smooth. In this case, for the rate of convergence we have $\|u - u_h\|_{L^\infty(\Omega)} \approx \mathcal{O}(h^2|\ln h|)$ if linear elements are employed (see [Rannacher, Scott]). The rate of convergence of the iteration error $\|u_h - \tilde{u}_h\|_{L^\infty(\Omega)}$ depends upon the condition number of A, which is typically $\mathcal{O}(h^{-2})$ for second order elliptic problems (see Remark 5.21).

If A is complex, it is not advisable to replace (5.1) by a two times bigger system of real and imaginary parts of (5.1). The width of the band of such system would be two times bigger than that of (5.1). Hence, Gaussian elimination would need twice as much memory as in solving (5.1) directly in complex arithmetic, since any complex number requires twice as much memory as a real number. Analogous arguments can also be made for iterative methods. Moreover, there are a lot of programming languages which enable us to employ complex arithmetic directly.

Systems of complex equations (5.1) arise in the approximation of the time harmonic Maxwell equations (cf. Chapters 12 and 13), Schrödinger equation, etc. The algorithms given in the following sections are applicable both to complex and real matrices.

5.2. Biconjugate gradient method

This method has been firstly described by [Fletcher, p. 80] for real nonsymmetric matrices and by [Jacobs, p. 447] for complex matrices. In Remark 5.8, we shall see that the standard conjugate gradient method for Hermitian matrices is a special case of the biconjugate gradient method.

Recall that α^C denotes the complex conjugate of α. Analogously A^C stands for the conjugate matrix to A. Further, let

$$A^H = (A^C)^T, \tag{5.2}$$

i.e., A^H is the conjugate transposed matrix to A. By $r^H = (r^C)^T$ we mean the conjugate row vector corresponding to a column vector r.

The *biconjugate gradient method* (BCG) for the system (5.1) is defined in the following way. Let x_0 be an initial guess for the solution of (5.1) and let $\tilde{r}_0$ be an arbitrary N-dimensional column vector such that $\tilde{p}_0^H A p_0 \neq 0$ and $\tilde{r}_0^H r_0 \neq 0$, where

$$p_0 = r_0 = b - Ax_0, \quad \tilde{p}_0 = \tilde{r}_0. \tag{5.3}$$

Then we set

$$\left.\begin{aligned}
\alpha_k &= \frac{\tilde{r}_k^H r_k}{\tilde{p}_k^H A p_k} && (5.4)\\
x_{k+1} &= x_k + \alpha_k p_k && (5.5)\\
r_{k+1} &= r_k - \alpha_k A p_k && (5.6)\\
\tilde{r}_{k+1} &= \tilde{r}_k - \alpha_k^C A^H \tilde{p}_k && (5.7)\\
\beta_k &= \frac{\tilde{r}_{k+1}^H r_{k+1}}{\tilde{r}_k^H r_k} && (5.8)\\
p_{k+1} &= r_{k+1} + \beta_k p_k && (5.9)\\
\tilde{p}_{k+1} &= \tilde{r}_{k+1} + \beta_k^C \tilde{p}_k && (5.10)
\end{aligned}\right\} \quad k = 0, 1, \ldots$$

Let

$$M = \inf\left\{k \in \{1, 2, \ldots,\} \mid \tilde{p}_k^H A p_k = 0 \quad \text{or } \tilde{r}_k^H r_k = 0\right\}, \tag{5.11}$$

i.e., the BCG algorithm (5.3)–(5.10) *does not break down* within M iterations.

Before we prove that the BCG algorithm terminates in at most N iterations, we prove several auxiliary assertions.

Lemma 5.1. *Let $x_0, \ldots, x_M$ and $r_0, \ldots, r_M$ be determined by (5.3)–(5.10). Then*

$$r_k = b - Ax_k, \quad k = 0, \ldots, M, \tag{5.12}$$

i.e., $r_0, \ldots, r_M$ represent a sequence of residual vectors.

Proof. We prove (5.12) by induction. From (5.3) we see that r_0 is the residual vector. Suppose now that $r_k = b - Ax_k$ holds for some $k \in \{0, ..., M-1\}$. Multiplying (5.5) by A, we have from (5.6) that

$$r_{k+1} = r_k - \alpha_k A p_k = b - Ax_k - \alpha_k A p_k = b - Ax_k - (Ax_{k+1} - Ax_k) = b - Ax_{k+1}.$$

□

Notice that $r_k = A(\overset{*}{x} - x_k)$, where $\overset{*}{x}$ is the true solution of (5.1). Hence, the matrix A multiplied by the error $\overset{*}{x} - x_k$ yields the residual vector r_k.

Remark 5.2. Let $\tilde{x}_0$ and $\tilde{b}$ be arbitrary N-dimensional column vectors, let $\tilde{r}_0 = \tilde{b} - A^H \tilde{x}_0$ and let

$$\tilde{x}_{k+1} = \tilde{x}_k + \alpha_k^C \tilde{p}_k \tag{5.13}$$

be included into the BCG algorithm just after the relation (5.5). Then, using an argument like in the proof of Lemma 5.1, we can prove that

$$\tilde{r}_k = \tilde{b} - A^H \tilde{x}_k, \quad k = 0, ..., M, \tag{5.14}$$

i.e., $\tilde{r}_0, ..., \tilde{r}_M$ are residual vectors associated to the problem

$$A^H \tilde{x} = \tilde{b}.$$

So by the BCG method we can solve simultaneously the two systems $Ax = b$ and $A^H \tilde{x} = \tilde{b}$.

In the next theorem we show that the sequences $\{r_k\}$, $\{\tilde{r}_k\}$ and $\{p_k\}$, $\{\tilde{p}_k\}$ satisfy the *biorthogonality and biconjugacy condition*, respectively.

Theorem 5.3. *Let (5.11) hold. Then for all $k, l \in \{0, ..., M\}$, $k \neq l$, we have*

$$\tilde{r}_k^H r_l = 0 \tag{5.15}$$

and

$$\tilde{p}_k^H A p_l = 0. \tag{5.16}$$

Proof. We prove the theorem by induction and only for $k < l$, since the case $k > l$ can treated analogously. By (5.6), (5.3) and (5.4), we have

$$\tilde{r}_0^H r_1 = \tilde{r}_0^H r_0 - \alpha_0 \tilde{p}_0^H A p_0 = 0, \tag{5.17}$$

and by (5.9), (5.7), (5.4), (5.17) and (5.8),

$$\begin{aligned}\tilde{p}_0^H A p_1 &= \tilde{p}_0^H A r_1 + \beta_0 \tilde{p}_0^H A p_0 = \frac{1}{\alpha_0}(\tilde{r}_0^H - \tilde{r}_1^H) r_1 + \frac{\beta_0}{\alpha_0} \tilde{r}_0^H r_0 \\ &= \frac{1}{\alpha_0}\big(- \tilde{r}_1^H r_1 + \beta_0 \tilde{r}_0^H r_0\big) = 0.\end{aligned}$$

Now let l be a fixed positive integer less than M. Assuming that both the relations (5.15) and (5.16) hold for all non-negative $k < l$, we prove that they remain valid also if l is replaced by $l+1$. Using (5.6), (5.15), (5.10) and (5.16) for $k < l$, we obtain

$$\tilde{r}_k^H r_{l+1} = \tilde{r}_k^H r_l - \alpha_l \tilde{r}_k^H A p_l = -\alpha_l (\tilde{p}_k^H - \beta_{k-1}\tilde{p}_{k-1}^H) A p_l = 0,$$

where $\beta_{-1} = 0$ and $\tilde{p}_{-1} = 0$ if necessary, and from (5.6), (5.4), (5.10) and (5.16),

$$\tilde{r}_l^H r_{l+1} = \tilde{r}_l^H r_l - \alpha_l \tilde{r}_l^H A p_l = \tilde{r}_l^H r_l - \frac{\tilde{r}_l^H r_l}{\tilde{p}_l^H A p_l} (\tilde{p}_l^H - \beta_{l-1}\tilde{p}_{k-1}^H) A p_l = 0.$$

Combining (5.9), (5.7), (5.16) and (5.15) for $k < l$, we get

$$\tilde{p}_k^H A p_{l+1} = \tilde{p}_k^H A r_{l+1} + \beta_l \tilde{p}_k^H A p_l = \frac{1}{\alpha_k} (\tilde{r}_k^H - \tilde{r}_{k+1}^H) r_{l+1} = 0,$$

and finally from (5.9), (5.7), (5.4), (5.15) and (5.8),

$$\begin{aligned} \tilde{p}_l^H A p_{l+1} &= \tilde{p}_l^H A r_{l+1} + \beta_l \tilde{p}_l^H A p_l = \frac{1}{\alpha_l} (\tilde{r}_l^H - \tilde{r}_{l+1}^H) r_{l+1} + \frac{\beta_l}{\alpha_l} \tilde{r}_l^H r_l \\ &= \frac{1}{\alpha_l} \big(- \tilde{r}_{l+1}^H r_{l+1} + \beta_l \tilde{r}_l^H r_l \big) = 0. \end{aligned}$$

□

Corollary 5.4. *If (5.11) holds then*

$$\tilde{r}_l^H p_k = \tilde{p}_k^H r_l = 0 \quad \textit{for } 0 \le k < l \le M.$$

Proof. Using (5.9) repeatedly, we find that

$$p_k = r_k + \beta_{k-1} r_{k-1} + \beta_{k-1}\beta_{k-2} r_{k-2} + \dots + \left(\prod_{j=0}^{k-1} \beta_j \right) r_0.$$

Thus (5.15) yields that $\tilde{r}_l^H p_k = 0$ for $k < l$. The second relation $\tilde{p}_k^H r_l = 0$ follows similarly from (5.10) and (5.15). □

Lemma 5.5. *If (5.11) holds then the vectors $\{r_l\}_{l=0}^{M-1}$ are linearly independent (i.e., $M \le N$ necessarily).*

Proof. By (5.11),

$$\tilde{r}_l^H r_l \neq 0 \quad \forall l \in \{0, \dots, M-1\}. \tag{5.18}$$

Suppose that $r_0, ..., r_{M-1}$ are linearly dependent. Then there exist complex numbers $c_0, ..., c_{M-1}$ and an integer $k \in \{0, ..., M-1\}$ so that

$$c_k \neq 0 \tag{5.19}$$

and

$$0 = \sum_{l=0}^{M-1} c_l r_l. \tag{5.20}$$

Now by (5.20) and (5.15),

$$0 = \sum_{l=0}^{M-1} c_l \tilde{r}_k^H r_l = c_k \tilde{r}_k^H r_k,$$

which contradicts (5.19) and (5.18). □

Remark 5.6. Under the assumption (5.11) the BCG algorithm terminates in at most M iterations and from Lemma 5.5 we know that $M \leq N$. If $r_M = 0$ then, by Lemma 5.1, x_M is the true solution of the system (5.1). If $r_M \neq 0$, then, by (5.11), $\tilde{p}_M^H A p_M = 0$ or $\tilde{r}_M^H r_M = 0$, which means that the algorithm has broken down. However, the latter case happens very rarely in practical computations. In this case we have to restart the algorithm with another initial value (e.g., $x_0 = x_M$). The theoretical probability of the breakdown is zero, since the measure of the set, where (5.11) does not hold, is zero. But as any computer has a finite arithmetic, the actual probability of a breakdown is a very small positive number. Note that the real computational difficulty may come when the iteration is close to breakdown. For real symmetric and indefinite matrices the algorithm (5.3)–(5.10) can be modified so that it never breaks down – see [Luenberger, p. 1264], (cf. also [Freund]).

Let us introduce the so-called *Krylov space*

$$\mathcal{K}^k = \text{span}(r_0, r_1, ..., r_{k-1}), \quad k \in \{1, ..., M\},$$

where span stands for the linear span with complex (or real when (5.1) is real) scalars. Sometimes $\mathcal{K}^k$ is called the right Krylov space whereas $\text{span}(\tilde{r}_0, \tilde{r}_1, ..., \tilde{r}_{k-1})$ is called the left Krylov space. The next lemma establishes several expressions of the right space. A similar lemma can be stated also for the left space.

Lemma 5.7. *Let (5.11) hold and let*

$$\begin{aligned}
\mathcal{K}_1^k &= \text{span}(p_0, ..., p_{k-1}), \\
\mathcal{K}_2^k &= \text{span}(r_0, ..., A^{k-1} r_0), \\
\mathcal{K}_3^k &= \text{span}(A(\overset{*}{x} - x_0), ..., A^k(\overset{*}{x} - x_0)),
\end{aligned}$$

where $\overset{*}{x}$ is the true solution of (5.1). Then $\mathcal{K}^k = \mathcal{K}_1^k = \mathcal{K}_2^k = \mathcal{K}_3^k$.

Proof. We prove the lemma by induction. The case $k = 1$ is evident due to (5.3). So let the assertion hold for some $k \in \{1, ..., M-1\}$. We divide the proof into the following four steps, showing successively that $\mathcal{K}^{k+1} \subset \mathcal{K}_1^{k+1} \subset \mathcal{K}_2^{k+1} \subset \mathcal{K}_3^{k+1} \subset \mathcal{K}^{k+1}$.

1. By (5.9), $r_k = p_k - \beta_{k-1}p_{k-1}$ and thus $\mathcal{K}^{k+1} \subset \mathcal{K}_1^{k+1}$.
2. We check that $p_k \in \mathcal{K}_2^{k+1}$. By the induction assumption $r_{k-1} \in \mathcal{K}^k = \mathcal{K}_2^k \subset \mathcal{K}_2^{k+1}$, $p_{k-1} \in \mathcal{K}_1^k = \mathcal{K}_2^k \subset \mathcal{K}_2^{k+1}$, and thus we obtain $Ap_{k-1} \in \mathcal{K}_2^{k+1}$. Using now (5.9) and (5.6), we find that

$$p_k = r_k + \beta_{k-1}p_{k-1} = r_{k-1} - \alpha_{k-1}Ap_{k-1} + \beta_{k-1}p_{k-1} \in \mathcal{K}_2^{k+1},$$

   Hence, $\mathcal{K}_1^{k+1} \subset \mathcal{K}_2^{k+1}$.
3. Since $r_0 = b - Ax_0 = A(\overset{*}{x} - x_0)$, we immediately see that $\mathcal{K}_2^{k+1} = \mathcal{K}_3^{k+1}$.
4. By Lemma 5.5, $\dim \mathcal{K}^{k+1} = k+1$. We already know that $\mathcal{K}^{k+1} \subset \mathcal{K}_3^{k+1}$. Hence, $\dim K_3^{k+1} = k+1$ and thus necessarily $\mathcal{K}^{k+1} = \mathcal{K}_3^{k+1}$. □

Remark 5.8. If A is a nonsingular Hermitian matrix, that is

$$A = A^H,$$

and $\tilde{r}_0 = r_0$, then the BCG method (5.3)–(5.10) leads to the *standard conjugate gradient method* with the real coefficients α_k, β_k and the vectors

$$\tilde{p}_k = p_k, \quad \tilde{r}_k = r_k, \quad k = 0, ..., M-1,$$

which are complex in general (they are real if A is real). The problem (5.1) is then equivalent to finding a stationary point of the real functional

$$J(x) = x^H Ax - b^H x - x^H b. \tag{5.21}$$

We will return to this case in Section 5.4.

Remark 5.9. As proposed, e.g., in [Beckman, p. 68], problem (5.1) can be transformed to

$$A^H Ax = A^H b, \tag{5.22}$$

where $A^H A$ is a positive definite Hermitian matrix. So the conjugate gradient algorithm cannot break down. However, its convergence is too slow (see Example 5.12), since it is sensitive to the condition number of $A^H A$ (cf. (5.82)) which is, in general, much worse than that of A.

Theorem 5.10. *Let A be a nonsingular symmetric matrix (i.e., $A = A^T$), $\tilde{r}_0 = r_0^C$ and let the sequences $\{r_k\}$, $\{\tilde{r}_k\}$, $\{p_k\}$, $\{\tilde{p}_k\}$ be generated by (5.3)–(5.10). Then*

$$\tilde{r}_k = r_k^C \quad \text{for } k = 0, ..., M-1, \tag{5.23}$$

and

$$\tilde{p}_k = p_k^C \quad \text{for } k = 0, ..., M-1. \tag{5.24}$$

Proof. By (5.3) and the assumption $\tilde{r}_0 = r_0^C$ we see that the theorem is valid for $k = 0$. Further we suppose its validity for some $k \in \{0, ..., M-2\}$ and prove its validity for $k+1$. By (5.7) and (5.6),

$$\tilde{r}_{k+1} = \tilde{r}_k - \alpha_k^C A^H \tilde{p}_k = r_k^C - \alpha_k^C (A^T)^C p_k^C = r_k^C - \alpha_k^C A^C p_k^C = r_{k+1}^C.$$

To verify (5.24) we employ (5.10), (5.23) and (5.9),

$$\tilde{p}_{k+1} = \tilde{r}_{k+1} + \beta_k^C \tilde{p}_k = r_{k+1}^C + \beta_k^C p_k^C = p_{k+1}^C.$$

□

Corollary 5.11. *For complex symmetric matrices the BCG algorithm (5.3)–(5.10) reduces to the form*

$$p_0 = r_0 = b - Ax_0, \tag{5.25}$$

$$\left.\begin{aligned}
\alpha_k &= \frac{r_k^T r_k}{p_k^T A p_k} && (5.26)\\
x_{k+1} &= x_k + \alpha_k p_k && (5.27)\\
r_{k+1} &= r_k - \alpha_k A p_k && (5.28)\\
\beta_k &= \frac{r_{k+1}^T r_{k+1}}{r_k^T r_k} && (5.29)\\
p_{k+1} &= r_{k+1} + \beta_k p_k && (5.30)
\end{aligned}\right\} \quad k = 0, 1, ...$$

Notice that only one matrix–vector multiplication is necessary in each iteration. Moreover, it is sufficient to store in computer memory only four vectors b, p, r, x (besides A). The vectors Ap_k, $k = 0, 1, ...$ occurring in (5.26) and (5.28) can be currently stored in b. Contrary to Remark 5.8 the coefficients α_k and β_k are complex in general and $r_k^T r_k$ does not represent a scalar product in complex case. Hence, there are still two sources (see (5.26), (5.29)) of possible breakdown in the above BCG algorithm.

Note that the popular conjugate gradient type methods like ORTHODIR, GMRES, ORTHOMIN, ORTHORES (see, e.g., [Van der Vorst, Dekker]) require much more memory cells than the BCG method. If we have troubles with the amount of memory, we usually repeatedly restart these methods from the last iteration.

For the numerical stability of GMRES, see [Drkošová, Rozložník, Strakoš, Greenbaum].

Example 5.12. We compared the BCG method with other iterative methods in solving the system (5.1) of $N = 740$ complex algebraic equations in an example from [Křížek, Neittaanmäki, 1990, p. 163]. This system has arisen by discretization of the equation

$$-\Delta u + i\sigma u = f \quad \text{in } \Omega \subset \mathbf{R}^2$$

with some boundary conditions, where σ is a real function having big jumps. Table 5.3 contains the minimum number of iterations necessary to achieve the prescribed tolerance of the error and residual vector. The true solution $\overset{*}{x}$ from Table 5.3 has been obtained by the Gaussian elimination and the norm $\|\cdot\|_\infty$ is defined by

$$\|s\|_\infty = \max_{j=1,\dots,N} |s^j|$$

for $s = (s^1, \dots, s^N)^T$. We have taken the initial guess $x_0 = 0$ for all considered cases.

Method	Jacobi	CG for (5.22)	Gauss-Seidel	SOR with $\omega = 1.4 - 0.2i$	BCG
$\Vert\overset{*}{x} - x_k\Vert_\infty < 5.10^{-3}$	∞	689	404	192	93
$\Vert r_k\Vert_\infty < 5.10^{-4}$	∞	696	338	170	108

Table 5.3.

From the second column we observe that the Jacobi method does not converge. (Sufficient conditions for its convergence stated in [Varga, p. 73] are not fulfilled in our case.) Further we see that the standard conjugate gradient method applied to the system (5.22) converges very slowly. Moreover, it was necessary to use the double precision in this case. The fourth and fifth column contain the results of the successive overrelaxation method for $\omega = 1$ and $\omega = 1.4 - 0.2i$ (almost optimal). Let us emphasize that the choice of an optimal overrelaxation complex parameter ω is a quite difficult task. In Figure 5.1 we see a domain of those ω for which the SOR method converges. However, this domain is different for another data. Finally, in the last column of Table 5.3 we see the results for the BCG method defined by (5.25)–(5.30). Note that the duration of one iteration by the BCG method was only 1.2 times longer than for the Gauss–Seidel method.

5.3. Preconditioned biconjugate gradient method

Let A be a nonsingular complex $N \times N$ matrix and let Z be its "easily invertible" approximation (cf. Remark 5.13). Suppose that Z is of the form

$$Z = LU, \tag{5.31}$$

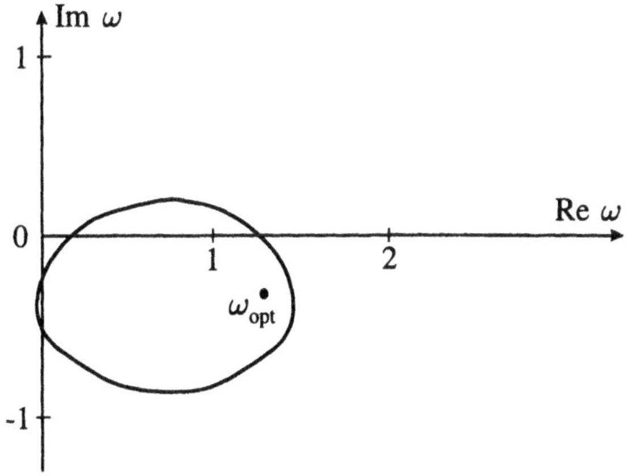

Figure 5.1.

where L and U are lower and upper triangular matrices, respectively. Multiplying (5.1) by L^{-1} and setting

$$\hat{x} = Ux, \tag{5.32}$$
$$\widehat{A} = L^{-1}AU^{-1}, \tag{5.33}$$
$$\hat{b} = L^{-1}b, \tag{5.34}$$

we may write

$$\widehat{A}\hat{x} = L^{-1}AU^{-1}Ux = L^{-1}Ax = L^{-1}b = \hat{b},$$

that is

$$\widehat{A}\hat{x} = \hat{b}. \tag{5.35}$$

It is known that the condition number of $\widehat{A}$ has a considerable influence on the rate of convergence of all iterative methods. An appropriate choice of Z enables us to reduce significantly the condition number of $\widehat{A}$. We call $\widehat{A}$ the *preconditioned matrix*, and $Z = LU$ the *preconditioning matrix.*

By the *preconditioned biconjugate gradient method* we shall mean (see [Křížek, Mlýnek]) the biconjugate gradient method (5.3)–(5.10) applied to the system (5.35). Thus formally we get:

$$\hat{p}_0 = \hat{r}_0 = \hat{b} - \widehat{A}\hat{x}_0, \quad \hat{\tilde{p}}_0 = \hat{\tilde{r}}_0, \tag{5.36}$$

$$\left.\begin{aligned}
\hat{\alpha}_k &= \frac{\hat{\tilde{r}}_k^H \hat{r}_k}{\hat{\tilde{p}}_k^H \widehat{A}\hat{p}_k} && (5.37)\\
\hat{x}_{k+1} &= \hat{x}_k + \hat{\alpha}_k \hat{p}_k && (5.38)\\
\hat{r}_{k+1} &= \hat{r}_k - \hat{\alpha}_k \widehat{A}\hat{p}_k && (5.39)\\
\hat{\tilde{r}}_{k+1} &= \hat{\tilde{r}}_k - \hat{\alpha}_k^C \widehat{A}^H \hat{\tilde{p}}_k && (5.40)\\
\hat{\beta}_k &= \frac{\hat{\tilde{r}}_{k+1}^H \hat{r}_{k+1}}{\hat{\tilde{r}}_k^H \hat{r}_k} && (5.41)\\
\hat{p}_{k+1} &= \hat{r}_{k+1} + \hat{\beta}_k \hat{p}_k && (5.42)\\
\hat{\tilde{p}}_{k+1} &= \hat{\tilde{r}}_{k+1} + \hat{\beta}_k^C \hat{\tilde{p}}_k && (5.43)
\end{aligned}\right\} \quad k = 0, 1, \ldots$$

where $\hat{x}_0$ is an initial guess for (5.35) and $\hat{\tilde{r}}_0$ is an arbitrary N-dimensional column vector. Throughout this section we again assume that the BCG algorithm (5.36)–(5.43) does not break down within M iterations. In this algorithm $\widehat{A}$ appears but $\widehat{A}$ is not suitable for computer implementation. That is why, for $k = 0, 1, \ldots$ we set (cf. (5.32))

$$x_k = U^{-1}\hat{x}_k \tag{5.44}$$

and introduce the following notation

$$\begin{aligned}
r_k &= L\hat{r}_k, & \tilde{r}_k &= U^H \hat{\tilde{r}}_k,\\
s_k &= U^{-1}\hat{r}_k, & \tilde{s}_k &= L^{-H}\hat{\tilde{r}}_k,\\
v_k &= U^{-1}\hat{p}_k, & \tilde{v}_k &= L^{-H}\hat{\tilde{p}}_k,\\
\alpha_k &= \hat{\alpha}_k, & \beta_k &= \hat{\beta}_k,
\end{aligned} \tag{5.45}$$

where $L^{-H} = (L^{-1})^H$. We will try now to eliminate from (5.36)–(5.43) all quantities with the symbol " ^ ". Thus by (5.36), (5.45), (5.34), (5.33) and (5.44) we get

$$L^{-1}r_0 = L^{-1}b - L^{-1}AU^{-1}Ux_0 = L^{-1}(b - Ax_0). \tag{5.46}$$

From (5.36) and (5.45) we further have

$$Uv_0 = Us_0 = L^{-1}r_0, \tag{5.47}$$

$$L^H\tilde{v}_0 = L^H\tilde{s}_0 = U^{-H}\tilde{r}_0. \tag{5.48}$$

The coefficient (5.37) can be, by (5.45) and (5.33), rewritten as follows

$$\alpha_k = \frac{(L^H\tilde{s}_k)^H L^{-1} r_k}{(L^H\tilde{v}_k)^H L^{-1}AU^{-1}Uv_k} = \frac{\tilde{s}_k^H r_k}{\tilde{v}_k^H A v_k} \tag{5.49}$$

and the equation (5.38), by (5.44) and (5.45),

$$Ux_{k+1} = Ux_k + \alpha_k U v_k. \tag{5.50}$$

To rearrange (5.39) and (5.40) we employ (5.45) and (5.33),

$$L^{-1}r_{k+1} = L^{-1}r_k - \alpha_k L^{-1}AU^{-1}Uv_k, \tag{5.51}$$
$$U^{-H}\tilde{r}_{k+1} = U^{-H}\tilde{r}_k - \alpha_k^C U^{-H}A^H L^{-H}L^H\tilde{v}_k. \tag{5.52}$$

Finally the coefficient (5.41) and the equations (5.42) and (5.43) can be transformed by means of (5.44) and (5.45) to the form

$$\beta_k = \frac{(L^H\tilde{s}_{k+1})^H L^{-1}r_{k+1}}{(L^H\tilde{s}_k)^H L^{-1}r_k} = \frac{\tilde{s}_{k+1}^H r_{k+1}}{\tilde{s}_k^H r_k}, \tag{5.53}$$
$$Uv_{k+1} = Us_{k+1} + \beta_k Uv_k, \tag{5.54}$$
$$L^H\tilde{v}_{k+1} = L^H\tilde{s}_{k+1} + \beta_k^C L^H\tilde{v}_k. \tag{5.55}$$

By (5.31) and (5.45)–(5.55), the *preconditioned biconjugate gradient method* for systems of complex algebraic equations may be written down in the following way: Let x_0 and $\tilde{r}_0$ be arbitrary N-dimensional column vectors. Then we set

$$r_0 = b - Ax_0, \tag{5.56}$$
$$v_0 = s_0 = Z^{-1}r_0, \tag{5.57}$$
$$\tilde{v}_0 = \tilde{s}_0 = Z^{-H}\tilde{r}_0, \tag{5.58}$$

$$\left.\begin{aligned}
\alpha_k &= \frac{\tilde{s}_k^H r_k}{\tilde{v}_k^H A v_k} && (5.59)\\
x_{k+1} &= x_k + \alpha_k v_k && (5.60)\\
r_{k+1} &= r_k - \alpha_k A v_k && (5.61)\\
\tilde{r}_{k+1} &= \tilde{r}_k - \alpha_k^C A^H \tilde{v}_k && (5.62)\\
s_{k+1} &= Z^{-1} r_{k+1} && (5.63)\\
\tilde{s}_{k+1} &= Z^{-H}\tilde{r}_{k+1} && (5.64)\\
\beta_k &= \frac{\tilde{s}_{k+1}^H r_{k+1}}{\tilde{s}_k^H r_k} && (5.65)\\
v_{k+1} &= s_{k+1} + \beta_k v_k && (5.66)\\
\tilde{v}_{k+1} &= \tilde{s}_{k+1} + \beta_k^C \tilde{v}_k && (5.67)
\end{aligned}\right\} \quad k = 0, 1, \ldots$$

Remark 5.13. If A is sparse then the matrices L and U in (5.31) may also be chosen so that they are sparse. The inversion of Z in (5.57) need not be performed, since the vector s_0 can be obtained more easily in the following two steps

$$\begin{aligned} Ly &= r_0,\\ Us_0 &= y, \end{aligned} \tag{5.68}$$

which are not expensive. Analogously we carry on in (5.58), (5.63) and (5.64).

Remark 5.14. It follows from (5.56), (5.60) and (5.61) that

$$r_k = b - Ax_k, \quad k = 0, 1, \ldots \tag{5.69}$$

i.e., the sequence $\{r_k\}$ represents the residual vectors of the original equation (5.1). The proof is the same as in Lemma 5.1.

Example 5.15. Note that if A is Hermitian (symmetric) then Z can be chosen in the form $Z = LL^H$ ($Z = LL^T$), where L is a lower triangular matrix. The algorithm (5.56)–(5.67) may be then simplified as in Remark 5.8 or Corollary 5.11. In particular, the preconditioned matrix of a real symmetric matrix A is, by (5.33), of the form

$$\widehat{A} = L^{-1} A L^{-T}.$$

The effect of three common preconditioning techniques is sketched in Figure 5.2 for

$$A = \left(\int_\Omega \left(\frac{\partial v^i}{\partial r} \frac{\partial v^j}{\partial r} + \frac{\partial v^i}{\partial z} \frac{\partial v^j}{\partial z} \right) r \, dr \, dz \right)_{i,j=1}^{25},$$

where Ω is a rectangular domain and v^i are the Courant basis functions corresponding to 5×5 degrees of freedom. The matrix A arises by discretization of the Laplace equation (with the homogeneous Dirichlet boundary conditions) in cylindrical coordinates. For a graphical illustration of entries of A we use bilinear elements over a square mesh with 25×25 nodal points (see Figure 5.2).

For the first preconditioner we take $L = D^{\frac{1}{2}}$ (see [Van der Sluis, p. 20]), where D is a diagonal matrix containing the main diagonal of A. Each diagonal entry of the associated preconditioned matrix $\widehat{A} = D^{-\frac{1}{2}} A D^{-\frac{1}{2}}$ is thus equal to 1 (cf. Figure 5.2).

The second preconditioner is constructed by means of the incomplete Choleski factorization (see [Axelsson, Barker, p. 41]). In this case we decompose A so that $A = \tilde{L}\tilde{L}^T + R$, where $\tilde{L}_{ij} = 0$ whenever $A_{ij} = 0$, and R is a small residual matrix. Then we set $L = \tilde{L}$. We observe from Figure 5.2 that the preconditioned matrix $\widehat{A} = \tilde{L}^{-1} A \tilde{L}^{-T}$ is almost the identity matrix.

Finally, in the third case we set

$$L = \left(\frac{1}{\omega} D + \overline{L} \right) \left(\frac{1}{\omega} D \right)^{-\frac{1}{2}},$$

where ω is an overrelaxation parameter ($\omega = 1$ in Figure 5.2) and $\overline{L}$ is a lower triangular matrix such that $A = \overline{L} + D + \overline{L}^T$. Such a choice of L leads to the so-called SSOR preconditioning technique (see [Axelsson, Barker, p. 51]).

For more details about preconditioning we refer to [Axelsson], [Axelsson, Kolotilina], [Axelsson, Lindskog], [Bruaset], [Cullum, Willoughby], [George, Liu], [Sarkar].

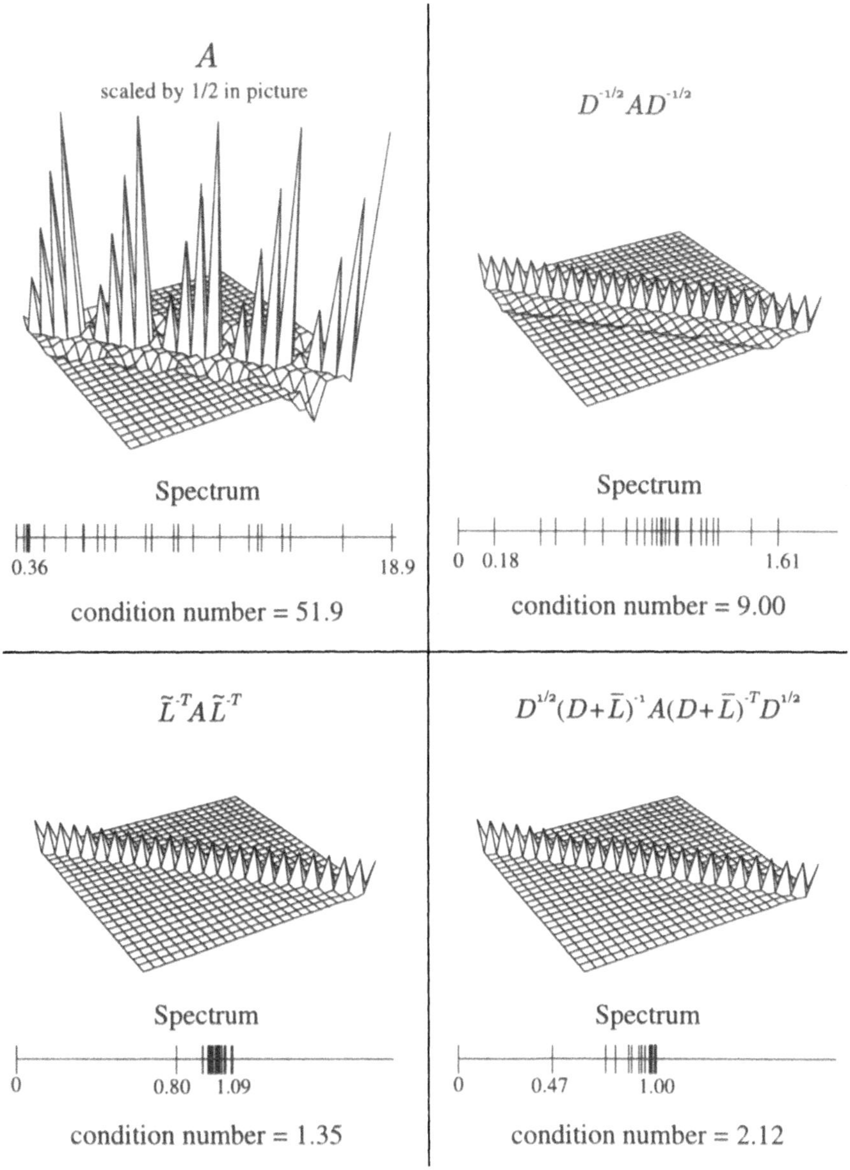

Figure 5.2.

5.4. Rate of convergence of the conjugate gradient method

From Theorem 5.3 and Remark 5.6 we know that the BCG method converges (under some assumptions) to the true solution in a finite number of iterations. The aim of this section is to show that if A is also positive definite then we have a linear rate of convergence (or even superlinear, which means that the conjugate gradient method converges faster than a geometrical sequence, see [Jennings, Malik]). This enables us to stop the iteration process much earlier than after N iterations.

By Remark 5.8 the conjugate gradient method for system (5.1) with a Hermitian matrix A ($A = A^H$) is defined as follows. Let x_0 be an initial guess for the solution of (5.1) such that $Ax_0 \neq b$. Then we set

$$p_0 = r_0 = b - Ax_0, \tag{5.70}$$

$$\left.\begin{aligned} \alpha_k &= \frac{r_k^H r_k}{p_k^H A p_k} && (5.71)\\ x_{k+1} &= x_k + \alpha_k p_k && (5.72)\\ r_{k+1} &= r_k - \alpha_k A p_k && (5.73)\\ \beta_k &= \frac{r_{k+1}^H r_{k+1}}{r_k^H r_k} && (5.74)\\ p_{k+1} &= r_{k+1} + \beta_k p_k && (5.75) \end{aligned}\right\} \quad k = 0, 1, \ldots, M-1$$

where (cf. (5.11))

$$M = \inf\left\{k \in \{1, 2, \ldots, \} \mid p_k^H A p_k = 0\right\}.$$

Note that α_k and β_k are real now.

We will investigate the rate of convergence of the above algorithm only for A positive definite, that is

$$x^H A x > 0 \quad \text{for } x \neq 0. \tag{5.76}$$

In this case the algorithm (5.70)–(5.75) cannot break down if x_k is different from the true solution. Due to Remark 5.6 the true solution of (5.1) is equal to x_M. Note that, by induction and (5.72), we have

$$x_M = x_0 + \sum_{k=0}^{M-1} \alpha_k p_k.$$

Moreover, according to (5.16) and (5.15), the vectors p_k are A-conjugate and r_k are orthogonal, i.e.,

$$r_k^H r_l = 0, \quad p_k^H A p_l = 0 \quad \text{for } k \neq l.$$

Lemma 5.16. *Let A be a Hermitian positive definite matrix and let $\overset{*}{x}$ be the true solution of the system (5.1). Then the real functional*

$$J(x) = x^H A x - b^H x - x^H b \tag{5.77}$$

has a global minimum at $\overset{*}{x}$ and there are no other local minima, maxima or saddle points.

Proof. For any complex vectors x, p and any complex parameter α we have

$$\begin{aligned} J(x + \alpha p) - J(x) &= \alpha\alpha^C p^H A p + \alpha x^H A p + \alpha^C p^H A x - \alpha b^H p - \alpha^C p^H b \\ &= |\alpha|^2 p^H A p - \alpha r^H p - \alpha^C p^H r\ , \end{aligned} \tag{5.78}$$

where $r = b - Ax$ is the residual vector. Thus the increment of J at the point $x = \overset{*}{x}$,

$$J(\overset{*}{x} + \alpha p) - J(\overset{*}{x}) = |\alpha|^2 p^H Ap ,$$

is by (5.76) always positive for any $\alpha \neq 0$ and $p \neq 0$.

On the other hand, if $x \neq \overset{*}{x}$ then by (5.78) the increment at x in the direction $p = r \neq 0$ is of the form

$$J(x + \alpha r) - J(x) = |\alpha|^2 r^H Ar - (\alpha + \alpha^C) r^H r.$$

We observe that it changes the sign if a sufficiently small α is replaced by $-\alpha$, i.e., there is no a stationary point at x. □

Remark 5.17. Under the assumptions of Lemma 5.16 the functional (5.77) is strictly convex and its contour lines are hyperellipsoids (see, e.g., [Křížek, Neittaanmäki, 1990, p. 79]). In each iteration of the CG algorithm (5.70)–(5.75), J is minimized in the plane given by x_{k+1} and by the vectors p_k and r_{k+1} for establishing x_{k+2} (see Figure 5.3 for $k = 0$, where p_0 and p_1 form conjugate directions).

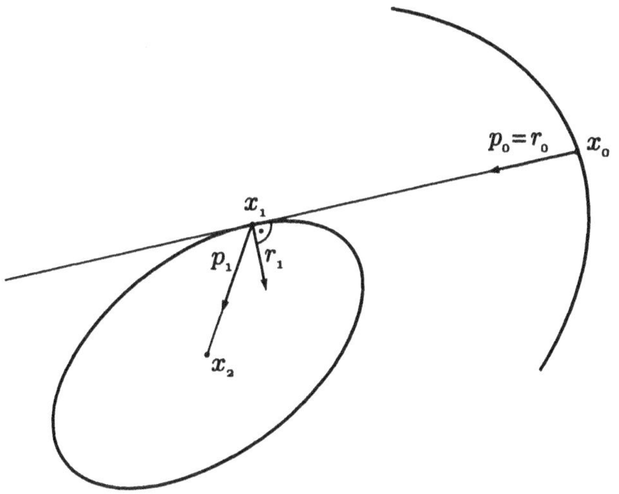

Figure 5.3.

Example 5.18. Let us illustrate a geometrical interpretation of a possible break down of the CG algorithm when (5.76) is not fulfilled. For the real symmetric matrix

$$A = \begin{pmatrix} 1 & 2 & 2 \\ 2 & 4 & -2 \\ 2 & -2 & 0 \end{pmatrix}$$

and $b = 0$ the associated quadratic functional (5.77) is of the form

$$J(x_1, x_2, x_3) = (x_1 + 2x_2)^2 + 4x_1x_3 - 4x_2x_3. \tag{5.79}$$

Taking $x_0 = (4,4,3)^T$, we easily find that $p_0 = r_0 = (-18,-18,0)^T$, $x_1 = (0,0,3)^T$, $r_1 = (-6,6,0)^T$, $p_1 = (-8,4,0)^T$ and $p_1^T A p_1 = 0$, i.e., the algorithm breaks down. In Figure 5.4 we see parabolic contour lines of the functional (5.79) in the plane given by x_1 and the vectors p_0 and r_1. We observe that there is no minimum of J in the direction p_1.

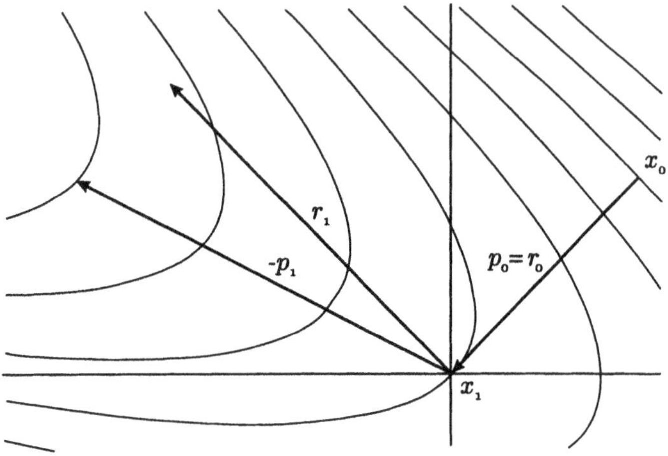

Figure 5.4.

Recall that any Hermitian positive definite matrix A has positive eigenvalues $\lambda_N \geq ... \geq \lambda_1 > 0$, even though A is complex. Its *condition number* is defined as

$$\kappa(A) = \max_j \lambda_j / \min_j \lambda_j = \lambda_N / \lambda_1. \tag{5.80}$$

We will establish now the rate of convergence of the conjugate gradient method in the so-called *energy norm*

$$\|x\|_A = \sqrt{x^H A x}. \tag{5.81}$$

Note that the points x for which $\|\overset{*}{x} - x\|_A$ is constant lie on a hyperellipsoid (see Figure 5.3 for the real case), since from (5.77) and (5.81) we have $J(x) + \overset{*}{x}{}^H A \overset{*}{x} = \|x - \overset{*}{x}\|_A^2$, where $\overset{*}{x}$ is the true solution of (5.1).

Theorem 5.19. *Let A be a Hermitian positive definite matrix and let x_0 be the initial approximation of the true solution $\overset{*}{x}$ of (5.1). Then for the CG algorithm (5.70)–(5.75) we have*

$$\|\overset{*}{x} - x_k\|_A \leq 2 \left(\frac{\sqrt{\kappa(A)} - 1}{\sqrt{\kappa(A)} + 1} \right)^k \|\overset{*}{x} - x_0\|_A \quad \forall k \in \{0, 1, ..., M\}\,, \tag{5.82}$$

where $\kappa(A) \geq 1$ is the condition number of A. Moreover,

$$\|\overset{*}{x} - x_k\|_A = \min_{x \in x_0 + \mathcal{K}^k} \|\overset{*}{x} - x\|_A \quad \forall k \in \{0, 1, ..., M\}.$$

Proof. If $\lambda_1 = \lambda_N$ then A is a multiple of the identity matrix and thus the CG algorithm converges after one iteration, i.e., (5.82) holds with $\kappa(A) = 1$.

So let us assume that $\lambda_1 < \lambda_N$ and $k \in \{1, ..., M\}$. According to (5.72),

$$x_k - x_0 = \sum_{j=0}^{k-1} \alpha_j p_j.$$

From here, the A-conjugacy of p_j and the fact that $\overset{*}{x} = x_M$, we obtain

$$(x_k - x_0)^H A p_l = \left(\sum_{j=0}^{k-1} \alpha_j p_j\right)^H A p_l = (\alpha_l p_l)^H A p_l = \left(\sum_{j=0}^{M-1} \alpha_j p_j\right)^H A p_l$$
$$= (\overset{*}{x} - x_0)^H A p_l \quad \forall l \in \{0, ..., k-1\}\ ,$$

i.e., $x_k - x_0$ is the projection of the initial error $\overset{*}{x} - x_0$ onto the Krylov space $\mathrm{span}(p_0, ..., p_{k-1})$ with respect to the energy scalar product

$$(x, p)_A = x^H A p.$$

Therefore,

$$\|\overset{*}{x} - x_k\|_A = \|\overset{*}{x} - x_0 - (x_k - x_0)\|_A \leq \|\overset{*}{x} - x_0 - y\|_A \quad \forall y \in \mathrm{span}(p_0, ..., p_{k-1}).$$

However, by Lemma 5.7 we have that

$$\mathrm{span}(p_0, ..., p_{k-1}) = \mathrm{span}(A(\overset{*}{x} - x_0), ..., A^k(\overset{*}{x} - x_0)).$$

Let $q\colon \mathbf{R}^1 \to \mathbf{R}^1$ be a polynomial. Then according to a theorem on mappings of the spectrum (see [Taylor, p. 303]), $q(\lambda_1), ..., q(\lambda_N)$ are the eigenvalues of the matrix $q(A)$. From here we get

$$\begin{aligned}\|\overset{*}{x} - x_k\|_A &= \|(\overset{*}{x} - x_0) - (x_k - x_0)\|_A \leq \|q_k(A)(\overset{*}{x} - x_0)\|_A \\ &\leq \max_{j=1,...,N} |q_k(\lambda_j)| \|\overset{*}{x} - x_0\|_A \leq \max_{\lambda \in [\lambda_1, \lambda_N]} |q_k(\lambda)| \|\overset{*}{x} - x_0\|_A\end{aligned} \tag{5.83}$$

for any polynomial q_k of the kth degree such that

$$q_k(0) = 1. \tag{5.84}$$

Taking $\lambda = 0$, we easily check that the following polynomial is of the form (5.84)

$$q_k(\lambda) = \frac{T_k\left(\dfrac{\lambda_N + \lambda_1 - 2\lambda}{\lambda_N - \lambda_1}\right)}{T_k\left(\dfrac{\lambda_N + \lambda_1}{\lambda_N - \lambda_1}\right)}, \quad \lambda \in [\lambda_1, \lambda_N], \tag{5.85}$$

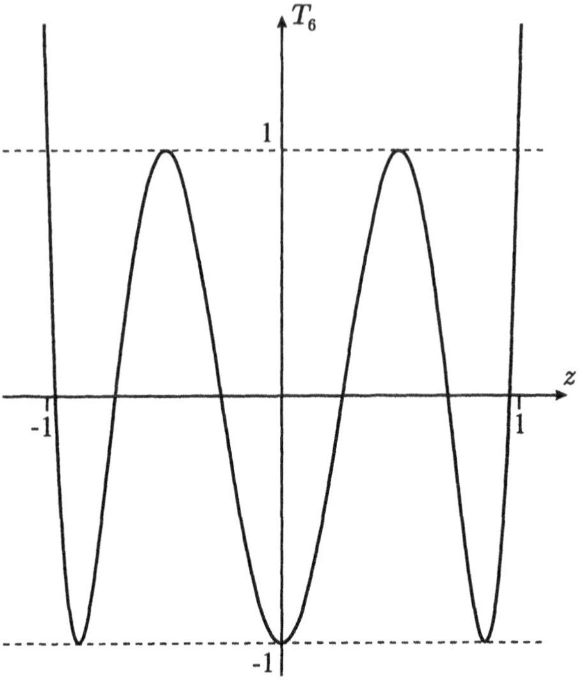

Figure 5.5.

where T_k is the Chebyshev polynomial of degree k (see Figure 5.5 for $k = 6$),

$$T_k(z) = \frac{1}{2}\left((z+\sqrt{z^2-1})^k + (z-\sqrt{z^2-1})^k\right), \quad z \in \mathbf{R}^1 \setminus (-1,1), \quad (5.86)$$

$$T_k(z) = \cos(k \arccos z), \quad z \in [-1,1]. \quad (5.87)$$

(It is known that these expressions really define a polynomial.)

From (5.86) and (5.80) we have

$$\begin{aligned} 2T_k&\left(\frac{\lambda_N+\lambda_1}{\lambda_N-\lambda_1}\right) \\ &= \left(\frac{\lambda_N+\lambda_1}{\lambda_N-\lambda_1} + \sqrt{\left(\frac{\lambda_N+\lambda_1}{\lambda_N-\lambda_1}\right)^2 - 1}\right)^k + \left(\frac{\lambda_N+\lambda_1}{\lambda_N-\lambda_1} - \sqrt{\left(\frac{\lambda_N+\lambda_1}{\lambda_N-\lambda_1}\right)^2 - 1}\right)^k \\ &= \left(\frac{\lambda_N+\lambda_1+2\sqrt{\lambda_N\lambda_1}}{\lambda_N-\lambda_1}\right)^k + \left(\frac{\lambda_N+\lambda_1-2\sqrt{\lambda_N\lambda_1}}{\lambda_N-\lambda_1}\right)^k \\ &> \left(\frac{(\sqrt{\lambda_N}+\sqrt{\lambda_1})^2}{(\sqrt{\lambda_N}-\sqrt{\lambda_1})(\sqrt{\lambda_N}+\sqrt{\lambda_1})}\right)^k = \left(\frac{\sqrt{\lambda_N}+\sqrt{\lambda_1}}{\sqrt{\lambda_N}-\sqrt{\lambda_1}}\right)^k = \left(\frac{\sqrt{\kappa(A)}+1}{\sqrt{\kappa(A)}-1}\right)^k. \end{aligned} \quad (5.88)$$

If $\lambda \in [\lambda_1, \lambda_N]$ then obviously $z = \frac{\lambda_N+\lambda_1-2\lambda}{\lambda_N-\lambda_1}$ lies in the interval $[-1,1]$ and thus by (5.87) we have

$$|T_k(z)| \leq 1 \quad \text{for } z \in [-1,1].$$

This, (5.85) and (5.88) imply that

$$|q_k(\lambda)| < \frac{1}{\frac{1}{2}\left(\frac{\sqrt{\kappa(A)}+1}{\sqrt{\kappa(A)}-1}\right)^k} = 2\left(\frac{\sqrt{\kappa(A)}-1}{\sqrt{\kappa(A)}+1}\right)^k .$$

From here and (5.83) we get (5.82). □

Corollary 5.20. *Let the suppositions of Theorem 5.19 be fulfilled and $\varepsilon \in (0,1)$ be given. Then for any $k > \frac{1}{2}\sqrt{\kappa(A)}\ln(\frac{2}{\varepsilon})$ $(k \leq M)$ and any initial vector x_0 we have*

$$\|\overset{*}{x} - x_k\|_A \leq \varepsilon \|\overset{*}{x} - x_0\|_A. \tag{5.89}$$

Proof. For $\kappa(A) = 1$ the assertion is trivial, since $x_1 = \overset{*}{x}$. So let $\kappa \equiv \kappa(A) > 1$ and let $\varepsilon \in (0,1)$ be given. Comparing to (5.82) with (5.89), we see that it suffices to consider those $k \in \{0, 1, ..., M\}$ for which

$$2\left(\frac{\sqrt{\kappa}-1}{\sqrt{\kappa}+1}\right)^k \leq \varepsilon ,$$

i.e.,

$$\frac{2}{\varepsilon} \leq \left(\frac{1+\frac{1}{\sqrt{\kappa}}}{1-\frac{1}{\sqrt{\kappa}}}\right)^k . \tag{5.90}$$

Employing the well-known formula (see, e.g., [Ryžik, Gradštejn, p. 58])

$$\ln\frac{1+y}{1-y} = 2\sum_{j=1}^{\infty}\frac{1}{2j-1}y^{2j-1}, \quad y \in (-1,1),$$

we get from (5.90) that

$$\begin{aligned} \ln\left(\frac{2}{\varepsilon}\right) &\leq k\ln\left(\frac{1+\frac{1}{\sqrt{\kappa}}}{1-\frac{1}{\sqrt{\kappa}}}\right) = 2k\left(\frac{1}{\sqrt{\kappa}} + \frac{1}{3(\sqrt{\kappa})^3} + \frac{1}{5(\sqrt{\kappa})^5} + ...\right) \\ &< 2k\left(\frac{1}{\sqrt{\kappa}} + \frac{1}{(\sqrt{\kappa})^3} + \frac{1}{(\sqrt{\kappa})^5} + \cdots\right) = 2k\frac{\sqrt{\kappa}}{\kappa-1} , \end{aligned} \tag{5.91}$$

where the last equality expresses the sum of the geometrical sequence. Taking now k so that

$$k \geq \frac{1}{2}\ln\left(\frac{2}{\varepsilon}\right)\sqrt{\kappa} > \frac{1}{2}\ln\left(\frac{2}{\varepsilon}\right)\frac{\kappa-1}{\sqrt{\kappa}} ,$$

we see that (5.91) is valid. □

Remark 5.21. The discrete problem over a given triangulation is only an approximation of the continuous problem in question. So we need not solve the discrete problem exactly. It suffices to reduce the iteration error down to the level of the discretization error. In Corollary 5.20, we have just derived that the required number of iterations for the CG method is $\mathcal{O}(\sqrt{\kappa(A)})$. For the stiffness matrices corresponding to second order elliptic problems and strongly regular families of triangulations, we have $\kappa(A) = \mathcal{O}(h^{-2})$ independently of d (see Remark 11.4 or [Axelsson, Barker]). Consequently, since $N \approx h^{-d}$, we need to perform $\mathcal{O}(\sqrt[d]{N})$ iterations and each iteration takes $\mathcal{O}(N)$ arithmetic operations (cf. Table 5.2). From Section 5.3 we know that the preconditioned conjugate gradient (PCG) method is, in fact, the CG method for the system (5.35) and a good preconditioning yields $\kappa(\widehat{A}) \approx \sqrt{\kappa(A)}$. Hence, the rate of convergence of the PCG method can be also established from Theorem 5.19 and Corollary 5.20. For a geometrical interpretation of preconditioning we refer to [Křížek, Neittaanmäki, 1990, p. 91].

Theorem 5.22. *Let A be a Hermitian positive definite matrix and let J be given by (5.77). Then for the conjugate gradient method (5.70)–(5.75) we have*

$$J(x_{k+1}) < J(x_k) \quad \forall k \in \{0, 1, ..., M-1\}.$$

Proof. According to Corollary 5.4 and Remark 5.8, we get

$$r_l^H p_k = 0 \quad \text{if} \quad 0 \le k < l \le M. \tag{5.92}$$

From here and (5.75)

$$r_k^H p_k = r_k^H r_k + \beta_{k-1} r_k^H p_{k-1} = r_k^H r_k, \quad p_k^H r_k = r_k^H r_k \in \mathbf{R}^1. \tag{5.93}$$

Finally, by (5.72), (5.78), (5.93) and (5.71) we come to

$$\begin{aligned} J(x_{k+1}) - J(x_k) &= \alpha_k^2 p_k^H A p_k - \alpha_k (r_k^H p_k + p_k^H r_k) = \alpha_k^2 p_k^H A p_k - 2\alpha_k r_k^H r_k \\ &= \frac{(r_k^H r_k)^2}{p_k^H A p_k} - 2\frac{(r_k^H r_k)^2}{p_k^H A p_k} < 0. \end{aligned}$$

□

In Theorem 5.19 we have derived that the *error*

$$e_k = \overset{*}{x} - x_k \tag{5.94}$$

decreases in the energy norm (5.81) when k increases. In the next theorem we show that the error e_k decreases also in the Euclidean norm.

Theorem 5.23. *Let the assumptions of Theorem 5.19 hold and let $\{x_k\}$ be the sequence defined by the conjugate gradient algorithm (5.70)–(5.75). Then there exists a constant $C \in (0,1)$ such that*

$$\|\overset{*}{x} - x_{k+1}\| < C\|\overset{*}{x} - x_k\| \quad \forall k \in \{0, 1, ..., M-1\}. \tag{5.95}$$

Proof. From (5.75) and (5.92) we find that

$$p_k^H p_j = r_k^H p_j + \beta_{k-1} p_{k-1}^H p_j = \beta_{k-1} p_{k-1}^H p_j \quad \text{for } j < k,$$

where $\beta_{-1} = 0$ and $p_{-1} = 0$ if necessary. Thus by induction,

$$p_k^H p_j = \beta_{k-1} \ldots \beta_j \|p_j\|^2 > 0. \tag{5.96}$$

Moreover, using (5.94), (5.72) and (5.96), we get

$$\begin{aligned} p_k^H e_{k+1} &= p_k^H (\overset{*}{x} - x_{k+1}) = p_k^H \left(x_0 + \sum_{j=0}^{M-1} \alpha_j p_j - x_0 - \sum_{j=0}^{k} \alpha_j p_j \right) \\ &= p_k^H \sum_{j=k+1}^{M-1} \alpha_j p_j = \sum_{j=k+1}^{M-1} \alpha_j p_k^H p_j > 0. \end{aligned} \tag{5.97}$$

Employing again (5.94) and (5.72), we have

$$e_k = \overset{*}{x} - x_k = \overset{*}{x} - x_{k+1} + \alpha_k p_k = e_{k+1} + \alpha_k p_k.$$

From here and (5.97) we see that

$$\|e_k\|^2 = \|e_{k+1}\|^2 + 2\alpha_k p_k^H e_{k+1} + \alpha_k^2 \|p_k\|^2$$

and that the last two terms are positive. Hence,

$$\|\overset{*}{x} - x_{k+1}\| < C_k \|\overset{*}{x} - x_k\|$$

for some $C_k \in (0,1)$. Since the sequence $\{x_k\}$ terminates in a finite number of iterations, we may set $C = \max_k C_k$. □

Remark 5.24. A generalization of (5.82) or (5.95) to the biconjugate gradient method for non-Hermitian matrices is up to now an open problem. The BCG algorithm sometimes exhibits a rather erratic behaviour of iterates. This problem can be overcome by the quasi-minimal residual method, the convergence of which is more smooth (see [Freund], [Freund, Nachtigal]).

Remark 5.25. Having enough computer time and memory, it is recommended (see [Duff, Erisman, Reid], [Golub, Van Loan], [Irons]) to apply direct methods for solving the system (5.1). Then we have no problems with the choice of the initial approximation and the stopping criterion as in iterative methods. Recall one of the most popular direct method – the Choleski factorization $A = LDL^H$ (or $A = LL^H$) for Hermitian matrices. Having L, the solution x^* can be computed using two backward substitutions and one diagonal scaling: $Ly = b$, $z = D^{-1}y$, $L^H x = z$.

Remark 5.26. Assume that

$$A = A_1 + \mathrm{i}A_2$$

is a nonsingular matrix, where A_i are real square matrices. If A_1 is also nonsingular then we have

$$(A_1 + \mathrm{i}A_2)(I - \mathrm{i}A_1^{-1}A_2) = A_1 + A_2A_1^{-1}A_2.$$

From here we easily find that

$$A^{-1} = (A_1 + A_2A_1^{-1}A_2)^{-1} - \mathrm{i}A_1^{-1}A_2(A_1 + A_2A_1^{-1}A_2)^{-1}.$$

6. Magnetic potential of transformer window

6.1. Dimensional reduction

We describe how to calculate the magnetic potential in the window of an ideal transformer. The knowledge of this potential is a starting point for the determination of some other quantities of practical importance (such as leakage field, overheating of windings, circular currents, additional losses, short circuit forces in windings, etc.).

Let us begin with the linear stationary Maxwell equations in $\mathbb{R}^3$ (cf. (2.5)) in a bounded simply connected domain (for a nonlinear case see Section 7.2)

$$\operatorname{rot} H = J, \tag{6.1}$$
$$B = \mu H, \tag{6.2}$$
$$\operatorname{div} B = 0. \tag{6.3}$$

Due to (6.3) there exists, by Theorem 3.35, a vector magnetic potential $A = (A_1, A_2, A_3)$ such that

$$B = \operatorname{rot} A. \tag{6.4}$$

Hence, by (6.1), (6.2) and (6.4),

$$\operatorname{rot}(\frac{1}{\mu} \operatorname{rot} A) = J. \tag{6.5}$$

To solve (6.5) in a transformer (with appropriate boundary conditions), we have to introduce several simplifications.

Since windings are axially symmetric, we will transform the problem into the cylindrical coordinates (r, φ, z) for which

$$\begin{aligned} x_1 &= r \cos \varphi, \\ x_2 &= r \sin \varphi, \\ x_3 &= z, \end{aligned} \tag{6.6}$$

and we shall assume that the axis $x_3 = z$ is identical with the axis of the inner leg. Later this enables us to consider only a two-dimensional problem in the rectangular window Ω of a transformer (see Figures 6.1 and 9.1). Since the permeability of copper (or aluminium) wires, insulation, oil, air, etc., is almost equal to vacuum permeability $\mu_0 = 4\pi \cdot 10^{-7}$ [H/m], we set

$$\mu = \mu_0 \quad \text{in } \Omega. \tag{6.7}$$

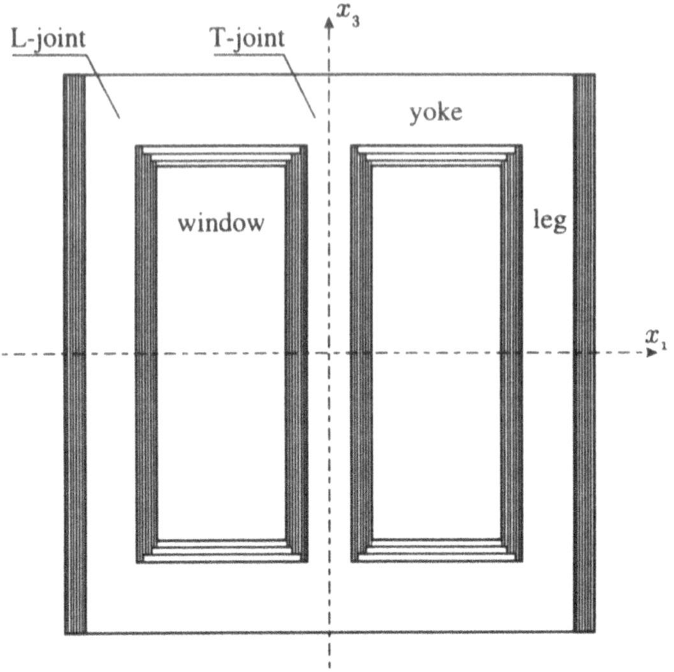

Figure 6.1.

It is known that the magnetic field outside iron reaches its maximum in the window. For that reason no great error will be made if the magnetic circuit is supposed to be axisymmetric with respect to the axis z (see [Doležel, p. 566]). Recall (see, e.g., [Rektorys, 1969, p. 274]) that the components of the vector rot A in the directions of r, φ, z are:

$$\frac{1}{r}\frac{\partial A_z}{\partial \varphi} - \frac{\partial A_\varphi}{\partial z}, \quad \frac{\partial A_r}{\partial z} - \frac{\partial A_z}{\partial r}, \quad \frac{1}{r}\frac{\partial}{\partial r}(rA_\varphi) - \frac{1}{r}\frac{\partial A_r}{\partial \varphi}. \tag{6.8}$$

If we further neglect the winding pitch then the current density J and the vector potential A has nonzero components in the direction φ only and does not depend on φ, i.e.,

$$J = (0, J_\varphi(r,z), 0)^T, \quad A = (0, A_\varphi(r,z), 0)^T. \tag{6.9}$$

Therefore, (6.8) can be reduced as follows

$$-\frac{\partial A_\varphi}{\partial z}, \ 0, \ \frac{1}{r}\frac{\partial}{\partial r}(rA_\varphi). \tag{6.10}$$

From here, (6.5), (6.7), (6.8) and (6.9) we easily find that

$$-\frac{\partial}{\partial r}\left(\frac{1}{r}\frac{\partial(rA_\varphi)}{\partial r}\right) - \frac{\partial^2 A_\varphi}{\partial z^2} = \mu_0 J_\varphi,$$

and thus

$$-\frac{\partial}{\partial r}\left(\frac{1}{r}\frac{\partial(rA_\varphi)}{\partial r}\right)-\frac{\partial}{\partial z}\left(\frac{1}{r}\frac{\partial(rA_\varphi)}{\partial z}\right)=\mu_0 J_\varphi \quad \text{in } \Omega. \tag{6.11}$$

Since $\mu_{\mathrm{Fe}} \gg \mu_0$, the entry of force lines from the window into iron is almost perpendicular (cf. Figure 6.5). That is why we will further assume that

$$\frac{\partial(rA_\varphi)}{\partial n}=0 \text{ on } \partial\Omega, \tag{6.12}$$

where n is the unit outward normal to $\partial\Omega$. Note that the components of the magnetic field corresponding to A are given by

$$H_r=\frac{1}{\mu_0}\frac{\partial A}{\partial z}, \quad H_z=\frac{1}{\mu_0}\left(\frac{\partial A}{\partial r}+\frac{A}{r}\right).$$

Introducing now a modified magnetic potential

$$u=rA_\varphi, \tag{6.13}$$

we get from (6.11) and (6.12) the following problem with variable coefficients:

$$-\frac{\partial}{\partial r}\left(\frac{1}{r}\frac{\partial u}{\partial r}\right)-\frac{\partial}{\partial z}\left(\frac{1}{r}\frac{\partial u}{\partial z}\right)=\mu_0 J_\varphi \quad \text{in } \Omega, \tag{6.14}$$

$$\frac{\partial u}{\partial n}=0 \qquad \text{on } \partial\Omega. \tag{6.15}$$

Finally we assume the balance of magnetomotive forces which can be expressed by the formula

$$\sum_{k=1}^{p} N_k I_k = 0, \tag{6.16}$$

where p is the number of windings, N_k is the number of threads in the kth winding and I_k is the corresponding current. The formula (6.16) is now equivalent to the equilibrium condition

$$\int_\Omega J_\varphi \, dr\, dz = 0, \tag{6.17}$$

which is necessary for solvability of the Neumann problem (6.14)–(6.15) (cf. (3.64)). To employ the variational formulation (3.70) and Theorems 3.19 and 3.20, we set (see [Křížek, Neittaanmäki, Vondrák])

$$a(v,w)=\int_\Omega\left(\frac{\partial v}{\partial r}\frac{\partial w}{\partial r}+\frac{\partial v}{\partial z}\frac{\partial w}{\partial z}\right)\frac{1}{r}\,dr\,dz, \tag{6.18}$$

$$\tilde{a}(v,w)=a(v,w)+\lambda\int_\Gamma v\,ds\int_\Gamma w\,ds, \tag{6.19}$$

$$F(v)=\mu_0\int_\Omega J_\varphi v\,dr\,dz \tag{6.20}$$

for $v, w \in V$, where $V = H^1(\Omega)$, $\lambda > 0$ is a given number and Γ is, e.g., the upper side of $\overline{\Omega}$ (see Figure 6.1).

Let us approximate V by bilinear elements, i.e., we set

$$V_h = \{v_h \in V \mid v_h\big|_K \in Q_1(K) \quad \forall K \in \mathcal{T}_h\},$$

where $\mathcal{T}_h$ consists of rectangles. We shall deal with the discrete problem: Find $u_h \in V_h$ such that

$$\tilde{a}(v_h, u_h) = F(v_h) \quad \forall v_h \in V_h. \tag{6.21}$$

Let $N = \dim V_h$ and let $\{v^i\}$ be the standard basis in V_h satisfying (4.8).

In the first place we describe how to compute analytically the matrix

$$A = (a(v^i, v^j))_{i,j=1}^N, \tag{6.22}$$

whose entries are the integrals over Ω of piecewise rational functions. The bilinear form (6.18) may be rewritten as the sum

$$a(v, w) = \sum_{K \in \mathcal{T}_h} a_K(v, w), \quad v, w \in V,$$

where

$$a_K(v, w) = \int_K \left(\frac{\partial v}{\partial r}\frac{\partial w}{\partial r} + \frac{\partial v}{\partial z}\frac{\partial w}{\partial z} \right) \frac{1}{r}\, dr\, dz. \tag{6.23}$$

The computation of the last integral over K will be transformed to the reference unit square $\widehat{K}$ with the vertices $(0,0)^T$, $(1,0)^T$, $(1,1)^T$ and $(0,1)^T$. Let $(R, Z)^T$ be coordinates of the left lower vertex of some $K \in \mathcal{T}_h$ and let b and c be the lengths of its sides as marked in Figure 6.2.

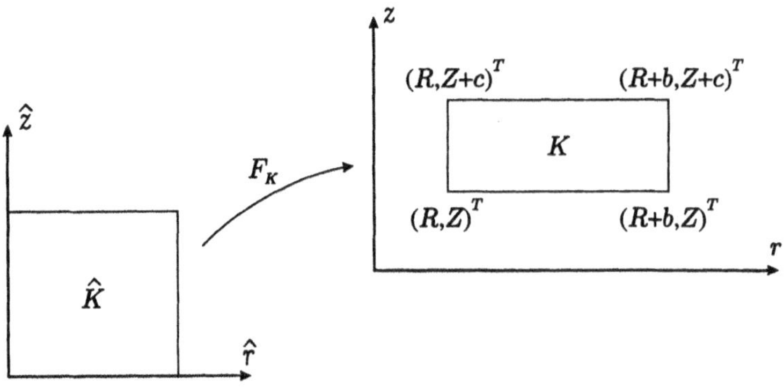

Figure 6.2.

Set

$$B_K = \begin{bmatrix} b & 0 \\ 0 & c \end{bmatrix}. \tag{6.24}$$

Then

$$F_K(\hat{r},\hat{z}) = B_K \begin{bmatrix} \hat{r} \\ \hat{z} \end{bmatrix} + \begin{bmatrix} R \\ Z \end{bmatrix} \tag{6.25}$$

is an affine one-to-one mapping between $\widehat{K}$ and K. For every $v \in H^1(K)$ and a.e. $(\hat{r},\hat{z})^T \in \widehat{K}$ we further set

$$\hat{v}(\hat{r},\hat{z}) = v(r,z), \tag{6.26}$$

where (see (6.25))

$$\begin{aligned} r &= b\hat{r} + R, \\ z &= c\hat{z} + Z. \end{aligned} \tag{6.27}$$

Hence,

$$\frac{\partial v}{\partial r} = b^{-1}\frac{\partial \hat{v}}{\partial \hat{r}}, \quad \frac{\partial v}{\partial z} = c^{-1}\frac{\partial \hat{v}}{\partial \hat{z}}, \tag{6.28}$$

and by the substitution theorem, (6.23), (6.28) and (6.27), we obtain

$$\begin{aligned} a_K(v,w) &= |\det B_K| \int_{\widehat{K}} \left(b^{-2}\frac{\partial \hat{v}}{\partial \hat{r}}\frac{\partial \hat{w}}{\partial \hat{r}} + c^{-2}\frac{\partial \hat{v}}{\partial \hat{z}}\frac{\partial \hat{w}}{\partial \hat{z}} \right) \frac{1}{b\hat{r}+R}\,d\hat{r}\,d\hat{z} \\ &= \int_{\widehat{K}} \left(\frac{c}{b}\frac{\partial \hat{v}}{\partial \hat{r}}\frac{\partial \hat{w}}{\partial \hat{r}} + \frac{b}{c}\frac{\partial \hat{v}}{\partial \hat{z}}\frac{\partial \hat{w}}{\partial \hat{z}} \right) \frac{1}{b\hat{r}+R}\,d\hat{r}\,d\hat{z}, \end{aligned} \tag{6.29}$$

which is independent of Z. The bilinear basis functions for the reference square $\widehat{K}$ are

$$\begin{aligned} \hat{v}_1(\hat{r},\hat{z}) &= (1-\hat{r})(1-\hat{z}), \\ \hat{v}_2(\hat{r},\hat{z}) &= \hat{r}(1-\hat{z}), \\ \hat{v}_3(\hat{r},\hat{z}) &= \hat{r}\hat{z}, \\ \hat{v}_4(\hat{r},\hat{z}) &= (1-\hat{r})\hat{z}. \end{aligned} \tag{6.30}$$

The local basis function v_i of the given rectangle K are defined by (cf. (6.26))

$$v_i(r,z) = \hat{v}_i(\hat{r},\hat{z}), \quad i = 1,2,3,4. \tag{6.31}$$

From (6.29) and (6.31) we obtain the entries

$$A_{ij}^K = a_K(v_i,v_j) = \int_{\widehat{K}} \left(\frac{c}{b}\frac{\partial \hat{v}_i}{\partial \hat{r}}\frac{\partial \hat{v}_j}{\partial \hat{r}} + \frac{b}{c}\frac{\partial \hat{v}_i}{\partial \hat{z}}\frac{\partial \hat{v}_j}{\partial \hat{z}} \right) \frac{1}{b\hat{r}+R}\,d\hat{r}\,d\hat{z} \tag{6.32}$$

of the element matrix

$$A^K = (A_{ij}^K)_{i,j=1}^4 \tag{6.33}$$

corresponding to the element K. To continue in (6.32) we have to evaluate the next three integrals:

$$I_0 = \int_{\widehat{K}} \frac{d\hat{r}\,d\hat{z}}{b\hat{r}+R} = \int_0^1 \frac{d\hat{r}}{b\hat{r}+R} = \frac{1}{b}\Big[\ln|b\hat{r}+R|\Big]_0^1 = \frac{1}{b}\ln\frac{b+R}{R},$$

$$I_1 = \int_{\widehat{K}} \frac{\hat{r}\,d\hat{r}\,d\hat{z}}{b\hat{r}+R} = \int_0^1 \frac{\hat{r}\,d\hat{r}}{b\hat{r}+R} = \left[\frac{\hat{r}}{b} - \frac{R}{b^2}\ln|b\hat{r}+R|\right]_0^1 = \frac{1}{b} - \frac{R}{b^2}\ln\frac{b+R}{R},$$

$$\begin{aligned} I_2 &= \int_{\widehat{K}} \frac{\hat{r}^2\,d\hat{r}\,d\hat{z}}{b\hat{r}+R} = \int_0^1 \frac{\hat{r}^2\,d\hat{r}}{b\hat{r}+R} = \frac{1}{b^3}\left[\frac{1}{2}(b\hat{r}+R)^2 - 2R(b\hat{r}+R) + R^2\ln|b\hat{r}+R|\right]_0^1 \\ &= \frac{1}{2b} - \frac{R}{b^2} + \frac{R^2}{b^3}\ln\frac{b+R}{R}. \end{aligned}$$

From here, (6.30) and (6.32) we get for instance that

$$A_{12}^K = \int_{\widehat{K}} \left(-\frac{c}{b}(1-\hat{z})^2 + \frac{b}{c}(1-\hat{r})\hat{r} \right) \frac{d\hat{r}\, d\hat{z}}{b\hat{r}+R} = -\frac{cI_0}{3b} + \frac{b}{c}(I_1 - I_2),$$

and in the same way we obtain the other entries of the element matrix (6.33):

$$A^K = \frac{cI_0}{6b} \begin{bmatrix} 2 & & \text{sym.} & \\ -2 & 2 & & \\ -1 & 1 & 2 & \\ 1 & -1 & -2 & 2 \end{bmatrix} + \frac{b}{c} \begin{bmatrix} I_2 - 2I_1 + I_0 & & & \text{sym.} \\ I_1 - I_2 & I_2 & & \\ I_2 - I_1 & -I_2 & I_2 & \\ 2I_1 - I_2 - I_0 & I_2 - I_1 & I_1 - I_2 & I_2 - 2I_1 + I_0 \end{bmatrix}.$$

We thus see that no numerical integration is necessary to get the matrix (6.22). To obtain the whole stiffness matrix

$$\widetilde{A} = \widetilde{A}(\gamma) = (\tilde{a}(v^i, v^j))_{i,j=1}^N, \tag{6.34}$$

where $\tilde{a}$ is given by (6.19), we still have to evaluate some one-dimensional integrals what is a trivial task. The structure of $\widetilde{A}$ for $N = 5 \times 5$ and for a row-wise enumeration of nodes is sketched in Figure 6.3 (x stands for non-zero entries). The left upper 5×5 submatrix of Figure 6.3 is full due to the terms

$$\lambda \int_\Gamma v^i\, ds \int_\Gamma v^j\, ds$$

occurring in $\widetilde{A}$. Nevertheless the half bandwidth of $\widetilde{A}$ is the same as that of A. Note that $\widetilde{A}$ is symmetric positive definite whereas A is only a singular positive semidefinite matrix.

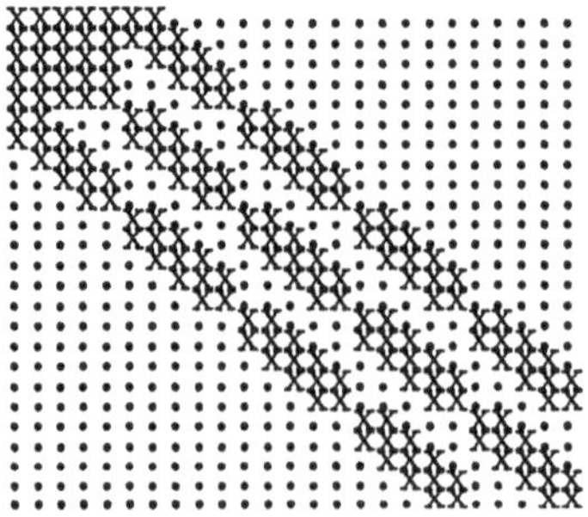

Figure 6.3.

If J_φ from (6.20) is constant on each element K (see Figure 6.2), the associated load vector $(F(v^i))_{i=1}^N$ can be obtained with the help of the formula

$$\int_K v^i\,dx = \frac{bc}{4}.$$

Remark 6.1. Problem (6.21) may be equivalently formulated as follows: Find $u_h \in W_h$ such that

$$a(w_h, u_h) = F(w_h) \quad \forall w_h \in W_h,$$

where

$$W_h = \{w_h \in V_h \mid \int_\Gamma w_h\,ds = 0\}.$$

Figure 6.4 illustrates the trace of a basis function $w^i \in W_h$ corresponding to a node lying on $\overline{\Gamma}$. Basis functions from W_h corresponding to nodes which do not lie on $\overline{\Gamma}$ have the standard shape.

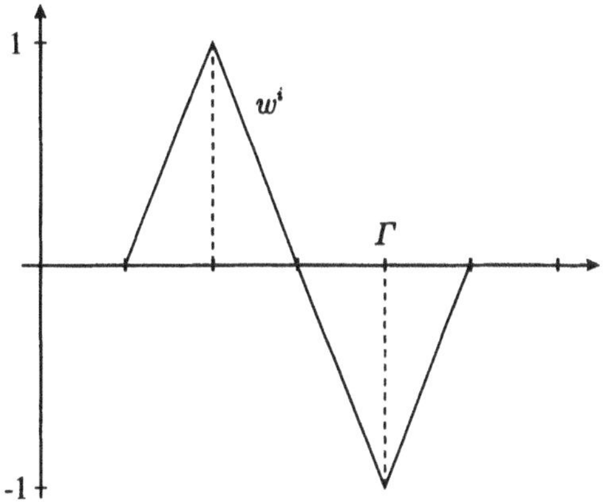

Figure 6.4.

6.2. Example

Let $\Omega = (0.3, 0.65) \times (-0.5, 0.5)$ and suppose that the problem (6.14)–(6.15) is symmetric with respect to the axis r (see Figure 6.1). Then we may restrict ourselves, e.g., to the upper half of Ω. Assume that this half is decomposed into three subdomains Ω_0, Ω_1, Ω_2 as marked in Figure 6.5 a), where Ω_1 and Ω_2 correspond to two windings.

For simplicity we assume further that the current density J_φ is piecewise constant,

$$J_\varphi(r,z) = \begin{cases} 0 & \text{in } \Omega_0, \\ 10^6 & \text{in } \Omega_1, \\ -10^6 & \text{in } \Omega_2, \end{cases}$$

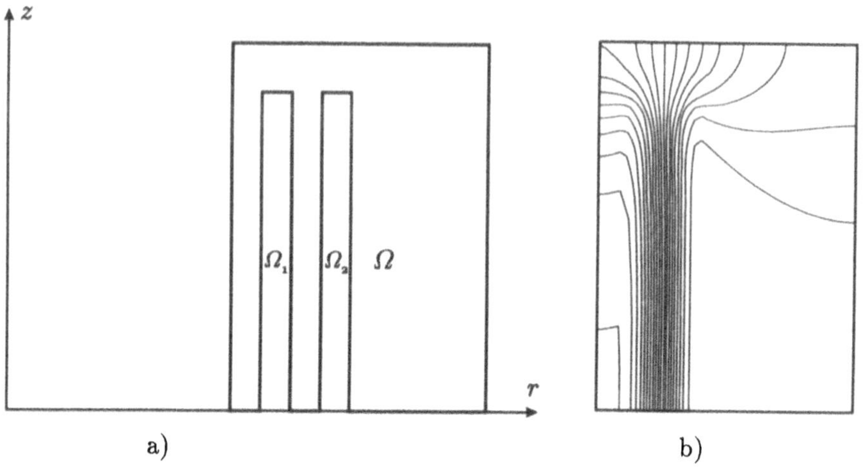

Figure 6.5.

(the values are given in [A/m^2]). In Figure 6.5 b) we see the contour of the corresponding finite element solution u_h over a triangulation containing 35×50 elements.

The components of the magnetic induction (which are more important than the potential u) can be derived from (6.4), (6.10) and (6.13)

$$B_r = -\frac{\partial A_\varphi}{\partial z} = -\frac{1}{r}\frac{\partial u}{\partial z}, \tag{6.35}$$

$$B_z = \frac{1}{r}\frac{\partial}{\partial r}(rA_\varphi) = \frac{1}{r}\frac{\partial u}{\partial r}. \tag{6.36}$$

The behaviour of their finite element approximations B_r^h and B_z^h can be seen in Figures 6.6 and 6.7.

For another real-life technical problem based on dimensional reduction, see [Křížek, Neittaanmäki, 1990, Chapter 19].

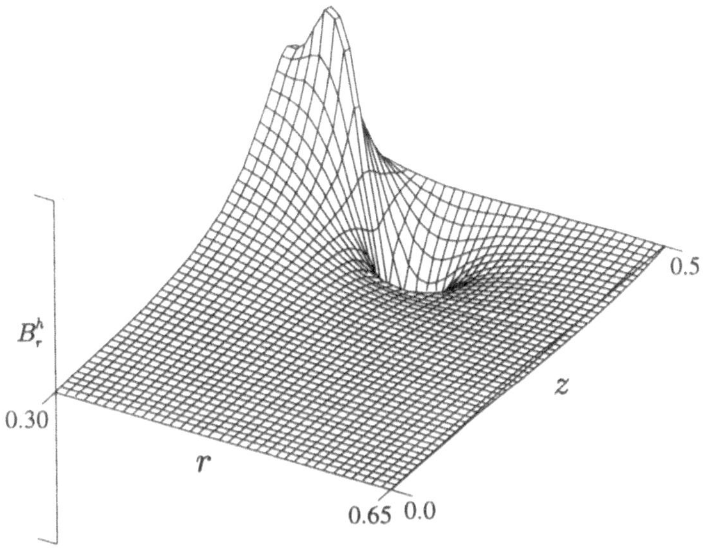

Figure 6.6.

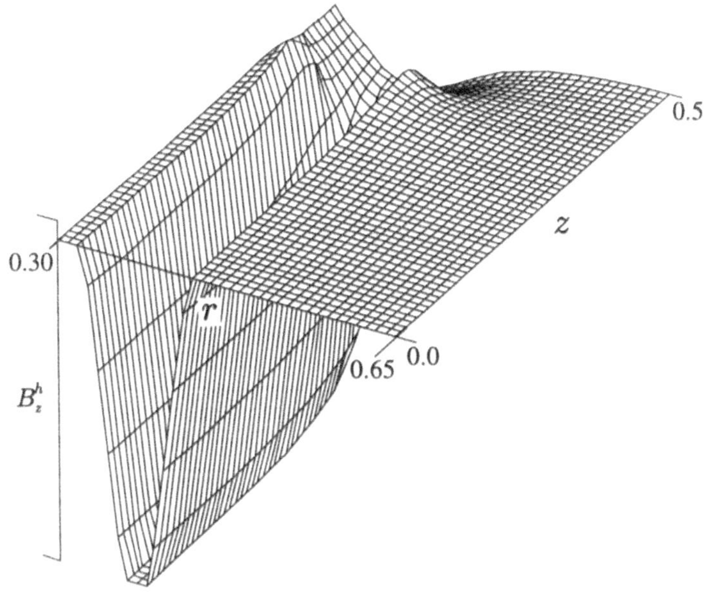

Figure 6.7.

7. Calculation of nonlinear stationary magnetic field

7.1. Introduction to theory of monotone operators

Currently, linear models of various physical fields can successfully be implemented numerically. Efficient numerical methods have been developed during last two or three decades and sufficiently capable computers are available. The situation is different with nonlinear models. There is no general numerical method for solving all nonlinear problems, and consequently every class of problems has to be investigated individually. The specific features of the class are taken into account in this process [Berger].

Here we introduce the theory of monotone operators which represents a useful tool for investigating a wide class of nonlinear problems.

Throughout this section, let V be a Hilbert space with a scalar product $\langle \cdot , \cdot \rangle$ (cf. Section 3.1). Consider an operator $\mathsf{A}: V \to V$ (in general nonlinear). We will treat the problem of finding $u \in V$ such that

$$\mathsf{A}u = b \tag{7.1}$$

for a given $b \in V$.

Note that the equation (7.1) can be equivalently rewritten as

$$\langle \mathsf{A}u, v \rangle = \langle b, v \rangle \quad \forall v \in V. \tag{7.2}$$

Indeed, (7.1) immediately implies (7.2). Conversely, taking $v = \mathsf{A}u - b$, we see by (7.2) and (3.5) that $\langle \mathsf{A}u - b, \mathsf{A}u - b \rangle = 0$, and thus (3.8) yields $\mathsf{A}u = b$.

Definition 7.1. *An operator $\mathsf{A}: V \to V$ is said to be*

(i) monotone, if

$$\langle \mathsf{A}v_1 - \mathsf{A}v_2, v_1 - v_2 \rangle \geq 0 \quad \forall v_1, v_2 \in V, \tag{7.3}$$

(ii) strictly monotone, if

$$\langle \mathsf{A}v_1 - \mathsf{A}v_2, v_1 - v_2 \rangle > 0 \quad \forall v_1, v_2 \in V \quad v_1 \neq v_2, \tag{7.4}$$

(iii) strongly monotone, if there exists a constant $C > 0$ such that

$$\langle \mathsf{A}v_1 - \mathsf{A}v_2, v_1 - v_2 \rangle \geq C \|v_1 - v_2\|_V^2 \quad \forall v_1, v_2 \in V, \tag{7.5}$$

where $\| \cdot \|_V$ is the induced norm ($\|v\|_V = \sqrt{\langle v, v \rangle}$).

Notice that (iii) $\implies$ (ii) $\implies$ (i).

Examples 7.2.

a) Let $V = \mathbf{R}^1$ and let $f\colon \mathbf{R}^1 \to \mathbf{R}^1$ be a nondecreasing function. We see that

$$(f(x_1) - f(x_2))(x_1 - x_2) \geq 0 \quad x_1, x_2 \in \mathbf{R}^1,$$

i.e., f represents a monotone operator. For an increasing function f we find, by (7.4), that f is a strictly monotone operator.

b) If $V = \mathbf{R}^N$ and A is a symmetric positive semidefinite $N \times N$ matrix, then the linear mapping $\mathcal{A}\colon\ x \in \mathbf{R}^N \mapsto Ax \in \mathbf{R}^N$ is a monotone operator, since

$$\langle \mathcal{A}x_1 - \mathcal{A}x_2, x_1 - x_2\rangle = (x_1 - x_2)^T A(x_1 - x_2) \geq 0 \quad \forall x_1, x_2 \in \mathbf{R}^N.$$

Here $\langle \cdot\,,\cdot\rangle$ is the usual scalar product in $\mathbf{R}^N$. If A is positive definite, the associated operator $\mathcal{A}$ is obviously strongly monotone. But if A is a monotone matrix (i.e., A^{-1} exists and $A^{-1} \geq 0$) then the associated operator need not be monotone.

c) Let $a(\cdot\,,\cdot)$ be a continuous V-elliptic bilinear form corresponding, e.g., to some linear problem of Chapter 3 or 6. Then the associated operator $\mathcal{A}$ is well defined by

$$\langle \mathcal{A}w, v\rangle = a(w, v), \quad v, w \in V. \tag{7.6}$$

Indeed, for a given fixed $w \in V$ the right-hand side of (7.6) is a linear continuous functional, $\langle \cdot\,,\cdot\rangle$ is a scalar product in V and thus, by the Riesz Theorem 3.2, the existence of a unique element $\mathcal{A}w \in V$ follows. From the V-ellipticity condition (3.13), we find that $\mathcal{A}$ is strongly monotone,

$$\langle \mathcal{A}v_1 - \mathcal{A}v_2, v_1 - v_2\rangle = a(v_1, v_1 - v_2) - a(v_2, v_1 - v_2) \geq C\|v_1 - v_2\|_V^2 \quad \forall v_1, v_2 \in V$$

for some $C > 0$.

d) Assume that $f \in L^2((0,1))$ and consider the nonlinear problem

$$-u'' + \mathrm{e}^u = f \quad \text{in } (0,1), \tag{7.7}$$

$$u(0) = u(1) = 0. \tag{7.8}$$

This kind of problems arises, e.g., in semiconductor devices (cf. (10.1)). A weak formulation of (7.7) and (7.8) reads:

Find $u \in V = H_0^1((0,1))$ such that

$$\langle \mathcal{A}u, v\rangle = (f, v)_0 \quad \forall v \in V,$$

where

$$\langle \mathcal{A}u, v\rangle = (u', v')_0 + (\mathrm{e}^u, v)_0 \tag{7.9}$$

and where

$$\langle z, v\rangle = (z, v)_1$$

is the standard scalar product in $H^1((0,1))$ – see (3.23).

To ensure that $\mathcal{A}$ is well-defined, let us first observe that the term $(e^u, v)_0$ is finite for $u, v \in V$, since by (3.30) the functions u and v are continuous on $[0,1]$. The function $v \mapsto (u', v')_0 + (e^u, v)_0$ thus represents a continuous linear functional on V and the existence of an element $\mathcal{A}u \in V$ follows again from the Riesz Theorem.

Since the exponential function is increasing, we have

$$(e^{\xi_1} - e^{\xi_2})(\xi_1 - \xi_2) \geq 0 \quad \forall \xi_1, \xi_2 \in \mathbf{R}^1. \tag{7.10}$$

Due to (7.9), (7.10) and Friedrichs' inequality (3.43) there exists a constant $C > 0$ such that

$$\begin{aligned} \langle \mathcal{A}v_1 - \mathcal{A}v_2, v_1 - v_2 \rangle &= (v_1' - v_2', v_1' - v_2')_0 + (e^{v_1} - e^{v_2}, v_1 - v_2)_0 \\ &\geq \|v_1' - v_2'\|_0^2 \geq C\|v_1 - v_2\|_1^2 \quad \forall v_1, v_2 \in V, \end{aligned}$$

and by (7.5) we see that the operator $\mathcal{A}$ is strongly monotone.

Theorem 7.3. *The equation (7.1) has at most one solution $u \in V$ provided A is strictly (strongly) monotone.*

Proof. If $\mathcal{A}u_1 = \mathcal{A}u_2 = b$ for some $u_1, u_2 \in V$ then $\langle \mathcal{A}u_1 - \mathcal{A}u_2, u_1 - u_2 \rangle = \langle b - b, u_1 - u_2 \rangle = 0$. Thus by (7.4) (or (7.5)) we get $u_1 = u_2$. □

Theorem 7.4. *Let $\mathcal{A}: V \to V$ be a strongly monotone operator which is Lipschitz continuous, i.e., there exists a constant $C > 0$ such that*

$$\|\mathcal{A}v_1 - \mathcal{A}v_2\|_V \leq C\|v_1 - v_2\|_V \quad \forall v_1, v_2 \in V. \tag{7.11}$$

Then the equation (7.1) has precisely one solution.

The proof can be found, e.g., in [Franců, p. 264] or [Křížek, Neittaanmäki, 1990, p. 203]. It is based upon the well-known Banach fixed-point theorem.

Definition 7.5. *Let $V_h \subset V$ be a nonempty finite dimensional (finite element) space. A function $u_h \in V_h$ is said to be a Galerkin solution of the problem (7.2) if*

$$\langle \mathcal{A}u_h, v_h \rangle = \langle b, v_h \rangle \quad \forall v_h \in V_h. \tag{7.12}$$

Sufficient conditions for the existence and uniqueness of u_h are stated in Theorems 7.3 and 7.4. Further we introduce an analogue of Céa's Lemma 4.11 which enables us to transform the question of the convergence $\|u - u_h\|_V \to 0$ for $h \to 0$ into the study of approximation properties of a family $\{V_h\}$ in V (see Chapter 4).

Theorem 7.6. *Let* $\mathcal{A}: V \to V$ *be a strongly monotone and Lipschitz continuous operator. Let* $V_h \neq \emptyset$ *be finite dimensional subspace of* V. *Then there exists a constant* C *independent of the space* V_h *such that*

$$\|u - u_h\|_V \leq C \inf_{v_h \in V_h} \|u - v_h\|_V, \tag{7.13}$$

where u *and* u_h *are solutions of (7.1) and (7.12), respectively.*

Proof. By Theorem 7.4 the solutions u and u_h exist uniquely. Thus (7.1) and (7.12) implies

$$\langle \mathcal{A}u - \mathcal{A}u_h, u_h - v_h \rangle = 0 \quad \forall v_h \in V_h.$$

From here, (7.5) and (7.11) we find that

$$\begin{aligned} C\|u - u_h\|_V^2 &\leq \langle \mathcal{A}u - \mathcal{A}u_h, u - u_h\rangle + \langle \mathcal{A}u - \mathcal{A}u_h, u_h - v_h\rangle \\ &\leq \|\mathcal{A}u - \mathcal{A}u_h\|_V \|u - v_h\|_V \\ &\leq C'\|u - u_h\|_V \|u - v_h\|_V \qquad \forall v_h \in V_h. \end{aligned}$$

□

7.2. An application

The results of the previous Section 7.1 can be applied to calculation of a magnetic field in nonlinear media for electromagnetic devices, such as electromagnets, transformers and their magnetic cores or shieldings, synchronous and induction electrical machines, inductance coils, magnetic heads of tape recorders, magnetic lenses in electron microscopes, magnetic traps of tokamaks, linear accelerators, etc.

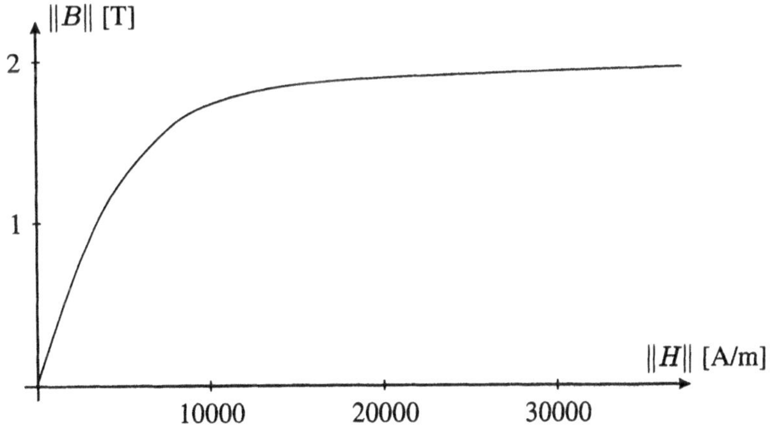

Figure 7.1.

Let $\Omega \in \mathcal{L}$ be a simply connected domain and assume for simplicity that $\Omega \subset \mathbf{R}^2$, i.e., we consider only the cross-section of a device. Recall Maxwell's system of equations describing a stationary magnetic field (cf. (2.5))

$$\operatorname{rot} H = f \quad \text{in } \Omega, \tag{7.14}$$
$$\operatorname{div} B = 0 \quad \text{in } \Omega, \tag{7.15}$$
$$n \cdot B = 0 \quad \text{on } \partial\Omega, \tag{7.16}$$

where $f \in L^2(\Omega)$ is an electric current density, the relation between the magnetic field $H = (H_1, H_2)^T$ and the magnetic induction $B = (B_1, B_2)^T$ is, in general, nonlinear (see Figure 7.1) through a scalar continuous function ν,

$$H(x) = \nu(x, \|B(x)\|^2)B(x) \quad \text{for a.e } x \in \Omega \tag{7.17}$$

($\|\cdot\|$ is the Euclidean norm), ν is the so-called *magnetic reluctance* which can be expressed in the following way:

For $\eta \geq 0$ set

$$\nu(x,\eta) = \begin{cases} \nu_0 & \text{for } x \in \Omega_0, \\ \nu_1(\eta) & \text{for } x \in \Omega_1, \end{cases} \tag{7.18}$$

where Ω_0 and Ω_1 are disjoint measurable parts of Ω, $\overline{\Omega} = \overline{\Omega}_0 \cup \overline{\Omega}_1$, Ω_1 corresponds to ferromagnetic media and Ω_0 to other media (copper or aluminium wires, insulation, air, ...), $\nu_0 = 1/\mu_0$, μ_0 is the vacuum permeability and $\nu_1 \in C^1([0,\infty))$ is a nondecreasing function such that

$$C_0 \leq \nu_1(\eta) \leq C_1, \tag{7.19}$$
$$|\vartheta \nu_1'(\eta)| \leq C_2, \tag{7.20}$$

for all $\eta \geq \vartheta \geq 0$ and some positive constants C_0, C_1 and C_2. The assumptions (7.19) and (7.20) are satisfied, e.g., by the function (see Figure 7.2)

$$\nu_1(\eta) = \frac{1}{\mu_0}\left(\alpha + (1-\alpha)\frac{\eta^8}{\eta^8 + \beta}\right),$$

which characterizes the reluctance of stator sheets in an electrical motor for $\alpha = 0.0003$ and $\beta = 16\,000$ (cf. [Glowinski, Marrocco, p. 59]). The situation above corresponds to isotropic media. (In the case of anisotropic media, ν is a matrix function.)

According to (7.15) and Theorem 3.30, there exists a stream function $u \in H^1(\Omega)$ (unique apart from an additive constant) such that

$$B = \operatorname{curl} u. \tag{7.21}$$

The function u is called a *scalar magnetic potential.* We see that

$$\|B(x)\| = \|\operatorname{curl} u(x)\| = \|\operatorname{grad} u(x)\|$$

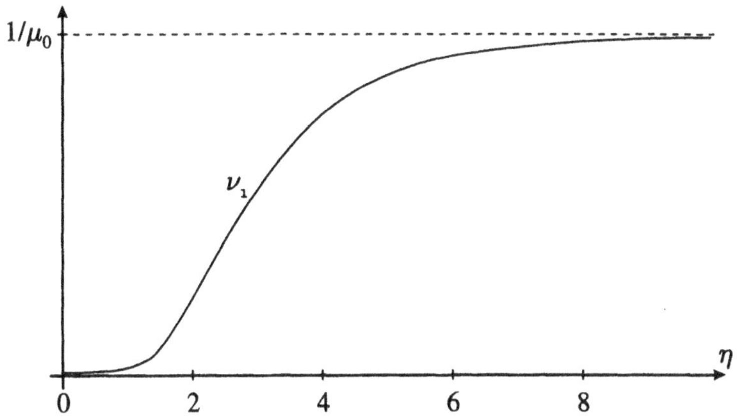

Figure 7.2.

and thus

$$\nu(x, \|B(x)\|^2) = \nu(x, \|\operatorname{grad} u(x)\|^2) \tag{7.22}$$

for a.a. $x \in \Omega$. Hence, by (7.14), (7.17), (7.22) and (7.21), we formally obtain

$$\begin{aligned} f = \operatorname{rot} H &= \operatorname{rot}(\nu(\cdot\,, \|B\|^2)B) = \operatorname{rot}(\nu(\cdot\,, \|\operatorname{curl} u\|^2) \operatorname{curl} u) \\ &= -\operatorname{div}(\nu(\cdot\,, \|\operatorname{grad} u\|^2) \operatorname{grad} u). \end{aligned}$$

For the unit tangent vector $t = (-n_2, n_1)$ to $\partial\Omega$ we get from (7.16) and (7.21) that

$$0 = n \cdot B = n \cdot \operatorname{curl} u = t \cdot \operatorname{grad} u = \frac{\partial u}{\partial t},$$

which means that u is constant on the boundary (we can take, e.g., $u = 0$ on $\partial\Omega$). Therefore, we arrive at the following nonlinear Dirichlet elliptic problem

$$\begin{aligned} -\sum_{i=1}^{2} \frac{\partial}{\partial x_i} \left(\nu(x, \|\operatorname{grad} u(x)\|^2) \frac{\partial u}{\partial x_i} \right) &= f \quad \text{in } \Omega, \\ u &= 0 \quad \text{on } \partial\Omega. \end{aligned} \tag{7.23}$$

For simplicity, we shall not usually explicitly mark a possible dependence of ν upon x. A weak formulation of the problem (7.23) reads:

Find $u \in V$ such that

$$a(u; u, v) = (f, v)_0 \quad \forall v \in V, \tag{7.24}$$

where

$$V = H_0^1(\Omega)$$

and

$$a(y; w, v) = \left(\nu(\|\operatorname{grad} y\|^2) \operatorname{grad} w, \operatorname{grad} v\right)_0, \quad v, w, y \in V. \tag{7.25}$$

Notice that the scalar product given by the integral over Ω is, by (7.18) and (7.19), finite.

In order to examine the solvability of problem (7.24) and its finite element approximation, we apply the results on monotone operators established in Theorems 7.4 and 7.6. To this end we define (as in (7.6)) an operator $\mathsf{A}\colon V \to V$ by

$$\langle \mathsf{A}w, v\rangle = a(w; w, v), \quad v \in V, \tag{7.26}$$

where $\langle \cdot\,,\cdot\rangle$ is the standard scalar product in $H^1(\Omega)$. For any $w \in V$ fixed, $a(w; w, \cdot)$ represents a linear continuous functional over V. Hence, $\mathsf{A}w \in V$ exists and is unique by the Riesz Theorem 3.2.

Now we verify whether A is Lipschitz continuous and strongly monotone. First we will check condition (7.11). Consider the function $F\colon \mathbf{R}^2 \to \mathbf{R}^2$ defined by

$$F(\xi) = (F_1(\xi), F_2(\xi))^T = \nu_1(\xi_1^2 + \xi_2^2)\begin{pmatrix}\xi_1\\ \xi_2\end{pmatrix}, \quad \xi = (\xi_1, \xi_2)^T \in \mathbf{R}^2. \tag{7.27}$$

The matrix of the first derivatives of F is

$$\frac{\partial F}{\partial \xi} = \begin{pmatrix} 2\xi_1^2\nu_1'(\xi_1^2+\xi_2^2) + \nu_1(\xi_1^2+\xi_2^2) & \text{sym.} \\ 2\xi_1\xi_2\nu_1'(\xi_1^2+\xi_2^2) & 2\xi_2^2\nu_1'(\xi_1^2+\xi_2^2) + \nu_1(\xi_1^2+\xi_2^2)\end{pmatrix}.$$

Setting $\eta = \xi_1^2 + \xi_2^2$ and $\vartheta = \xi_1^2$ in (7.19) and (7.20), we find that

$$\left|\frac{\partial F_1}{\partial \xi_1}\right| \le \left|2\xi_1^2\nu_1'(\xi_1^2+\xi_2^2)\right| + \left|\nu_1(\xi_1^2+\xi_2^2)\right| \le 2C_2 + C_1.$$

Similarly,

$$\left|\frac{\partial F_1}{\partial \xi_2}\right| = \left|2\xi_1\xi_2\nu_1'(\xi_1^2+\xi_2^2)\right| \le \left|\xi_1^2\nu_1'(\eta)\right| + \left|\xi_2^2\nu_1'(\eta)\right| \le 2C_2,$$

etc. Hence, the partial derivatives of F are continuous and bounded and thus F is Lipschitz continuous. Then, according to (7.27) and (7.18),

$$\left\|\nu(x, \|\xi\|^2)\xi - \nu(x, \|\zeta\|^2)\zeta\right\| \le C\|\xi - \zeta\| \quad \forall \xi, \zeta \in \mathbf{R}^2,\ x \in \Omega. \tag{7.28}$$

We again omit the variable x. Putting $\xi = \operatorname{grad} v_1(x)$, $\zeta = \operatorname{grad} v_2(x)$ for arbitrary $v_1, v_2 \in V$, squaring (7.28) and then integrating over Ω, we get

$$\left\|\nu(\|\operatorname{grad} v_1\|^2)\operatorname{grad} v_1 - \nu(\|\operatorname{grad} v_2\|^2)\operatorname{grad} v_2\right\|_0 \le C\left\|\operatorname{grad}(v_1 - v_2)\right\|_0.$$

Finally, from here, (7.26), (7.25) and the Cauchy-Schwarz inequality (3.10) we obtain

$$\begin{aligned}\|\mathsf{A}v_1 - \mathsf{A}v_2\|_1 &= \sup_{\substack{w\neq 0\\ w\in H^1(\Omega)}} \frac{|\langle \mathsf{A}v_1 - \mathsf{A}v_2, w\rangle|}{\|w\|_1} = \sup_{\substack{w\neq 0\\ w\in H^1(\Omega)}} \frac{|a(v_1; v_1, w) - a(v_2; v_2, w)|}{\|w\|_1} \\ &\le C \sup_{\substack{w\neq 0\\ w\in H^1(\Omega)}} \frac{\|\operatorname{grad}(v_1 - v_2)\|_0 \|\operatorname{grad} w\|_0}{\|w\|_1} \le C\|v_1 - v_2\|_1,\end{aligned}$$

i.e., (7.11) holds.

Next we show that $\mathcal{A}$ is strongly monotone. Let $v_1, v_2 \in V$ be arbitrary. Set

$$w = v_1 - v_2$$

and

$$\varphi(t) = \int_{\Omega_1} \nu_1(\|\operatorname{grad}(v_2 + tw)\|^2) \operatorname{grad}(v_2 + tw) \cdot \operatorname{grad} w \, dx, \quad t \in [0,1].$$

Since ν_1 is supposed to be differentiable and nondecreasing, its derivative is non-negative and by (7.19) we have

$$\begin{aligned}\varphi'(t) = \int_{\Omega_1} \Big[&\nu_1(\|\operatorname{grad}(v_2 + tw)\|^2)\|\operatorname{grad} w\|^2 \\ &+ 2\nu_1'(\|\operatorname{grad}(v_2 + tw)\|^2)(\operatorname{grad}(v_2 + tw), \operatorname{grad} w)^2 \Big] \, dx \\ \geq C_0 \|&\operatorname{grad} w\|_{0,\Omega_1}^2,\end{aligned} \tag{7.29}$$

where $C_0 > 0$ is independent of $t \in [0,1]$. Moreover, we see that φ' is continuous. Referring to (7.26), (7.25), (7.18) and (7.29), we obtain

$$\begin{aligned}\langle \mathcal{A}v_1 - \mathcal{A}v_2, w\rangle &= \big(\nu(\|\operatorname{grad} v_1\|^2)\operatorname{grad} v_1, \operatorname{grad} w\big)_{0,\Omega_0\cup\Omega_1} \\ &\quad - \big(\nu(\|\operatorname{grad} v_2\|^2)\operatorname{grad} v_2, \operatorname{grad} w\big)_{0,\Omega_0\cup\Omega_1} \\ &= (\nu_0 \operatorname{grad} v_1, \operatorname{grad} w)_{0,\Omega_0} - (\nu_0 \operatorname{grad} v_2, \operatorname{grad} w)_{0,\Omega_0} + \varphi(1) - \varphi(0) \\ &= \nu_0(\operatorname{grad} w, \operatorname{grad} w)_{0,\Omega_0} + \varphi'(t_0) \geq \min(\nu_0, C_0)\|\operatorname{grad} w\|_0^2 \\ &\geq \overline{C}\|v_1 - v_2\|_1^2,\end{aligned}$$

where the positiveness of $\overline{C}$ follows from Friedrichs' inequality (3.43) and the existence of $t_0 \in (0,1)$ is guaranteed by the mean value theorem. Hence, (7.5) is valid and we may apply Theorem 7.4.

Consequently, there exist precisely one weak solution $u \in V$ of the problem (7.23) and precisely one solution of the problem:

Find $u_h \in V_h$ such that

$$a(u_h; u_h, v_h) = (f, v_h)_0 \quad \forall v_h \in V_h. \tag{7.30}$$

The function u_h is a Galerkin approximation of the solution of (7.24) as follows from (7.12) and (7.25). The problem of convergence of u_h to u in the $\|\cdot\|_1$-norm can be now transformed, by Theorem 7.6, to the problem of analyzing the approximation properties of the family $\{V_h\}$ in V (cf. Theorem 4.16). If the solution u is smooth enough we may moreover expect some rate of convergence (as in Theorem 4.17).

There are several ways to solve the discrete nonlinear problem (7.30). One way is to reduce this problem to a sequence of linear problems by the *method of successive approximations*:

Let $y^0 \in V_h$ be arbitrary. Suppose that $y^k \in V_h$, $k \in \{0, 1, ...\}$, is known. Then $y^{k+1} \in V_h$ is uniquely determined, by the Lax–Milgram Lemma 3.3, as the solution of the linear problem

$$a(y^k; y^{k+1}, v_h) = (f, v_h)_0 \quad \forall v_h \in V_h. \tag{7.31}$$

For a geometrical interpretation of this method see [Křížek, Neittaanmäki, 1990, p. 180]. The method is sometimes called also Kačanov's method or the method of secant modules or the method of freezing coefficients. The convergence $y^k \to u_h$ for $k \to \infty$ is investigated in [Fučík, Kratochvíl, Nečas] (see also [Kačúr, Nečas, Polák, Souček], [Nečas, Hlaváček]).

In Figure 7.3 we see induction lines in the cross-section of a part of the stator of a synchronous machine (cf. Figure 11.2). The situation corresponds to ν_1 as sketched in Figure 7.2. The current density was $f = 4.10^6[\mathrm{A/m^2}]$ in copper wires and $f = 0$ otherwise. We have used standard linear elements and $y^0 = 0$ was the initial approximation for Kačanov's method (7.31). After 15 iterations of this method we have got a "numerical fixed-point", i.e., $y^k \doteq y^{k+1}$ for $k \geq 15$.

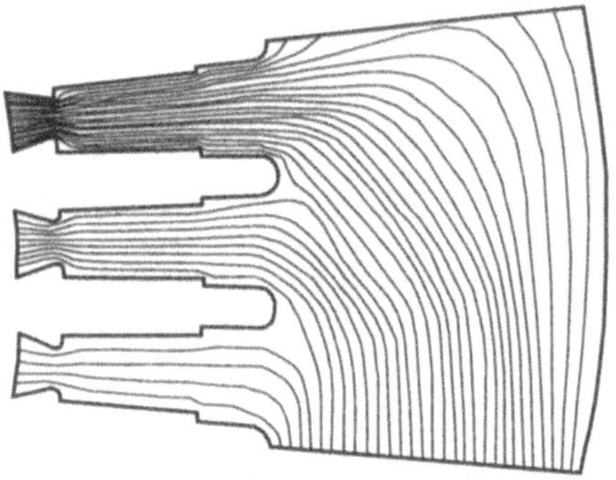

Figure 7.3.

In [Glowinski, Marrocco] or [Křížek, Neittaanmäki, 1990, p. 212], the problem (7.30) is transformed to the minimization of the nonquadratic convex functional

$$J(v) = \frac{1}{2}\int_\Omega \mathcal{N}(x, \|\operatorname{grad} v\|^2)\,dx - \int_\Omega fv\,dx$$

over the space V_h, where $\mathcal{N}(x, \cdot)$ is a primitive function to $\nu(x, \cdot)$ for any fixed $x \in \Omega$. The minimum is found by the Newton or relaxation method.

For various approaches to this kind of problems we further refer to [Bossavit], [Brebbia], [Feistauer, Ženíšek, 1990a,b], [Feistauer, Sobotíková], [Heise], [Kreisinger, Adam], [Křížek, Neittaanmäki, 1990], [Matyska], [Melkes, Zlámal], [Szabó], [Zlámal, 1983], [Ženíšek, 1990a,b] and the journal IEEE Transactions on Magnetics. The literature is very rich. More complicated models with hysteresis are investigated, e.g., in [Brokate], [Krasnoselskij, Pokrovskij], [Krejčí], [Mayergoyz], [Visintin].

7.3. The main theorem on monotone operators

We shall again deal with the solvability of the problem (7.1). However, this time we ignore the assumptions of the strong monotonicity and the Lipschitz continuity (cf. Theorem 7.4) which often cannot be fulfilled. Throughout this section we assume $\mathcal{A}$ to be a monotone operator over a Hilbert space V into itself. Let V be provided with the scalar product $\langle \cdot, \cdot \rangle$ and the induced norm $\| \cdot \|_V$. (Monotone operators in Banach spaces are treated, e.g., in [Fučík, Kufner], [Gajewski, Gröger, Zacharias].)

Definition 7.7. *A sequence $\{v_k\}_{k=1}^{\infty} \subset V$ is said to converge weakly to $v \in V$ (we write $v_k \rightharpoonup v$), if*

$$\langle v_k, w \rangle \to \langle v, w \rangle \quad \forall w \in V. \tag{7.32}$$

Remark 7.8. By the Riesz Theorem 3.2, the relation (7.32) is equivalent to

$$F(v_k) \to F(v) \quad \forall F \in V^*,$$

where V^* is the space of all continuous linear functionals defined on V. The space V^* is called the *dual space*. It is equipped with the norm

$$\|F\|_{V^*} = \sup_{\substack{v \in V \\ \|v\|_V = 1}} |F(v)|$$

and we have $\|F\|_{V^*} = \|w\|_V$, where $F(v) = \langle v, w \rangle$ for every $v \in V$ due to the Riesz Theorem 3.2. Moreover, $V = V^{**}$ which means that any Hilbert space is reflexive.

Obviously, if $V \subset L^2(\Omega)$ then $V^* \supset (L^2(\Omega))^*$ (in particular, for $V = H_0^1(\Omega)$ its dual space V^* is denoted by $H^{-1}(\Omega)$).

Further note that (strong) convergence $v_k \to v$ implies weak convergence $v_k \rightharpoonup v$. The converse is not true in general. Consider, e.g., the sequence $\{\sin kx\}_{k=1}^{\infty}$ in $L_2((0,\pi))$, which converges weakly to the zero element of $L_2((0,\pi))$ but does not converge in the $\| \cdot \|_{0,2}$-norm, since

$$\int_0^{\pi} (\sin kx - \sin jx)^2 \, dx = \pi \qquad \text{for } k \neq j.$$

However, in a finite dimensional space, weak convergence implies strong convergence.

Before we prove some auxiliary lemmas for monotone operators, we introduce two consequences of the Banach–Steinhaus theorem:

Theorem 7.9. *(The principle of uniform boundedness) Let $\{v_k\} \subset V$ be a sequence such that the sequence $\{\langle v_k, w \rangle\}$ is bounded for any $w \in V$. Then the sequence of norms $\{\|v_k\|_V\}$ is bounded.*

For the proof see [Taylor, Theorem 4.4-C].

Theorem 7.10. *The sequence $\{v_k\} \subset V$ converges weakly to the element $v \in V$ if and only if the sequence $\{\|v_k\|_V\}$ is bounded and*

$$\lim_{k\to\infty} \langle v_k, w\rangle = \langle v, w\rangle$$

for all w from a set which is dense in V.

For the proof see [Yosida, Theorem V.3].

Lemma 7.11. *Any monotone operator $\mathsf{A}\colon V \to V$ is locally bounded, i.e., for any $v \in V$ there exist constants $\delta > 0$ and $C > 0$ such that $\|\mathsf{A}z\|_V \le C$ when $z \in V$ and $\|z - v\|_V \le \delta$.*

Proof. Let A be a monotone operator which is not locally bounded. Then there exist $v \in V$ and a sequence $\{v_k\} \subset V$ such that

$$v_k \to v \quad \text{and} \quad \|\mathsf{A}v_k\|_V \to \infty. \tag{7.33}$$

From the monotonicity of A we have

$$\langle \mathsf{A}(v+w) - \mathsf{A}v_k, v + w - v_k\rangle \ge 0 \quad \forall w \in V. \tag{7.34}$$

Let us set for any $k = 1, 2, \ldots,$

$$\alpha_k = 1 + \|\mathsf{A}v_k\|_V \|v - v_k\|_V. \tag{7.35}$$

Then, by (7.34) and (7.35),

$$\begin{aligned}\frac{1}{\alpha_k}\langle \mathsf{A}v_k, w\rangle &\le \frac{1}{\alpha_k}(\langle \mathsf{A}v_k, v_k - v\rangle + \langle \mathsf{A}(v+w), w + v - v_k\rangle)\\ &\le 1 + \frac{1}{\alpha_k}\|\mathsf{A}(v+w)\|_V(\|w\|_V + \|v - v_k\|_V) \le C_1 \quad \forall w \in V,\end{aligned}$$

where C_1 depends on v, w, but is independent of k. The analogous estimate holds also for $(-w)$. Hence,

$$\limsup_{k\to\infty}\left|\frac{1}{\alpha_k}\langle \mathsf{A}v_k, w\rangle\right| < \infty \quad \forall w \in V.$$

According to Theorem 7.9, we arrive at

$$\frac{1}{\alpha_k}\|\mathsf{A}v_k\|_V \le C, \tag{7.36}$$

where $C > 0$ is a constant independent of k. Using now (7.36) together with (7.35), we obtain

$$\|\mathsf{A}v_k\|_V \le C\alpha_k = C(1 + \|\mathsf{A}v_k\|_V\|v - v_k\|_V). \tag{7.37}$$

Let us choose an integer $k_0 > 0$ such that for any $k > k_0$ it is

$$C\|v - v_k\|_V \le \tfrac{1}{2}.$$

Then, by (7.37),

$$\|\mathsf{A}v_k\|_V \le C + \tfrac{1}{2}\|\mathsf{A}v_k\|_V.$$

Consequently, $\|\mathsf{A}v_k\|_V \le 2C$ for all $k > k_0$ which contradicts (7.33). □

Lemma 7.12. *Let* $\mathsf{A}\colon V \to V$ *be a monotone operator and let* $\mathcal{U} \subset V$ *be such a subset that*

$$\|v\|_V \le C_1 \quad \text{and} \quad \langle \mathsf{A}v, v\rangle \le C_2 \quad \forall v \in \mathcal{U}. \tag{7.38}$$

Then there exists a constant $C > 0$ *such that*

$$\|\mathsf{A}v\|_V \le C \quad \forall v \in \mathcal{U}.$$

Proof. By the previous Lemma 7.11 the operator A is locally bounded. So it is locally bounded particularly at zero, i.e., there exist $\delta > 0$ and $C_3 > 0$ such that

$$\|\mathsf{A}v\|_V \le C_3 \quad \forall v \in V,\ \|v\|_V \le \delta. \tag{7.39}$$

Using the monotonicity of A and then (7.38) and (7.39), we find that

$$\begin{aligned}\|\mathsf{A}v\|_V &= \sup_{\|w\|_V=\delta} \frac{1}{\delta}\langle \mathsf{A}v, w\rangle \le \sup_{\|w\|_V=\delta} \frac{1}{\delta}(\langle \mathsf{A}v, v\rangle + \langle \mathsf{A}w, w\rangle - \langle \mathsf{A}w, v\rangle)\\ &\le \frac{1}{\delta}(C_2 + \delta C_3 + C_3 C_1).\end{aligned}$$

□

Lemma 7.13. *Let* $v_k, v, w_k, w \in V$ *and let* $v_k \rightharpoonup v$ *and* $w_k \to w$ *for* $k \to \infty$. *Then*

$$\langle v_k, w_k\rangle \to \langle v, w\rangle.$$

Proof. According to Theorem 7.10, the sequence $\|v_k\|_V$ is bounded as $v_h \rightharpoonup v$. The proof now follows directly from the triangle inequality, the Cauchy-Schwarz inequality (3.10) and (7.32)

$$\begin{aligned}|\langle v_k, w_k\rangle - \langle v, w\rangle| &\le |\langle v_k, w_k\rangle - \langle v_k, w\rangle| + |\langle v_k, w\rangle - \langle v, w\rangle|\\ &\le \|v_k\|_V\|w_k - w\|_V + |\langle v_k, w\rangle - \langle v, w\rangle| \to 0 \text{ as } k \to \infty.\end{aligned}$$

□

Note that scalar products of two weakly converging sequences need not converge to the scalar product of their weak limits. To see this, set for instance $v_k(x) = w_k(x) = \sin kx$ for $x \in (0, \pi)$. Then $v_k \rightharpoonup 0$, $w_k \rightharpoonup 0$ and

$$\langle v_k, w_k\rangle = \int_0^\pi v_k(x) w_k(x)\, dx = \frac{\pi}{2}.$$

Definition 7.14. *An operator* $\mathsf{A}\colon V \to V$ *is said to be demicontinuous, if*

$$v_k \to v \quad \Rightarrow \quad \mathsf{A}v_k \rightharpoonup \mathsf{A}v \quad (\textit{weakly}), \tag{7.40}$$

Theorem 7.15. *Let $\mathcal{A}: V \to V$ be a monotone and demicontinuous operator. Then the set of solutions of the equation $\mathcal{A}u = b$ is convex and closed for any $b \in V$.*

Proof. First of all we prove that

$$\langle b - \mathcal{A}v, u - v \rangle \geq 0 \quad \forall v \in V \quad \Rightarrow \quad \mathcal{A}u = b. \tag{7.41}$$

So let $w \in V$ be arbitrary and let us put

$$v_s = u - sw, \quad s > 0. \tag{7.42}$$

Then

$$\langle b - \mathcal{A}v_s, u - v_s \rangle \geq 0,$$

and (7.42) implies that $0 \leq s\langle b - \mathcal{A}v_s, w\rangle$, i.e.,

$$0 \leq \langle b - \mathcal{A}v_s, w\rangle = \langle b - \mathcal{A}(u - sw), w\rangle.$$

Since $\mathcal{A}$ is demicontinuous

$$\mathcal{A}(u - sw) \rightharpoonup \mathcal{A}u \quad \text{for } s \to 0.$$

Hence, by Lemma 7.13,

$$0 \leq \langle b - \mathcal{A}u, w\rangle \quad \forall w \in V.$$

Since w was arbitrary, $\mathcal{A}u = b$.

If the set of solutions is empty or contains only one point, then it is, of course, convex. So let $u_1, u_2 \in V$ be two different solutions of (7.1) and let

$$u_t = tu_1 + (1-t)u_2$$

for some $t \in (0, 1)$. Since $\mathcal{A}$ is monotone, we have

$$\langle b - \mathcal{A}v, u_t - v\rangle = t\langle \mathcal{A}u_1 - \mathcal{A}v, u_1 - v\rangle + (1-t)\langle \mathcal{A}u_2 - \mathcal{A}v, u_2 - v\rangle \geq 0$$

for every $v \in V$. Therefore, by (7.41), $\mathcal{A}u_t = b$.

Further we show that the set of solutions is closed. So let $\{u_k\}_{k=1}^{\infty} \subset V$ be a sequence of solutions of (7.1) and let $u_k \to u$ for some $u \in V$. Then, by Lemma 7.13, we obtain for any $v \in V$ that

$$\langle b - \mathcal{A}v, u - v\rangle = \lim_{k\to\infty} \langle b - \mathcal{A}v, u_k - v\rangle = \lim_{k\to\infty} \langle \mathcal{A}u_k - \mathcal{A}v, u_k - v\rangle \geq 0.$$

Using again the implication (7.41), we find that $\mathcal{A}u = b$. □

Definition 7.16. *An operator* $\mathsf{A}: V \to V$ *is said to be coercive, if*

$$\lim_{\|v\|_V \to \infty} \frac{\langle \mathsf{A}v, v\rangle}{\|v\|_V} = \infty. \tag{7.43}$$

Remark 7.17. If A is continuous (or Lipschitz continuous) then it is demicontinuous, but the converse implication does not hold, in general. We also see that any strongly monotone operator is coercive. Indeed, from (7.5), we have

$$\langle \mathsf{A}v - \mathsf{A}0, v\rangle \geq C\|v\|_V^2 \quad \forall v \in V,$$

where $C > 0$ is independent of v. Therefore,

$$\frac{\langle \mathsf{A}v, v\rangle}{\|v\|_V} \geq \frac{\langle \mathsf{A}0, v\rangle}{\|v\|_V} + C\|v\|_V \geq -\|\mathsf{A}0\|_V + C\|v\|_V \to \infty \text{ as } \|v\|_V \to \infty.$$

To illustrate the idea of a "coercive operator" we first restrict ourselves to finite dimensional spaces.

Theorem 7.18. *Let* $\mathsf{A}: \mathbf{R}^k \to \mathbf{R}^k$ *be a continuous coercive operator. Then the equation* $\mathsf{A}x = b$ *is solvable for any* $b \in \mathbf{R}^k$.

The proof will be given later. It is based on the following two theorems:

Theorem 7.19. *(Brouwer fixed-point theorem) Let* $\mathcal{B} \subset X$ *be a non-empty closed ball in a normed finite dimensional space* X. *Let* S *be a continuous mapping which maps the ball* $\mathcal{B}$ *into itself, i.e.,* $S(\mathcal{B}) \subset \mathcal{B}$. *Then there exists* $x \in S$ *such that* $Sx = x$.

For the proof see, e.g., [Fučík, Kufner, p. 252–253].

Theorem 7.20. *Let* $r > 0$ *be given, let* $(\cdot,\cdot)$ *and* $\|\cdot\|$ *be the standard scalar product and the Euclidean norm in* $\mathbf{R}^k$ *and let* $\mathsf{B}: \mathbf{R}^k \to \mathbf{R}^k$ *be a continuous operator such that*

$$(\mathsf{B}x, x) \geq 0 \quad \forall x \in \mathbf{R}^k,\ \|x\| = r. \tag{7.44}$$

Then there exists $x \in \mathbf{R}^k$ *such that* $\mathsf{B}x = 0$ *and* $\|x\| \leq r$.

Proof. Suppose, on the contrary, that $\mathsf{B}x \neq 0$ for all x from the closed ball

$$\mathcal{B}_r = \{x \in \mathbf{R}^k \mid \|x\| \leq r\}.$$

Then the operator

$$S: x \mapsto -r\frac{\mathsf{B}x}{\|\mathsf{B}x\|}$$

is a continuous mapping from $\mathcal{B}_r$ into $\mathcal{B}_r$. According to the Brouwer Theorem 7.19, there exists a point $x \in \mathcal{B}_r$ such that

$$x = -r \frac{\mathbf{B}x}{\|\mathbf{B}x\|}. \tag{7.45}$$

Obviously, $\|x\| = r$ and by (7.45),

$$(\mathbf{B}x, x) = (\mathbf{B}x, -r\mathbf{B}x/\|\mathbf{B}x\|) = -r\|\mathbf{B}x\| < 0,$$

which contradicts (7.44). □

Remark 7.21. A geometrical meaning of the condition (7.44) is that the angle between the vectors $\mathbf{B}x$ and x is not greater than a right angle on the surface of the ball $\mathcal{B}_r$, i.e., the vectors $\mathbf{B}x$ have "centrifugal orientations" on the surface (see Figure 7.4 for $k = 2$).

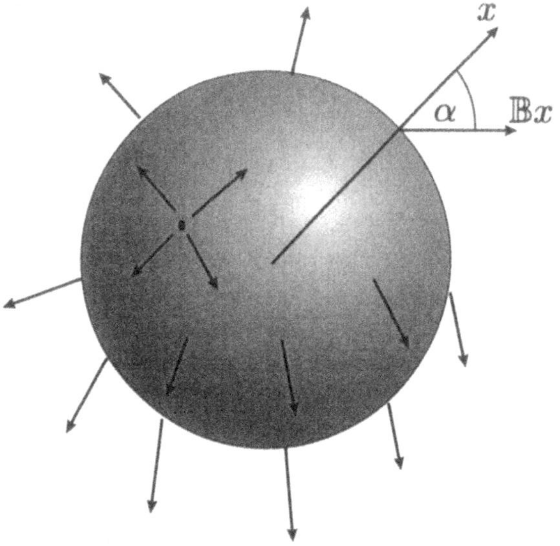

Figure 7.4.

Proof of Theorem 7.18. Given $b \in \mathbf{R}^k$ we find by the coercivity (7.43) that there exists sufficiently large $r > 0$ such that

$$(\mathbf{A}x - b, x) \geq (\mathbf{A}x, x) - \|b\|\|x\| \geq 0 \quad \forall x \in \mathbf{R}^k, \ \|x\| = r.$$

The rest of the proof follows from Theorem 7.20 for $\mathbf{B}x = \mathbf{A}x - b$. □

We apply now the above results in $\mathbf{R}^k$ to prove the main theorem of theory of monotone operators also in infinite dimensional separable spaces. Recall that a Hilbert space is called ***separable*** if it contains a countable dense subset. The following theorem is, in fact, a modification of the well-known Minty–Browder theorem (see, e.g., [Gajewski, Gröger, Zacharias, Chapter III.2.1]). Its proof is constructive using Galerkin approximations (cf. (7.52)). This represents a basis for many numerical methods.

Theorem 7.22. *(The main theorem on monotone operators) Let V be a separable Hilbert space and let $\mathsf{A}\colon V \to V$ be a monotone demicontinuous and coercive operator. Then the set of solutions of the equation $\mathsf{A}u = b$ is nonempty, convex and closed for any $b \in V$.*

Proof. According to Theorem 7.15, it suffices to prove that there exists at least one solution.

First we construct Galerkin approximations of the problem (7.1). Since V is separable, there exists a complete system (finite or infinite countable — see [Hirzebruch, Scharlau]) of linearly independent vectors $\{v^i\} \subset V$ (the so-called Schauder basis), i.e., the linear span of $\{v^i\}$ is dense in V. For $k = 1, 2, \ldots$, set

$$V_k = \operatorname{span}(v^1, \ldots, v^k).$$

Then

$$x = (x_1, \ldots, x_k) \mapsto u_k = \sum_{i=1}^{k} x_i v^i \tag{7.46}$$

defines a linear continuous one-to-one mapping F of the space $\mathbf{R}^k$ onto V_k,

$$Fx = u_k. \tag{7.47}$$

Using the norm equivalence on finite dimensional spaces, we find by (7.47) that there is a constant $C > 0$ such that

$$\|x\| \le C\|Fx\|_V = C\|u_k\|_V. \tag{7.48}$$

Let us define the operator $\mathsf{B}\colon \mathbf{R}^k \to \mathbf{R}^k$ by

$$\mathsf{B}x = (y_1, \ldots, y_k), \tag{7.49}$$

where

$$y_i = \langle \mathsf{A}Fx - b, v^i\rangle. \tag{7.50}$$

If $x^j \to x$ then, of course, $Fx^j \to Fx$ and from the demicontinuity of A we have that

$$\langle \mathsf{A}Fx^j - b, v^i\rangle \to \langle \mathsf{A}Fx - b, v^i\rangle, \quad i = 1, .., k.$$

Hence, B is continuous.

From the coercivity of A it follows that for sufficiently large $r_1 > 0$ we get

$$\Big(\frac{\langle \mathsf{A}u_k, u_k\rangle}{\|u_k\|_V} - \|b\|_V\Big)\|u_k\|_V \ge 0 \quad \text{for } \|u_k\|_V \ge r_1. \tag{7.51}$$

Using now (7.49), (7.50), (7.47), (7.46) and (7.51), we find that for any $x \in \mathbf{R}^k$ such that (see (7.48))

$$\|x\| = r = Cr_1,$$

we can calculate

$$(\mathsf{B}x, x) = \sum_{i=1}^{k} x_i y_i = \sum_{i=1}^{k} x_i \langle \mathsf{A}u_k - b, v^i \rangle = \langle \mathsf{A}u_k, u_k \rangle - \langle b, u_k \rangle$$
$$\geq \left(\frac{\langle \mathsf{A}u_k, u_k \rangle}{\|u_k\|_V} - \|b\|_V \right) \|u_k\|_V \geq 0.$$

Consequently, by virtue of Theorem 7.20, there exists $x \in \mathbf{R}^k$ such that

$$\mathsf{B}x = 0.$$

By (7.49) and (7.50) we have for $u_k = Fx$ the following Galerkin system

$$\langle \mathsf{A}u_k, v^i \rangle = \langle b, v^i \rangle, \quad i = 1, ..., k. \tag{7.52}$$

In particular,

$$\langle \mathsf{A}u_k, u_k \rangle = \langle b, u_k \rangle, \tag{7.53}$$

which yields

$$\frac{\langle \mathsf{A}u_k, u_k \rangle}{\|u_k\|_V} \leq \|b\|_V.$$

Thus from the coercivity of A we find that the sequence $\{u_k\}$ is bounded,

$$\|u_k\|_V \leq C_1. \tag{7.54}$$

Moreover, by (7.53) and (7.54)

$$\langle \mathsf{A}u_k, u_k \rangle \leq C_2 \tag{7.55}$$

independently of k. Now by Lemma 7.12, (7.54) and (7.55),

$$\|\mathsf{A}u_k\|_V \leq C \quad \text{for all } k = 1, 2, ... \tag{7.56}$$

Let $v \in \bigcup_k V_k$ be given, i.e., there exists a positive integer $k_0(v)$ such that $v \in V_k$ for all $k \geq k_0(v)$. According to (7.52),

$$\langle \mathsf{A}u_k, v \rangle = \langle b, v \rangle \quad \forall k \geq k_0(v)$$

and so we have

$$\lim_{k \to \infty} \langle \mathsf{A}u_k, v \rangle = \langle b, v \rangle \quad \forall v \in \bigcup_k V_k. \tag{7.57}$$

Therefore, by Lemma 7.10, (7.56) and (7.57),

$$\mathsf{A}u_k \rightharpoonup b \quad \text{in } V. \tag{7.58}$$

The relation (7.54) implies that there exists a subsequence $\{u_{k_j}\} \subset \{u_k\}$ which weakly converges to some element $u \in V$. For simplicity denote the subsequence again by $\{u_k\}$. We will show that this u is a solution of the problem $\mathcal{A}u = b$.

Using (7.53), we obtain

$$\lim_{k\to\infty} \langle \mathcal{A}u_k, u_k\rangle = \lim_{k\to\infty} \langle b, u_k\rangle = \langle b, u\rangle.$$

Thus from here and (7.58) we get for any $v \in V$

$$\begin{aligned}
\langle b - \mathcal{A}v, u - v\rangle &= \langle b, u\rangle - \langle b, v\rangle - \langle \mathcal{A}v, u - v\rangle \\
&= \limsup_{k\to\infty} \langle \mathcal{A}u_k, u_k\rangle - \langle b, v\rangle - \langle \mathcal{A}v, u - v\rangle \\
&= \limsup_{k\to\infty} \langle \mathcal{A}u_k, u_k\rangle - \lim_{k\to\infty} \langle \mathcal{A}u_k, v\rangle - \lim_{k\to\infty} \langle \mathcal{A}v, u_k - v\rangle \\
&= \limsup_{k\to\infty} \langle \mathcal{A}u_k, u_k\rangle + \limsup_{k\to\infty} \langle \mathcal{A}u_k, -v\rangle + \limsup_{k\to\infty} \langle -\mathcal{A}v, u_k - v\rangle \\
&\geq \limsup_{k\to\infty} \langle \mathcal{A}u_k - \mathcal{A}v, u_k - v\rangle \geq 0,
\end{aligned}$$

where the last inequality follows from the monotonicity of $\mathcal{A}$. Now (7.41) implies that $\mathcal{A}u = b$. □

Example 7.23. Let us return to the problem of Section 7.2. If, for instance, (7.20) is not satisfied, then we are not able to prove that $\mathcal{A}$ is Lipschitz continuous. Nevertheless, we can check that all the assumptions of Theorem 7.22 are satisfied.

The space $H^1(\Omega)$ is separable. An example of a countable set, dense in $H^1(\Omega)$, is the set of all polynomials with rational coefficients.

Books devoted to finite element approximation of elliptic problems of monotone type include [Feistauer, 1993], [Glowinski, 1984], [Křížek, Neittaanmäki, 1990], [Ženíšek, 1990].

8. Steady-state radiation heat transfer problem

8.1. Gâteaux and Fréchet differential

In Section 8.2, we shall see that the steady-state radiative heat transfer problem can be transformed to minimization of a smooth nonquadratic functional J over a convex and closed subset of a Banach space V. To this end we firstly shortly recall some basic definitions concerning differentiability of J, because these sometimes differ in the literature.

Let V be a Banach space, let $v, w \in V$ and let J be a functional defined on an open neighbourhood of the point v. Assume that the following limit

$$J'(v;w) \equiv \lim_{t \to 0} \frac{J(v+tw) - J(v)}{t}$$

exists and is finite. Then the number $J'(v;w)$ is called the (directional) *derivative of the functional J at the point v in the direction w*. Moreover, let the operator $J'(v;\cdot)\colon w \in V \mapsto J'(v;w) \in \mathbf{R}^1$ be linear and continuous. Then this operator is called the *Gâteaux differential* of the functional J at the point v. Note that some authors (see, e.g., [Ekeland, Temam, p. 23]) take the foregoing limit over $t \to 0+$ which yields a different definition of the directional derivative. If there exists a linear continuous operator $J'(v)(\cdot)\colon V \to \mathbf{R}^1$ such that

$$\lim_{\substack{w \to 0 \\ w \in V}} \frac{|J(v+w) - J(v) - J'(v)(w)|}{\|w\|_V} = 0,$$

then this operator is called the *Fréchet* (or *total*) *differential* of the functional J at the point v.

It is known that if the Fréchet differential exists at some point $v \in V$ then the Gâteaux differential at v also exists and is equal to the Fréchet differential, i.e., $J'(v)(w) = J'(v;w)$ for all directions $w \in V$. To see that the converse assertion is not true, it suffices to set $V = \mathbf{R}^2$,

$$J(x_1,x_2) = \begin{cases} 2|x_1| - \sqrt{x_2} & \text{if } x_2 < 4x_1^2 \text{ and } x_2 \geq x_1^2, \\ 2\sqrt{x_2} - |x_1| & \text{if } x_2 < x_1^2 \text{ and } x_2 \geq \frac{1}{4}x_1^2, \\ 0 & \text{otherwise.} \end{cases}$$

The graph of this functional is sketched in Figure 8.1.

We observe that J is continuous and that all directional derivatives of J at the origin are vanishing (cf. Figure 8.2). Thus the Gâteaux differential at 0 exists and $J'(0;\cdot) = 0$. On the other hand, the Fréchet differential at 0 does not exist, since for $w_k = (\frac{1}{k}, \frac{1}{k^2})$, $k = 1, 2, \ldots$, we get

$$w_k \to 0 \quad \text{as } k \to \infty$$

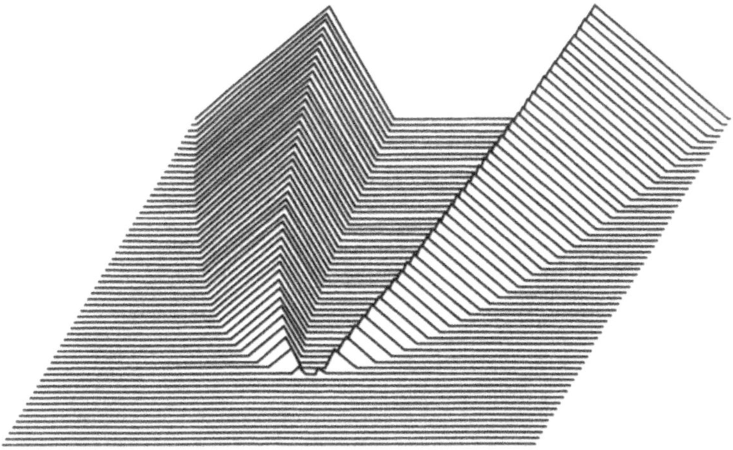

Figure 8.1.

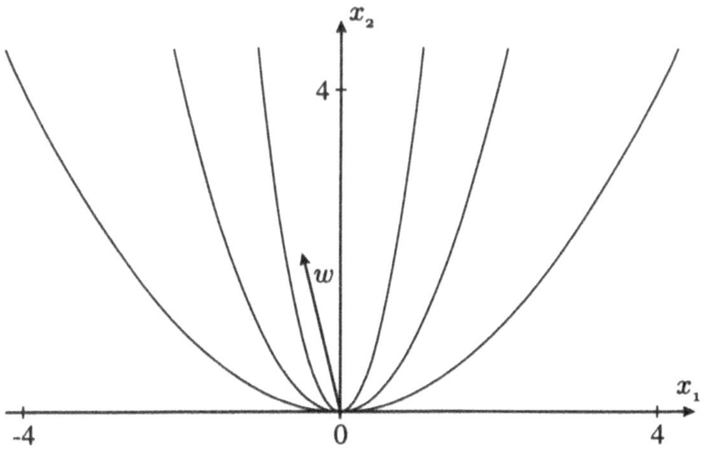

Figure 8.2.

and

$$\frac{|J(w_k) - J(0) - J'(0; w_k)|}{\|w_k\|} = \frac{|k^{-1} - 0 - 0|}{\sqrt{k^{-2} + k^{-4}}} \to 1 \quad \text{as } k \to \infty.$$

It is easy to see that if the Fréchet differential of J exists at $v \in V$ then J is continuous at v. This need not be true for the Gâteaux differential. To see it, we modify slightly the previous example as follows. Setting

$$\widetilde{J}(x_1, x_2) = \begin{cases} J(x_1, x_2)/|x_1| & \text{for } x_1 \neq 0, \\ 0 & \text{for } x_1 = 0, \end{cases}$$

we find that the functional $\widetilde{J}$ is equal to 1 on the parabola $x_2 = x_1^2$ except the point 0, where is discontinuity of $\widetilde{J}$, since $\widetilde{J}(0) = 0$. However, all directional derivatives of $\widetilde{J}$ at the point 0 exist and $\widetilde{J}'(0; \cdot) = 0$.

8.2. Classical and variational formulation

From physics we know that a body loses heat energy from its surface by electromagnetic waves. This phenomenon is called radiation. Losses of energy are proportional to the fourth power of the surface temperature (the Kirchhoff law). This effect is small at room temperature. But the radiation cannot be neglected when the surface temperature is high (e.g., in computation of temperature distribution in large dry transformers, electrical rotary machines, ...). It is represented by the nonlinear boundary condition

$$\alpha(u - u_0) + n^T \mathcal{A} \operatorname{grad} u + \beta(u^4 - u_0^4) = \tilde{g},$$

where $\alpha \geq 0$ is the coefficient of convective heat transfer, u is the temperature of the body, u_0 is the surrounding temperature, n is the outward unit normal to the surface, $\mathcal{A}$ is a symmetric uniformly positive definite matrix (cf. (3.47)) of heat conductivities, $\beta = \sigma f_{\text{em}}$, $\sigma = 5.669 \times 10^{-8}[\,\text{Wm}^{-2}\text{K}^{-4}]$ is the Stefan–Boltzmann constant, $0 \leq f_{\text{em}} \leq 1$ is the relative emissivity function and $\tilde{g}$ is the density of surface heat sources.

Consider the following classical formulation of the radiative problem with mixed boundary conditions: Find $u \in C^2(\overline{\Omega})$, $u \geq 0$, such that

$$\begin{aligned} -\operatorname{div}(\mathcal{A} \operatorname{grad} u) &= f \quad \text{in } \Omega, \\ u &= \overline{u} \quad \text{on } \Gamma_1, \\ \alpha u + n^T \mathcal{A} \operatorname{grad} u + \beta u^4 &= g \quad \text{on } \Gamma_2, \end{aligned} \tag{8.1}$$

where all functions are for simplicity assumed sufficiently smooth, $\Omega \in \mathcal{L}$ is a two-dimensional domain, Γ_1 and Γ_2 are non-empty sets satisfying (3.48), f is the density of body heat sources, $\overline{u} \geq 0$ is the prescribed temperature and

$$g = \tilde{g} + \alpha u_0 + \beta u_0^4.$$

The condition on Γ_2 is sometimes called the *Stefan–Boltzmann boundary condition.*

Since the classical solution of (8.1) need not exist, we introduce a weak formulation of (8.1). To this end we suppose that the entries of $\mathcal{A}$ belong to $L^\infty(\Omega)$, $f \in L^2(\Omega)$, $\overline{u} \in H^1(\Omega)$, $\alpha, \beta \in L^\infty(\Gamma_2)$ and $g \in L^2(\Gamma_2)$. Introduce the space of test functions

$$V = \{v \in H^1(\Omega) \mid v = 0 \text{ on } \Gamma_1\},$$

the convex set

$$U = \{v \in H^1(\Omega) \mid v \geq 0 \text{ in } \Omega,\ v = \overline{u} \text{ on } \Gamma_1\},$$

the symmetric bilinear continuous form

$$a(v, w) = \int_\Omega (\operatorname{grad} v)^T \mathcal{A} \operatorname{grad} w \, dx + \int_{\Gamma_2} \alpha v w \, ds, \quad v, w \in H^1(\Omega),$$

and the linear continuous form

$$F(v) = \int_\Omega fv\,dx + \int_{\Gamma_2} gv\,ds, \quad v \in H^1(\Omega).$$

Recall that the V-ellipticity of $a(\cdot,\cdot)$,

$$a(v,v) \geq C\|v\|_1^2 \quad \forall v \in V, \tag{8.2}$$

follows from Friedrichs' inequality (3.42).

Let $v \in U$ be arbitrary and suppose that a solution $u \in U$ of (8.1) exists. Multiplying the first equation of (8.1) by the function $v - u \in V$ and then integrating over Ω, we get by Green's Theorem 3.10 the following variational equality

$$\int_\Omega (\operatorname{grad} u)^T \mathcal{A} \operatorname{grad}(v-u)\,dx - \int_{\partial\Omega} n^T \mathcal{A} \operatorname{grad} u(v-u)\,ds = \int_\Omega f(v-u)\,dx \quad \forall v \in U,$$

and thus

$$a(u, v-u) + \int_{\Gamma_2} \beta u^4 (v-u)\,ds = F(v-u) \quad \forall\, v \in U.$$

From here we obviously get the variational inequality

$$a(u, v-u) + \int_{\Gamma_2} \beta u^4 (v-u)\,ds \geq F(v-u) \quad \forall v \in U. \tag{8.3}$$

A function $u \in U$ satisfying (8.3) is called the *weak solution* (or variational inequality solution) of the classical problem (8.1). It is clear that the classical solution is also the weak solution. The relationship between the classical and weak solution is discussed in [Milka]. Using potential theory, we show later that there is precisely one weak solution $u \in U$ of (8.3).

Define the functional of potential energy

$$J(v) = \frac{1}{2}a(v,v) + \frac{1}{5}\int_{\Gamma_2} \beta v^5\,ds - F(v), \quad v \in H^1(\Omega). \tag{8.4}$$

According to the trace theorem of [Nečas, 1967, p. 84, 86], for $d = 2$ and $q \in [1, \infty)$ there exists a unique linear continuous mapping $Z\colon H^1(\Omega) \to L^q(\partial\Omega)$ such that for any $v \in C^\infty(\overline{\Omega})$ we have $Zv = v$ on $\partial\Omega$, i.e.,

$$\|v\|_{0,q,\partial\Omega} \leq C_q \|v\|_{1,2,\Omega} \quad \forall v \in H^1(\Omega). \tag{8.5}$$

Therefore, the boundary integral in (8.4) is finite and J is thus correctly defined.

By (8.4), we can easily compute the directional derivative of J at the point $v \in H^1(\Omega)$ in the direction $w \in H^1(\Omega)$,

$$\begin{aligned} J'(v;w) &= \lim_{t\to 0} \frac{1}{t}(J(v+tw) - J(v)) \qquad (8.6)\\ &= \lim_{t\to 0} \frac{1}{t}\Big(ta(v,w) + \frac{1}{2}t^2 a(w,w) \\ &\quad + \int_{\Gamma_2} \beta\Big(tv^4 w + 2t^2 v^3 w^2 + 2t^3 v^2 w^3 + t^4 v w^4 + \frac{1}{5}t^5 w^5\Big)\,ds - tF(w)\Big) \\ &= a(v,w) + \int_{\Gamma_2} \beta v^4 w\,ds - F(w). \end{aligned}$$

Analogously we find the expression for the second directional derivative of J,

$$J''(v; w, w) = a(w, w) + 4 \int_{\Gamma_2} \beta v^3 w^2 \, ds, \quad v, w \in H^1(\Omega). \tag{8.7}$$

Lemma 8.1. *The functional J is continuous on $H^1(\Omega)$ with respect to the $\|\cdot\|_1$-norm and is strictly convex over the set U.*

Proof. To check the continuity of J it suffices to prove that the Gâteaux differential of J is the Fréchet differential, i.e., we show that for any $v \in H^1(\Omega)$

$$\lim_{\substack{w \to 0 \\ w \in H^1(\Omega)}} \frac{|J(v+w) - J(v) - J'(v; w)|}{\|w\|_1} = 0. \tag{8.8}$$

By (8.4), (8.6), the continuity of $a(\cdot, \cdot)$, the Cauchy–Schwarz inequality, the imbedding $L^5(\partial\Omega) \subset L^2(\partial\Omega)$ and (8.5) we obtain

$$\begin{aligned} &|J(v+w) - J(v) - J'(v; w)| \\ &\le \frac{1}{2} a(w, w) + \int_{\Gamma_2} \beta |2v^3 w^2 + 2v^2 w^3 + v w^4 + \frac{1}{5} w^5| \, ds \\ &\le C_1 \|w\|_1^2 + C_2 (\|v\|_{0,5,\partial\Omega}^3 \|w\|_{0,5,\partial\Omega}^2 + \|v\|_{0,5,\partial\Omega}^2 \|w\|_{0,5,\partial\Omega}^3 \\ &\quad + \|v\|_{0,5,\partial\Omega} \|w\|_{0,5,\partial\Omega}^4 + \|w\|_{0,5,\partial\Omega}^5) \\ &\le C(\|w\|_1^2 + \|v\|_1^3 \|w\|_1^2 + \|v\|_1^2 \|w\|_1^3 + \|v\|_1 \|w\|_1^4 + \|w\|_1^5) \\ &\le C(v) \|w\|_1^2 \end{aligned} \tag{8.9}$$

whenever $\|w\|_1 \le 1$. Thus (8.8) is valid and J is continuous.

From (8.7) and (8.2) we see that there exists a constant $C > 0$ such that

$$J''(v; w, w) \ge C \|w\|_1^2 \quad \forall v \in U \ \forall w \in V. \tag{8.10}$$

This means that the functional J is strictly convex over the set U which completes the proof. □

Remark 8.2. The functional J is not convex over the whole space $H^1(\Omega)$ due to the term $\frac{1}{5} \int_{\Gamma_2} \beta v^5 \, ds$. That is why we have restricted ourselves to sets of non-negative functions. This is in accordance with physics because the temperature cannot decrease below 0 [K]. Therefore, we have two nonlinearities. The first one is due to the Stefan–Boltzmann boundary condition and the second one is due to the constraint $v \ge 0$ contained in the definition of the convex set U.

Theorem 8.3. *There exists a unique solution of the variational problem: Find $u \in U$ such that*

$$J(u) = \inf_{v \in U} J(v). \tag{8.11}$$

Proof. By (8.4), (8.2) and the continuity of F, there exist positive constants C_1, C_2 such that

$$J(v) = \frac{1}{2}a(v,v) + \frac{1}{5}\int_{\Gamma_2} \beta v^5 \, ds - F(v) \geq C_1\|v\|_1^2 - C_2\|v\|_1 \quad \forall v \in U,$$

since the integral over Γ_2 is non-negative. Hence, the functional J is coercive on U, i.e.,

$$J(v) \to \infty \qquad \text{as } \|v\|_1 \to \infty, \; v \in U. \tag{8.12}$$

Let $\tilde{v} \in U$ be arbitrary. Then by (8.12) there exists $r > 0$ such that for any $v \in U$ for which $\|v\|_1 > r$, we have

$$J(\tilde{v}) < J(v).$$

Thus the minimization problem (8.11) over the unbounded set U is equivalent to the minimization of J over the bounded closed set

$$\tilde{U} = \{v \in U \mid \|v\|_1 \leq r\}.$$

The rest of the proof follows from Lemma 8.1 and the next theorem (see [Fučík, Kufner, p. 199]). □

Theorem 8.4. *Let $\tilde{U}$ be a non-empty convex closed and bounded subset of $H^1(\Omega)$. Let J be a continuous and convex functional defined on $\tilde{U}$. Then*

(a) $\inf_{v\in\tilde{U}} J(v) > -\infty$;

(b) there exists at least one $u \in \tilde{U}$ such that

$$J(u) = \inf_{v\in\tilde{U}} J(v);$$

(c) if, moreover, J is strictly convex on $\tilde{U}$ then there exists precisely one u with the property (b).

Remark 8.5. The unique solution of (8.11) is called the *variational (generalized) solution.* It is known (see [Céa, 1971, p. 118]) that the minimization of a convex Gâteaux-differentiable functional J over a convex subset $U \subset H^1(\Omega)$ is equivalent to the following variational inequality problem: Find $u \in U$ such that

$$J'(u; v-u) \geq 0 \quad \forall v \in U. \tag{8.13}$$

By (8.6), the inequality (8.13) takes just the form (8.3). Hence, there is also precisely one solution of the variational inequality problem (8.3).

8.3. Convergence of finite element approximations

Throughout this section we suppose that $\Omega \in \mathcal{L}$ is polygonal. We shall employ standard linear triangular elements. Triangulations $\mathcal{T}_h$ are supposed to satisfy the standard consistency condition, i.e., those points, where one type of the boundary condition changes into another, belong to the set of vertices of all $K \in \mathcal{T}_h$ (see Section 4.2). Let $\mathcal{F}$ be a family of triangulations of $\overline{\Omega}$. For a given triangulation $\mathcal{T}_h \in \mathcal{F}$ set

$$W_h = \{v_h \in H^1(\Omega) \mid v_h|_K \in P_1(K)\ \forall K \in \mathcal{T}_h\}.$$

Recall that the function $\overline{u}$ (see (8.1)) is nonnegative and that functions from W_h are continuous (see Theorem 4.1). For simplicity assume, moreover, that

$$\exists h_0 > 0\ \forall h \in (0, h_0) : \overline{u} \in W_h. \tag{8.14}$$

Then the following set will be non-empty

$$U_h = U \cap W_h.$$

(In Sections 9.6 and 9.7 we show how to treat nonhomogeneous Dirichlet boundary conditions when (8.14) is not satisfied.)

Analogously to Theorem 8.3 we can prove that for a sufficiently small $h > 0$ there exists a unique solution of the discrete problem: Find $u_h \in U_h$ such that

$$J(u_h) = \inf_{v_h \in U_h} J(v_h), \tag{8.15}$$

where J is defined by (8.4).

To prove the convergence $u_h \to u$ for $h \to 0$ without any smoothness assumptions upon u, we shall need the following lemma.

Lemma 8.6. *Let $\mathcal{F}$ be a family of triangulations of $\overline{\Omega}$ satisfying the maximum angle condition (4.33). Then $\bigcup\limits_{h \in (0,h_0)} U_h$ is dense in U with respect to the $\|\cdot\|_1$-norm.*

For the proof see [Milka].

Theorem 8.7. *Let $\mathcal{F}$ be a family of triangulations of $\overline{\Omega}$ satisfying the maximum angle condition (4.33). Then*

$$\|u - u_h\|_1 \to 0 \quad \text{as } h \to 0,$$

where u and u_h are, respectively, the unique solutions of (8.11) and (8.15).

Proof. According to Lemma 8.6 there exists a sequence $\{v_h\}$ such that $v_h \in U_h$ and

$$\|u - v_h\|_1 \to 0 \quad \text{as } h \to 0. \tag{8.16}$$

Obviously,

$$J(u) \le J(u_h) \le J(v_h). \tag{8.17}$$

By Lemma 8.1 the functional J is continuous and, therefore, the relations (8.16) and (8.17) yield

$$\lim_{h\to 0} J(u_h) = J(u). \tag{8.18}$$

Introduce the Taylor expansion of J at the point u (see [Céa, 1971, p. 52]),

$$J(u_h) = J(u) + J'(u; u_h - u) + \frac{1}{2} J''(u + \theta_h(u_h - u); u_h - u, u_h - u), \tag{8.19}$$

where $\theta_h \in (0,1)$. From (8.13), (8.10) and (8.19), we have

$$\begin{aligned} C\|u - u_h\|_1^2 &\le C\|u - u_h\|_1^2 + J'(u; u_h - u) \\ &\le \frac{1}{2} J''(u + \theta_h(u_h - u); u_h - u, u_h - u) + J'(u; u_h - u) \\ &= J(u_h) - J(u), \end{aligned} \tag{8.20}$$

where $C > 0$ is independent of h. From here and (8.18) we see that $\|u - u_h\|_1 \to 0$ as $h \to 0$. □

If the variational solution u is sufficiently smooth we can derive even a linear rate of convergence in the $\|\cdot\|_1$-norm.

Theorem 8.8. *Let the assumptions of Theorem 8.7 be fulfilled, let $u \in H^2(\Omega)$ and let there exist a constant C_u such that $u \ge C_u > 0$ in Ω. Then*

$$\|u - u_h\|_1 \le Ch \quad \text{as } h \to 0.$$

Proof. Let $w \in V \cap C^\infty(\overline{\Omega})$ be an arbitrary function such that $\|w\|_{C(\overline{\Omega})} < C_u$. Then clearly $u + w,\ u - w \in U$. Setting $v = u \pm w$ in (8.13), we find that $J'(u; w) \ge 0$ and $J'(u; -w) \ge 0$. Now from the linearity of $J'(u; \cdot)$ and the density of $V \cap C^\infty(\overline{\Omega})$ in V (see [Doktor, p. 618]), we obtain the well-known necessary Euler extremum condition:

$$J'(u; v) = 0 \quad \forall v \in V. \tag{8.21}$$

Since $u \in H^2(\Omega)$, the function u is continuous by the Sobolev Imbedding Theorem 3.11. We see by (8.14) that $\pi_h u \in U_h$, where $\pi_h u$ is the standard linear interpolant of u. Therefore,

$$J(u_h) \le J(\pi_h u), \tag{8.22}$$

due to (8.15). Using now (8.20), (8.22), (8.21), (8.9) and (4.62), we arrive at

$$\begin{aligned} C_1\|u - u_h\|_1^2 &\le J(u_h) - J(u) \le J(\pi_h u) - J(u) \\ &= J(\pi_h u) - J(u) - J'(u; \pi_h u - u) \le C(u)\|u - \pi_h u\|_1^2 \le C_2 h^2, \end{aligned}$$

where C_1, $C_2 > 0$ are independent of h (and C_2 depends on u). □

All results of this chapter can be modified to the case $\Gamma_1 = \emptyset$ if $\alpha > 0$ or $\beta > 0$ on some set of a positive measure. As an example we compute the temperature distribution in the cross section of a transformer yoke shown in Figure 8.3 (cf. also Figure 9.1). In (8.1) we have taken $f = 15000\,[\mathrm{W/m^3}]$, $u_0 = 70\,[°\mathrm{C}]$, $\tilde{g} = 0$ and

$$A = \begin{pmatrix} 1 & 0 \\ 0 & 10 \end{pmatrix}.$$

Due to the symmetry the temperature is computed only in the right half of the section. In Figure 8.4, the triangulation and the values of the coefficient α used in the calculation are illustrated. Figure 8.5 contains contour lines of the temperature for $\beta = 0$ and $\beta = \sigma$, respectively. We see that the difference between numerical results for the linear problem ($\beta = 0$) and the nonlinear problem ($\beta = \sigma$) is rather large.

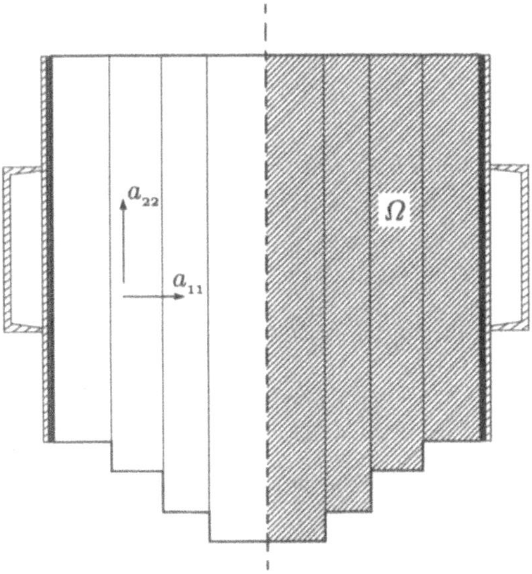

Figure 8.3.

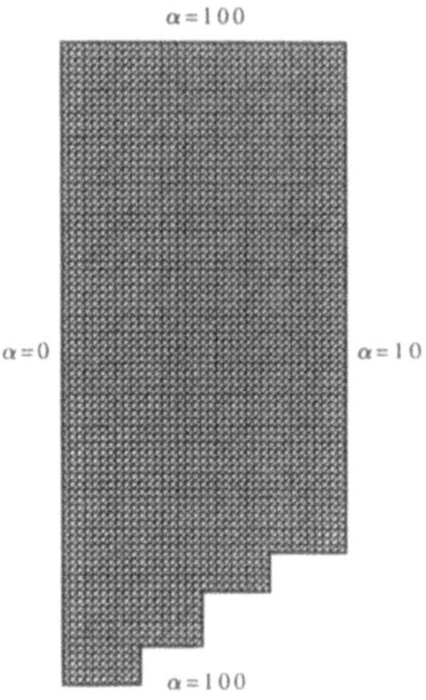

Figure 8.4.

Various approaches to the solution of the radiative heat transfer problem can be further find, e.g., in [Delfour, Payre, Zolésio], [Hottel, Sarofim], [Lewis, Morgan], [Liu, Křížek], [Ohayon, Gorge], [Szabó, Babuška], [Vaško] and [Vujanovič, Strauss]. A condition which is similar to the Stefan–Boltzmann boundary condition arises in mathematical modelling of electrolysis processes – see [Feistauer, Kalis, Rokyta].

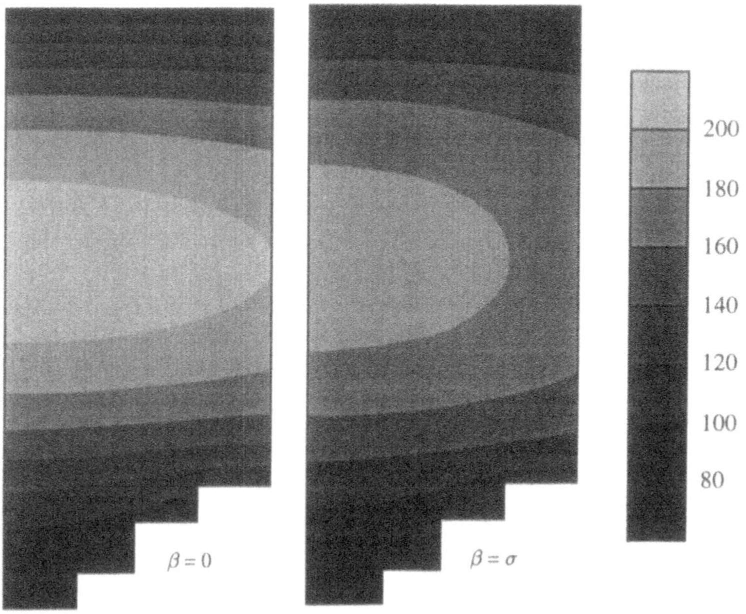

Figure 8.5.

8.4. Three-dimensional radiation problem

Finite element approximations of an axially symmetric heat radiation problem are studied in [Delfour, Payre, Zolésio]. In this section, we need not assume that Ω is axially symmetric. Then, however, we encounter some troubles with the definition domain of the functional J. Namely, the traces of functions from $H^1(\Omega)$ need not be in $L^5(\partial\Omega)$, in general (see the following example).

Example 8.9. Let us use standard spherical coordinates (r, φ, θ) to describe the set

$$\Omega = \{(x_1, x_2, x_3) \in \mathbf{R}^3 \mid x_1^2 + x_2^2 + x_3^2 < 1,\ x_i > 0,\ i = 1, 2, 3\}.$$

Consider the function r^α, where α is a real parameter. Then for $\alpha > -\frac{1}{2}$ we have

$$\begin{aligned} \|r^\alpha\|_1^2 &= \int_0^1 \int_0^{\frac{\pi}{2}} \int_0^{\frac{\pi}{2}} (r^{2\alpha} + \alpha^2 r^{2\alpha - 2}) r^2 \sin\theta \, d\theta \, d\varphi \, dr \\ &= \frac{\pi}{2}\Big(\frac{1}{2\alpha + 3} + \frac{\alpha^2}{2\alpha + 1}\Big) \in (0, \infty). \end{aligned}$$

The triple integral is not finite whenever $\alpha \in (-\infty, -\frac{1}{2}]$. Now we show that the trace of r^α is in $L^5(\partial\Omega)$ only if $\alpha > -\frac{2}{5}$.

It suffices to investigate traces on one of the three sectors of a circle which are contained in $\partial\Omega$. Denote by S that sector for which $\theta = \frac{\pi}{2}$, i.e.,

$$S = \{(x_1, x_2, 0) \in \mathbf{R}^3 \mid x_1^2 + x_2^2 < 1,\ x_1 > 0,\ x_2 > 0\}.$$

Then

$$\|r^\alpha\|_{0,5,S}^5 = \int_0^1 \int_0^{\frac{\pi}{2}} r^{5\alpha} r \, d\varphi \, dr = \frac{\pi}{2(5\alpha+2)} \in (0,\infty)$$

if and only if $\alpha > -\frac{2}{5}$. Hence, if $\alpha \in (-\frac{1}{2}, -\frac{2}{5}]$ then $r^\alpha \in H^1(\Omega)$, but its trace is not in $L^5(\partial\Omega)$. □

For $d = 3$ we can guarantee that the traces of functions from $H^1(\Omega)$ are only in $L^4(\partial\Omega)$, i.e., there exists a constant $C > 0$ such that (see [Nečas, 1967, p. 84])

$$\|v\|_{0,4,\partial\Omega} \le C\|v\|_{1,2,\Omega} \quad \forall v \in H^1(\Omega). \tag{8.23}$$

If we would restrict the definition domain of the functional (8.4) to the space $H^2(\Omega)$ (or $W_3^1(\Omega)$) then J would be not coercive on nonnegative functions from these spaces with respect to the norm $\|\cdot\|_2$ (or $\|\cdot\|_{1,3}$). Therefore, we shall consider the functional J over the linear space

$$Z = \{v \in H^1(\Omega) \mid v|_{\Gamma_2} \in L^5(\Gamma_2)\}$$

equipped with the norm

$$\|v\|_Z = (\|v\|_{1,2,\Omega}^2 + \|v\|_{0,5,\Gamma_2}^2)^{\frac{1}{2}}.$$

Lemma 8.10. *The space Z is a Banach space.*

Proof. Let $\{v_k\} \subset Z$ be a sequence such that

$$\lim_{k,m\to\infty} \|v_k - v_m\|_Z = 0.$$

Since $H^1(\Omega)$ and $L^5(\Gamma_2)$ are Banach spaces, there exist $v \in H^1(\Omega)$ and $w \in L^5(\Gamma_2)$ such that $v_k \to v$ in $H^1(\Omega)$ and $v_k|_{\Gamma_2} \to w$ in $L^5(\Gamma_2)$ for $k \to \infty$. By (8.23) we find that $v_k|_{\Gamma_2} \to v|_{\Gamma_2}$ in $L^4(\Gamma_2)$ and thus $v|_{\Gamma_2} = w$. Hence, the Cauchy sequence $\{v_k\}$ converges in the space Z, which means that Z is complete. □

Lemma 8.11. *The space Z is reflexive.*

Proof. We shall proceed similarly to [Delfour, Payre, Zolésio, p. 1079]. By [Fučík, Kufner, p. 187] the Cartesian product of two reflexive spaces is also reflexive. Hence, the space

$$T \equiv H^1(\Omega) \times L^5(\Gamma_2) = \{(v,w) \mid v \in H^1(\Omega),\ w \in L^5(\Gamma_2)\},$$

equipped with the norm

$$\|(v,w)\|_T = (\|v\|_{1,2,\Omega}^2 + \|w\|_{0,5,\Gamma_2}^2)^{\frac{1}{2}},$$

is a reflexive normed space. The mapping

$$v \in Z \mapsto (v, v|_{\Gamma_2}) \in T \tag{8.24}$$

is obviously an isomorphism between Z and some closed subspace T_1 of T. According to [Fučík, Kufner, p. 187], T_1 is reflexive since T is reflexive. Therefore, due to the isomorphism (8.24), the space Z is also reflexive. □

Lemma 8.12. *The functional J given by (8.4) is continuous on Z.*

Proof. For $v, w \in Z$ we have

$$\begin{aligned} |J(v+w) - J(v)| &\le |\tfrac{1}{2}a(v+w, v+w) - \tfrac{1}{2}a(v,v)| \\ &\quad + \left|\frac{1}{5}\int_{\Gamma_2} \beta((v+w)^5 - v^5)\, ds\right| + |F(v+w) - F(v)| \\ &\le |a(v,w)| + \tfrac{1}{2}|a(w,w)| + |F(w)| \\ &\quad + \int_{\Gamma_2} |\beta(v^4 w + 2v^3 w^2 + 2v^2 w^3 + v w^4 + \tfrac{1}{5}w^5)|\, ds \\ &\le C_1(\|v\|_1\|w\|_1 + \|w\|_1^2 + \|w\|_1) + C_2(\|v\|_{0,5,\Gamma_2}^4\|w\|_{0,5,\Gamma_2} \\ &\quad + \|v\|_{0,5,\Gamma_2}^3\|w\|_{0,5,\Gamma_2}^2 + \|v\|_{0,5,\Gamma_2}^2\|w\|_{0,5,\Gamma_2}^3 \\ &\quad + \|v\|_{0,5,\Gamma_2}\|w\|_{0,5,\Gamma_2}^4 + \|w\|_{0,5,\Gamma_2}^5), \end{aligned}$$

where C_1, C_2 are positive constants, since $\|w\|_1 \le \|w\|_Z$ and $\|w\|_{0,5,\Gamma_2} \le \|w\|_Z$, we find that

$$\lim_{\substack{w \in Z \\ \|w\|_Z \to 0}} |J(v+w) - J(v)| = 0.$$

□

Now we define

$$U = \{v \in Z \mid v \ge 0 \text{ in } \Omega,\ v = \overline{u} \text{ on } \Gamma_1\}. \tag{8.25}$$

Lemma 8.13. *The functional J is strictly convex over the set U given by (8.25).*

Proof. For any $v, w, z \in U$ we have by (8.7) and (8.2) that

$$\begin{aligned} J''(v; z-w\, z-w) &= a(z-w,\ z-w) + 4\int_{\Gamma_2} \beta v^3 (z-w)^2\, ds \\ &\ge C\|z-w\|_1^2, \end{aligned} \tag{8.26}$$

where C is a positive constant. Since $\|z-w\|_Z = 0$ if and only if $\|z-w\|_1 = 0$, we see that J is strictly convex on U. □

Lemma 8.14. *Let there exist a positive constant C_β such that*

$$\beta \ge C_\beta \text{ on } \Gamma_2. \tag{8.27}$$

Then the functional J is coercive on U.

Proof. For any $v \in U$ we find that

$$J(v) = \tfrac{1}{2}a(v,v) + \tfrac{1}{5}\int_{\Gamma_2} \beta v^5 \, ds - F(v)$$
$$\geq C_1\|v\|_1^2 + \tfrac{1}{5}C_\beta\|v\|_{0,5,\Gamma_2}^5 - C_2\|v\|_1.$$

Obviously, if $\|v\|_Z \to \infty$ then $\|v\|_1 \to \infty$ or $\|v\|_{0,5,\Gamma_2} \to \infty$, and thus

$$\lim_{\substack{v \in Z \\ \|v\|_Z \to \infty}} J(v) = \infty.$$

□

Theorem 8.15. *Let $\overline{u} \in Z$ and let (8.27) hold. Then there exists a unique function u which minimizes the functional J over the set U. The function u is a variational solution of the problem (8.1).*

Proof. Due to the coercivity of J (see Lemma 8.14), the minimization of J over the unbounded set U can be transformed to the minimization of J over the bounded subset

$$\tilde{U} = \{v \in U \mid \|v\|_Z \leq r\}$$

like in Theorem 8.3. The existence of a unique minimizer u now follows from Theorem 8.4 and Lemmas 8.10–8.13.

By (8.13), we again derive the associated variational inequality of the form (8.3). □

Now we briefly describe a finite element approximation of the three-dimensional problem. Let Ω be a bounded polyhedral domain with a Lipschitz continuous boundary (an example of a polyhedron whose boundary is not Lipschitz continuous is given in [Křížek, 1982, p. 48]). Let the one-dimensional boundary of the two-dimensional set $\Gamma_1 \subset \partial\Omega$ consist of a finite number of straight line segments. Denote by $\mathcal{T}_h$ a decomposition of $\overline{\Omega}$ into closed tetrahedra K in the usual sense, i.e., the union of all $K \in \mathcal{T}_h$ is $\overline{\Omega}$, the interiors of all tetrahedra are mutually disjoint, and any face of any tetrahedron $K \in \mathcal{T}_h$ is either a subset of $\overline{\Gamma}_1$ or $\overline{\Gamma}_2$ or a face of another $K' \in \mathcal{T}_h$. A constructive proof of the existence of such a decomposition is given in [Křížek, 1982, p. 61]. The space W_h is defined in the same way as for $d = 2$. Let us assume that $\overline{u} \in W_h$ for all h sufficiently small (cf. (8.14)). Then from Theorem 8.4 we again see that for any $\mathcal{T}_h$ (h small) there exists one and only one solution of the minimization problem: Find $u_h \in U_h$ such that

$$J(u_h) = \min_{v_h \in U_h} J(v_h),$$

where

$$U_h = U \cap W_h.$$

Recall that (see (4.88)) if $\mathcal{F}$ satisfies the maximum angle condition then

$$\|v - \pi_h v\|_{1,\infty} = \mathcal{O}(h) \tag{8.28}$$

for sufficiently smooth v. Here π_h is the standard linear interpolation operator. From property (8.28) we can prove the convergence of u_h to u in the $\|\cdot\|_Z$-norm as in Section 8.3 using the inequality $\|v\|_Z \leq C\|v\|_{1,\infty}$ for any v from the Sobolev space $W^1_\infty(\Omega)$.

9. Nonlinear anisotropic heat conduction in a transformer magnetic core

9.1. Non-monotonicity and non-potentionality of the problem

In this chapter we deal with a quasilinear elliptic problem whose classical formulation reads:

Find $u \in C^1(\overline{\Omega})$ such that $u|_\Omega \in C^2(\Omega)$ and

$$-\operatorname{div}(\mathcal{A}(\cdot\,,u)\operatorname{grad}u) = f \quad \text{in } \Omega, \tag{9.1}$$

$$u = \overline{u} \quad \text{on } \Gamma_1, \tag{9.2}$$

$$\alpha u + n^T\mathcal{A}(\cdot\,,u)\operatorname{grad}u = g \quad \text{on } \Gamma_2, \tag{9.3}$$

where $\Omega \in \mathcal{L}$, $n = (n_1, ..., n_d)^T$ is the outward unit normal to $\partial\Omega$, $d \in \{1, 2, ...\}$, Γ_1 and Γ_2 are relatively open sets in the boundary $\partial\Omega$, $\overline{\Gamma}_1 \cup \overline{\Gamma}_2 = \partial\Omega$, $\Gamma_1 \cap \Gamma_2 = \emptyset$, $\mathcal{A} = (a_{ij})_{i,j=1}^d$ is a uniformly positive definite matrix, $\alpha \geq 0$. Let the functions $\mathcal{A}$, α, f, $\overline{u}$ and g be sufficiently smooth for the time being (precise assumptions on these functions will be given later). The boundary condition (9.3) is called the *Newton boundary condition.*

This chapter is devoted to a steady-state heat conduction problem defined by (9.1)–(9.3) which describes a temperature distribution in large transformers. Their magnetic cores (consisting of iron sheets) are nonlinear orthotropic media the heat conductivities of which can be represented by a diagonal matrix $\mathcal{A} = \mathcal{A}(u)$.

First let us consider the following special case of (9.1)–(9.3) which describes a stationary heat conduction in a homogeneous and isotropic medium Ω,

$$-\operatorname{div}(\lambda(u)\operatorname{grad}u) = f \quad \text{in } \Omega,$$

$$u = 0 \quad \text{on } \Gamma_1,$$

$$n^T\lambda(u)\operatorname{grad}u = 0 \quad \text{on } \Gamma_2,$$

where $\lambda\colon \mathbf{R}^1 \to \mathbf{R}^1$ is a measurable bounded function such that

$$\lambda(\xi) \geq C > 0 \quad \forall \xi \in \mathbf{R}^1. \tag{9.4}$$

Note that this nonlinear problem can be converted by the well-known Kirchhoff transformation (cf. [Čermák, Zlámal], [Larsson, Thomée, Zhang])

$$\mathcal{K}(U) = \int_0^U \lambda(\xi)\,d\xi, \quad U \in \mathbf{R}^1,$$

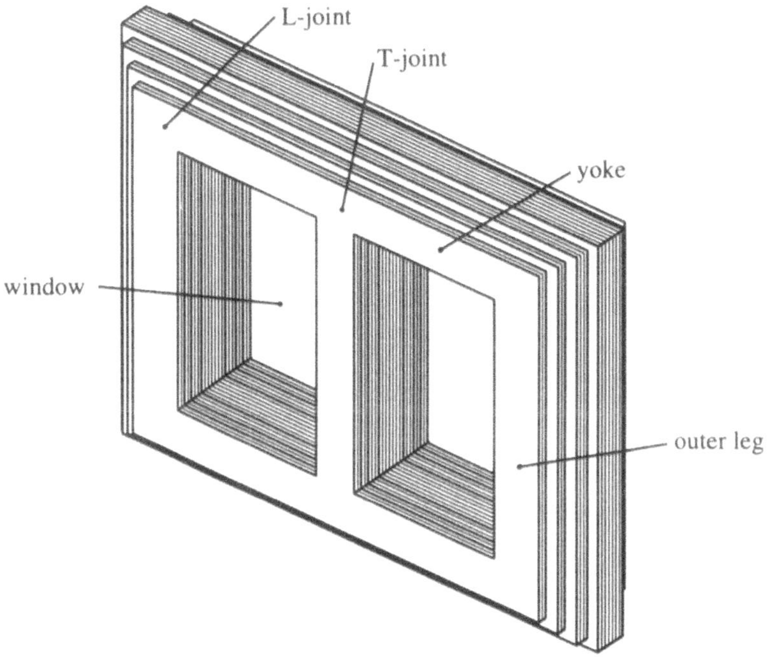

Figure 9.1.

to the linear problem

$$\begin{aligned} -\Delta z &= f \quad &&\text{in } \Omega, \\ z &= 0 \quad &&\text{on } \Gamma_1, \\ n^T \operatorname{grad} z &= 0 \quad &&\text{on } \Gamma_2, \end{aligned}$$

where $z(x) = \mathcal{K}(u(x))$. From (9.4) we observe that $\mathcal{K}$ is an increasing function, i.e., its inverse $\mathcal{K}^{-1}$ exists and we have $u(x) = \mathcal{K}^{-1}(z(x))$.

The Kirchhoff transformation, however, cannot be applied in the case of anisotropic nonlinear media. For instance, in examining a temperature field in the magnetic circuit of a transformer (see Figure 9.1), nonlinear temperature dependencies of the heat conductivities across and along the lamination differ. This is the case of an orthotropic material. The associated 3×3 matrix $\mathcal{A}$ of heat conductivities is diagonal and such that $a_{11} \neq a_{22} = a_{33}$. The temperature dependencies of the diagonal entries is illustrated in Figures 9.2 and 9.3. We see that the type of nonlinearity is different in each direction (it, moreover, depends upon the type of iron used to construct the sheets).

Since the Kirchhoff transformation cannot be used, we have to find another approach to solve the problem. In Remark 9.2, we show that the problem (9.1)–(9.3) does not lead to a problem with a monotone operator, in general (cf. Chapter 7). We also demonstrate that the potential operator theory of Chapter 8 cannot be applied.

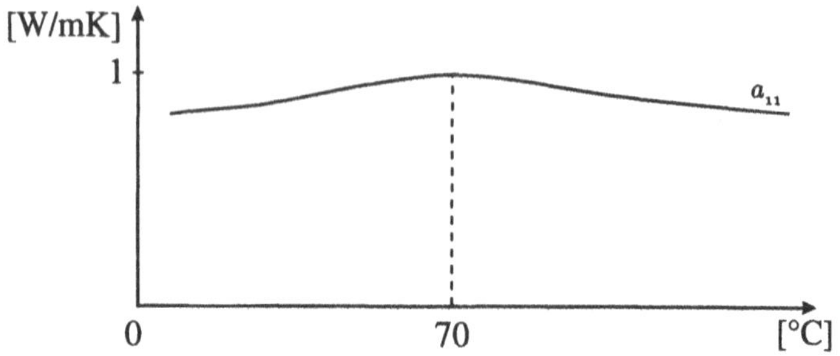

Figure 9.2.

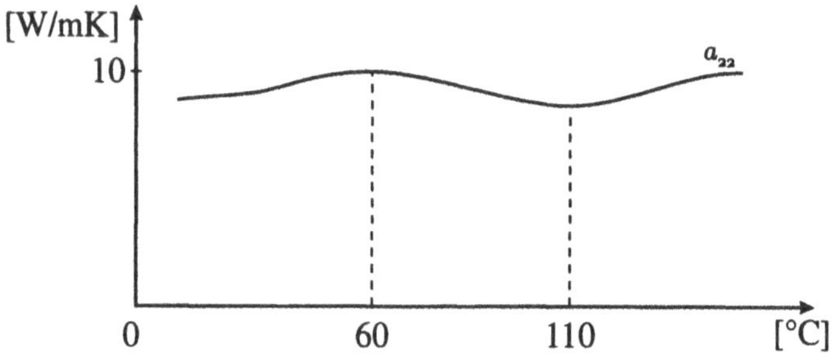

Figure 9.3.

To state a weak formulation of problem (9.1)–(9.3) we assume that $\mathcal{A} = \mathcal{A}(\cdot,\cdot)$ and $\alpha = \alpha(\cdot)$ are bounded measurable functions,

$$\operatorname*{ess\,sup}_{x,\xi,i,j} |a_{ij}(x,\xi)| \le C, \quad \operatorname*{ess\,sup}_{s} |\alpha(s)| \le C, \tag{9.5}$$

where $x \in \Omega$, $\xi \in \mathbf{R}^1$, $i,j \in \{1,...,d\}$ and $s \in \Gamma_2$. The components a_{ij} are assumed to be Lipschitz continuous with respect to the second variable, i.e., there exists $C_L > 0$ such that for all $\zeta, \xi \in \mathbf{R}^1$ and almost all $x \in \Omega$ we have

$$|a_{ij}(x,\zeta) - a_{ij}(x,\xi)| \le C_L|\zeta - \xi|, \quad i,j = 1,...,d. \tag{9.6}$$

Moreover, let there exist $C_0 > 0$ such that for almost all $x \in \Omega$

$$C_0\eta^T\eta \le \eta^T\mathcal{A}(x,\xi)\eta \quad \forall\xi \in \mathbf{R}^1\ \forall\eta \in \mathbf{R}^d \tag{9.7}$$

and let

$$0 \le \alpha(s)$$

for almost all $s \in \Gamma_2$. Finally, let $f \in L^2(\Omega)$, $\overline{u} \in H^1(\Omega)$, $g \in L^2(\Gamma_2)$ and

$$V = \{v \in H^1(\Omega) \mid v = 0 \text{ on } \Gamma_1\}.$$

For simplicity, a possible dependence of $\mathcal{A}$ on x will usually not be explicitly indicated in what follows. Set

$$a(y;w,v) = (\mathcal{A}(y)\operatorname{grad} w, \operatorname{grad} v)_{0,\Omega} + \langle\alpha w, v\rangle_{0,\Gamma_2}, \quad y,w,v \in H^1(\Omega), \tag{9.8}$$

$$F(v) = (f,v)_{0,\Omega} + \langle g, v\rangle_{0,\Gamma_2}, \quad v \in H^1(\Omega), \tag{9.9}$$

where $\langle\cdot,\cdot\rangle_{0,\Gamma_2}$ stands for the usual scalar product in $L^2(\Gamma_2)$. Since $\mathcal{A}$ and α are bounded, we observe that both the terms in (9.8) are finite, i.e., $a(\cdot;\cdot,\cdot)$ is well-defined.

Definition 9.1. *A function $u \in H^1(\Omega)$ is said to be a weak solution of the problem (9.1)–(9.3) if $u - \overline{u} \in V$ and*

$$a(u;u,v) = F(v) \quad \forall v \in V.$$

Remark 9.2. To prove the existence of a weak solution $u \in V$ we cannot apply the main theorem for monotone operators (cf. Theorem 7.22), since our problem does not lead to a monotone operator, in general. To see this we put, e.g., $d = 1$, $\Omega = (0,1)$, $\Gamma_1 = \emptyset$, $\alpha = 1/50$ and define (a nondecreasing function $\mathcal{A}$ – see Figure 9.4)

$$\mathcal{A}(x,v) = \mathcal{A}(v) = \begin{cases} 1 & \text{if } v \le 2, \\ v-1 & \text{if } v \in (2,4), \\ 3 & \text{if } v \ge 4. \end{cases}$$

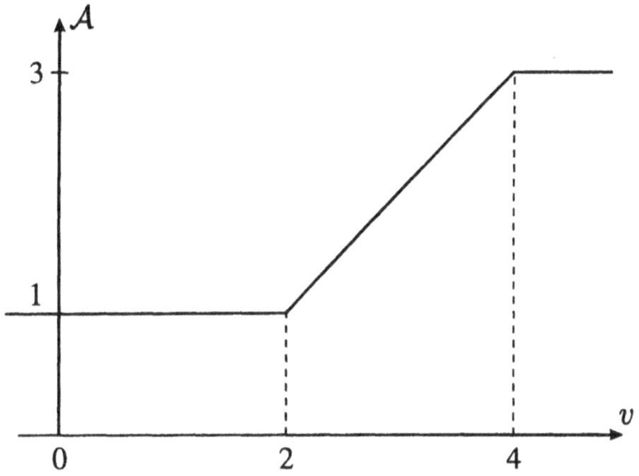

Figure 9.4.

Define the nonlinear operator $\mathsf{A}: V \to V$ by

$$\langle \mathsf{A}w, v\rangle = a(w;w,v), \quad v,w \in V, \tag{9.10}$$

where $\langle\cdot,\cdot\rangle$ is the $H^1(\Omega)$-scalar product. We see that the functions $v(x) = 2x$ and $w(x) = x+4$ violate the monotonicity condition (7.3) for the operator A, since

$$\begin{aligned}
&\langle \mathsf{A}v - \mathsf{A}w, v-w\rangle = a(v;v,v-w) - a(w;w,v-w)\\
&= \int_0^1 (\mathcal{A}(v)v' - \mathcal{A}(w)w')(v'-w')\,dx + \frac{1}{50}\int_{\Gamma_2}(v-w)^2\,ds\\
&= \int_0^1 (1\times 2 - 3\times 1)(2-1)\,dx + \frac{1}{50}\left((v(1)-w(1))^2 + (v(0)-w(0))^2\right)\\
&= -1 + \frac{1}{2} < 0,
\end{aligned}$$

where v' denotes the derivative with respect to x. To show that the problem (9.1)–(9.3) can lead to a non-monotone operator also for $d > 1$ it suffices to put $\Omega = [(0,1)]^d$, $a_{11}(v) = \mathcal{A}(v)$, $a_{ij} = 0$ for $i \neq j$, $a_{ii} = 1$ for $i = 2, ..., d$, and employ the functions $v(x_1, ..., x_d) = v(x_1)$ and $w(w_1, ..., w_d) = w(x_1)$ for $(x_1, ..., x_d) \in \Omega$, $\alpha(s) = \alpha_0$ for sufficiently small $\alpha_0 \in (0,1)$ and again $\Gamma_1 = \emptyset$.

Remark 9.3. The problem of Definition 9.1 cannot be transformed to the minimization of some functional, since the associated operator A is not potential in general. To see this we put $d = 1$, $\Omega = (0,1)$, $\Gamma_1 = \{0\}$, $\alpha \equiv 0$,

$$\mathcal{A}(x,v) = \mathcal{A}(v) = v + 1 \quad \text{for } v \in (0,1)$$

and $\mathcal{A}(v)$ may be defined in an arbitrary way outside of $(0,1)$ so that (9.6), (9.7) and the boundedness of $\mathcal{A}$ hold. By [Gajewski, Gröger, Zacharias, Lemma III.4.1], a demicontinuous operator A (see Definition 7.14) is potential if and only if

$$\int_0^1 \langle \mathsf{A}tv, v\rangle \, dt - \int_0^1 \langle \mathsf{A}tw, w\rangle \, dt = \int_0^1 \langle \mathsf{A}(w + t(v-w)), v - w\rangle \, dt \tag{9.11}$$

for all $v, w \in V$.

So let $w_k \to w$ in V as $k \to \infty$. Then by the boundedness and Lipschitz continuity of $\mathcal{A}$ (see (9.6)), the Cauchy–Schwarz inequality and (3.31) we get

$$\begin{aligned} |a(w_k; w_k, v) - a(w; w, v)| &\le |a(w_k; w_k - w, v)| + |a(w_k; w, v) - a(w; w, v)| \\ &\le C\|w_k - w\|_1 \|v\|_1 + C_L \|w_k - w\|_0 \|w'\|_0 \|v'\|_{0,\infty} \to 0 \quad \text{as } k \to \infty \end{aligned}$$

for any $v \in V \cap C^1([0,1])$. Now from (9.10) and the density $V = \overline{V \cap C^1([0,1])}$ we see that $\mathsf{A}w_k \rightharpoonup \mathsf{A}w$, which means that A is demicontinuous.

Setting $v(x) = x^2$ and $w(x) = x$, we arrive at

$$\begin{aligned} \int_0^1 \langle \mathsf{A}tv, v\rangle \, dt &= \int_0^1 \int_0^1 \mathcal{A}(tv)(tv)'v' \, dx \, dt \\ &= \int_0^1 \int_0^1 (1 + tx^2)(2tx)2x \, dx \, dt = \frac{14}{15}, \end{aligned}$$

$$\int_0^1 \langle \mathsf{A}tw, w\rangle \, dt = \int_0^1 \int_0^1 \mathcal{A}(tw)(tw)'w' \, dx \, dt = \int_0^1 \int_0^1 (1 + tx)t \, dx \, dt = \frac{2}{3},$$

$$\begin{aligned} &\int_0^1 \langle \mathsf{A}(w + t(v-w)), v - w\rangle \, dt \\ &= \int_0^1 \int_0^1 \mathcal{A}(w + t(v-w))(w + t(v-w))'(v-w)' \, dx \, dt \\ &= \int_0^1 \int_0^1 (x + t(x^2 - x) + 1)(1 + t(2x-1))(2x-1) \, dx \, dt = \frac{13}{180}, \end{aligned}$$

i.e., (9.11) does not hold and thus A is not potential. Another way would be to check that the well-known necessary symmetry conditions from [Nečas, 1983, p. 41] are not fulfilled. The fact that there is no potential for A means that there is no functional $J\colon V \to \mathsf{R}^1$, the Gâteaux differential of which satisfies the relation

$$\langle \mathsf{A}v, w\rangle = J'(v;w) \quad \forall v, w \in V.$$

9.2. Existence of the weak and discrete solutions

From now until Section 9.5 we shall assume for simplicity that

$$d \in \{2,3\} \quad \text{and} \quad \Gamma_1 = \emptyset$$

which is an interesting case in practice (see Section 9.5). For the Dirichlet boundary conditions (see Section 9.6) and for the mixed boundary conditions (9.2)–(9.3) (which can be possibly nonlinear) we refer to [Hlaváček, Křížek, Malý]. So we shall look for the weak solution $u \in H^1(\Omega)$ such that (cf. Definition 9.1)

$$a(u;u,v) = F(v) \quad \forall v \in H^1(\Omega). \tag{9.12}$$

To guarantee the existence of such a u, we moreover assume that there exist a constant $\alpha_0 > 0$ and a non-empty relatively open subset $\Gamma_3 \subset \Gamma_2 = \partial\Omega$ such that

$$\alpha(s) \geq \alpha_0$$

for almost all $s \in \Gamma_3$. Then there exists a constant $C_0 > 0$ such that

$$C_0\|v\|_1^2 \leq a(y;v,v) \quad \forall y, v \in H^1(\Omega). \tag{9.13}$$

This inequality is a direct consequence of (9.7), (9.8) and the following Friedrichs' inequality (3.42),

$$\|v\|_{1,\Omega}^2 \leq C(\|\operatorname{grad} v\|_{0,\Omega}^2 + \|v\|_{0,\Gamma_3}^2) \quad \forall v \in H^1(\Omega).$$

Remark 9.4. Using (9.8), (9.9), the boundedness of $\mathcal{A}$, α (see (9.5)) and the Trace Theorem 3.8, it is not difficult to verify that

$$|a(y;w,v)| \leq C\|w\|_1\|v\|_1 \quad \forall y, w, v \in H^1(\Omega), \tag{9.14}$$

$$|F(v)| \leq C\|v\|_1 \qquad \forall v \in H^1(\Omega). \tag{9.15}$$

Before we prove that the problem (9.12) has a solution, we introduce one theorem.

Theorem 9.5. *Let $V_h \subset C(\overline{\Omega})$ be a non-empty finite-dimensional subspace. Then*

(i) there exists a Galerkin approximation $u_h \in V_h$ such that

$$a(u_h; u_h, v_h) = F(v_h) \quad \forall v_h \in V_h, \tag{9.16}$$

(ii) if the constant C_L from (9.6) is sufficiently small, there exists a unique Galerkin approximation u_h. Moreover, u_h can be calculated by means of the method of successive approximations (Kačanov's method) as follows: Let $y^0 \in V_h$ be arbitrary. If $y^k \in V_h$ is known, $y^{k+1} \in V_h$ is defined by the relation

$$a(y^k; y^{k+1}, v_h) = F(v_h) \quad \forall v_h \in V_h \tag{9.17}$$

and

$$\|u_h - y^k\|_1 \to 0 \qquad \text{as } k \to \infty.$$

Proof. (i) Let us consider a mapping

$$S\colon V_h \to V_h$$

defined by the following relation

$$a(y; Sy, v) = F(v) \quad \forall v \in V_h. \tag{9.18}$$

The mapping is uniquely determined. This follows from (9.13)–(9.15) and the Lax–Milgram Lemma 3.3. Setting $v = Sy$ in (9.18), we may write by (9.13) and (9.15)

$$C_0\|Sy\|_1^2 \le a(y; Sy, Sy) = F(Sy) \le C\|Sy\|_1.$$

Consequently,

$$\|Sy\|_1 \le C/C_0 \quad \forall y \in V_h. \tag{9.19}$$

Next we show that the mapping S is Lipschitz continuous, i.e., a constant $C_S > 0$ exists such that C_S is independent of C_L and

$$\|Sy - Sz\|_1 \le C_L C_S \|y - z\|_1 \quad \forall y, z \in V_h. \tag{9.20}$$

So let $y, z \in V_h$ be arbitrary and let $v = Sy - Sz \in V_h$. Using (9.13), (9.18), (9.19), (9.6) and the equivalence of all norms in V_h, we obtain

$$\begin{aligned} C_0\|v\|_1^2 &\le a(y; v, v) = a(y; Sy, v) - a(y; Sz, v) = F(v) - a(y; Sz, v) \\ &= a(z; Sz, v) - a(y; Sz, v) = ((\mathcal{A}(z) - \mathcal{A}(y))\operatorname{grad}(Sz), \operatorname{grad} v)_0 \\ &\le C_L d^2 \|z - y\|_{0,\infty} \|Sz\|_1 \|v\|_1 \le C_L C \|z - y\|_1 \|v\|_1, \end{aligned}$$

where C is independent of C_L, i.e., (9.20) is valid.

The existence of a fixed point $y = Sy$ follows now from (9.19), the continuity of S and the Brouwer fixed-point Theorem 7.19. Setting $u_h = y$, we obtain the Galerkin approximation satisfying (9.16).

(ii) If $C_LC_S < 1$, the mapping S is, by (9.20), contractive and the well-known iterative procedure

$$y^0 \in V_h, \qquad y^{k+1} = Sy^k$$

leads to the unique fixed point y, since

$$\|y^k - y\|_1 \le (1 - C_LC_S)^{-1}(C_LC_S)^k\|y^0 - Sy^0\|_1$$

(see, e.g., [Nečas, Hlaváček, Sec. 11.7]). Consequently, (ii) is proved. □

In Theorem 9.6, we obtain the existence of a solution u of the problem (9.12) as a weak limit of the Galerkin approximations u_h under the following assumption:

Let $\{V_h\}_{h\to 0}$ be a family of finite-dimensional subspaces of $H^1(\Omega)\cap C(\overline{\Omega})$ such that

$$\forall v \in C^\infty(\overline{\Omega})\ \exists\{v_h\}_{h\to 0}\ :\ v_h \in V_h,\ \|v - v_h\|_1 \to 0 \ \text{ as } h \to 0. \tag{9.21}$$

This condition is easy to satisfy for many families of finite element spaces (see, e.g., Theorem 4.15 or Remarks 4.34, 4.35).

Theorem 9.6. *Let (9.21) hold and let $\{u_h\}_{h\to 0}$ be a sequence of Galerkin approximations satisfying (9.16). Then there exist a subsequence $\{u_{\hat{h}}\} \subset \{u_h\}$ and an element $u \in H^1(\Omega)$ such that*

$$u_{\hat{h}} \rightharpoonup u \quad \text{(weakly) in } H^1(\Omega) \quad \text{as } \hat{h} \to 0, \tag{9.22}$$

and u is a solution of problem (9.12). Moreover, any weak cluster (accumulation) point of the sequence $\{u_h\}$ is a solution of (9.12).

Proof. First we show the boundedness of $\{u_h\}$ in $H^1(\Omega)$. Using (9.13) and (9.15), we may write

$$C_0\|u_h\|_1^2 \le a(u_h; u_h, u_h) = F(u_h) \le C_1\|u_h\|_1. \tag{9.23}$$

Consequently,

$$\|u_h\|_1 \le C \tag{9.24}$$

and thus, as a consequence of the Eberlein–Schmulyan theorem (see [Yosida, Chap. V]), there exist an element $u \in H^1(\Omega)$ and a subsequence $\{u_{\hat{h}}\} \subset \{u_h\}$ such that (9.22) holds. For simplicity we write h instead of $\hat{h}$ in what follows.

We show that u is a solution of (9.12). Since $u_h \in H^1(\Omega)$, the weak limit u also belongs to $H^1(\Omega)$. So it remains to prove the equality (9.12).

Let $v \in C^\infty(\overline{\Omega})$ be arbitrary but fixed. Then, by (9.6),

$$\begin{aligned} |a(u; u_h, v) - a(u_h; u_h, v)| &\le \sum_{i,j} \int_\Omega |a_{ij}(u) - a_{ij}(u_h)||\partial_i u_h||\partial_j v|\, dx \\ &\le C_L C(v)\|u - u_h\|_0\|u_h\|_1. \end{aligned} \tag{9.25}$$

Furthermore, let $\{v_h\}_{h\to 0}$ be a sequence satisfying (9.21). Then from (9.16), (9.25), (9.14), (9.15), (9.22), (9.24), the Rellich Theorem 3.9 and (9.21), we get

$$\begin{aligned}&|a(u;u,v)-F(v)|\leq\\&\leq|a(u;u-u_h,v)|+|a(u;u_h,v)-a(u_h;u_h,v)|+|a(u_h;u_h,v-v_h)|+|F(v_h-v)|\\&\leq|a(u;u-u_h,v)|+C_1\|u-u_h\|_0\|u_h\|_1+C_2\|u_h\|_1\|v-v_h\|_1+C_3\|v-v_h\|_1\to 0\end{aligned}$$

as $h\to 0$. Hence,

$$a(u;u,v)=F(v)\quad\forall v\in C^\infty(\overline{\Omega}).$$

By the density of $C^\infty(\overline{\Omega})$ in $H^1(\Omega)$, (9.14) and (9.15), we find that u is a solution of the problem (9.12). □

The next corollary enables us to estimate the norm $\|u\|_1$ by the "data".

Corollary 9.7. *There exists a constant $C>0$ (independent of the data f,g) such that*

$$\|u\|_1\leq C(\|f\|_{0,\Omega}+\|g\|_{0,\partial\Omega}).$$

Proof. From (9.13), (9.12), (9.9) and the Trace Theorem 3.8 we have

$$C_0\|u\|_1^2\leq a(u;u,u)=(f,u)_{0,\Omega}+\langle g,u\rangle_{0,\partial\Omega}\leq C(\|f\|_{0,\Omega}+\|g\|_{0,\partial\Omega})\|u\|_1.$$

□

9.3. Uniqueness and nonuniqueness

First of all note that if an elliptic equation is not in the divergence form there exist examples of nonunique solutions (see, e.g., [Gilbarg, Trudinger, p. 209], [Meyers, p. 178]). We can also get non-unique solutions of our problem (9.1)–(9.3) if condition (9.6) is violated. To see this, we introduce the following example due to J. Malý. Let $d=1$, $\Omega=(0,1)$ and consider two fixed real smooth functions u_1, u_2 such that $u_1<u_2$ on $(0,1)$, $u_1(0)=u_2(0)$, $u_1'(0)=u_2'(0)$, $u_1(1)=u_2(1)$, $u_1'(1)=u_2'(1)$, $u_1'\geq 1$ and $u_2'\geq 1$ (see Figure 9.5).

Let us define a real function $\mathcal{A}$ on the graphs of u_1 and u_2 as follows

$$\mathcal{A}(x,\xi)=\frac{1}{u_i'(x)}\qquad\text{for } x\in[0,1],\quad \xi=u_i(x),\quad i=1,2.$$

Then by Tietze's extension theorem (see, e.g., [Rudin, p. 422]) there exists a continuous extension (still denoted by $\mathcal{A}$) so that $\mathcal{A}(\cdot,\cdot)\colon\Omega\times\mathbf{R}^1\to\mathbf{R}^1$ and relations (9.5) and (9.7) hold. We see that

$$-(\mathcal{A}(x,u_i)u_i')'=0\quad\text{for } i=1,2,$$

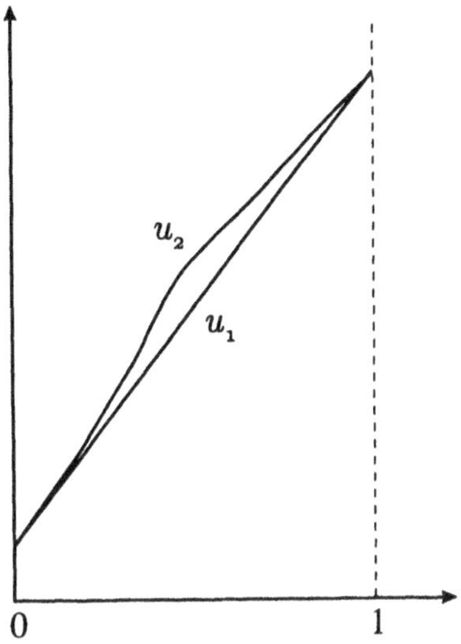

Figure 9.5.

i.e., u_1 and u_2 are solutions of (9.1) with nonhomogeneous boundary conditions and $f = 0$. However, in this case it is not difficult to check that $\mathcal{A}$ is not Lipschitz continuous (with respect to the second variable) near those points, where u_1 and u_2 bifurcate. The condition (9.6) is thus essential.

Note that a proof of the uniqueness of the classical solution of the problem (9.1)–(9.3) is given by [Douglas, Dupont, Serrin] for the Dirichlet boundary conditions and by [Hlaváček, Křížek, 1993a] for mixed conditions. The uniqueness of the weak solution for the mixed nonlinear boundary conditions is proved in [Hlaváček, Křížek, Malý].

Throughout this section assume for simplicity again that $\Gamma_1 = \emptyset$, $d \in \{1, 2, ...\}$ and that there exists a constant $\alpha_0 > 0$ and a non-empty relatively open subset $\Gamma_3 \subset \Gamma_2 = \partial\Omega$ such that

$$\alpha \geq \alpha_0 \quad \text{on } \Gamma_3. \tag{9.26}$$

Theorem 9.8. *Let (9.5)–(9.7) and (9.26) hold and let u_1, $u_2 \in H^1(\Omega)$ be two weak solutions of the problem (9.12). Then $u_1 = u_2$ a.e. in Ω.*

Proof. Introduce the following set

$$\Omega_0 = \{x \in \Omega \mid u_1(x) < u_2(x)\}$$

and assume that

$$\operatorname{meas} \Omega_0 > 0. \tag{9.27}$$

Let us consider an arbitrary $\varepsilon > 0$ and define (see Figure 9.6)

$$\Omega_\varepsilon = \{x \in \Omega_0 \mid u_2 - u_1 > \varepsilon\}, \tag{9.28}$$

$$v_\varepsilon = \begin{cases} \min(\varepsilon, u_2 - u_1) & \text{in } \Omega_0, \\ 0 & \text{in } \mathbf{R}^d \setminus \Omega_0. \end{cases} \tag{9.29}$$

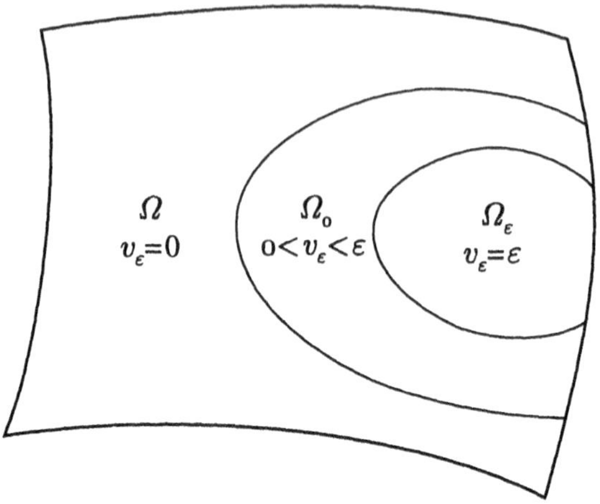

Figure 9.6.

As $u_2 - u_1 \in H^1(\Omega)$, the positive part $(u_2 - u_1)^+$ also lies in $H^1(\Omega)$. This is due to the fact that $v \mapsto v^+$ represents a continuous mapping from $H^1(\Omega)$ to $H^1(\Omega)$ – see, e.g., [Glowinski, p. 29]. The mapping $v \mapsto |v| = v^+ + v^-$ is continuous as well. Therefore, the equality $\min(a,b) = \frac{1}{2}(a + b - |a-b|)$ implies that

$$v_\varepsilon = \min\left(\varepsilon, (u_2 - u_1)^+\right) \in H^1(\Omega). \tag{9.30}$$

Thus, by (9.12), we may write

$$(\mathcal{A}(u_i)\,\mathrm{grad}\, u_i, \mathrm{grad}\, v_\varepsilon)_{0,\Omega} + \langle \alpha u_i, v_\varepsilon \rangle_{0,\partial\Omega} = F(v_\varepsilon), \ \ i = 1,2. \tag{9.31}$$

Since $v_\varepsilon \geq 0$ and $\alpha \geq 0$, we have

$$\alpha(u_1 - u_2)v_\varepsilon \leq 0 \quad \text{on } \partial\Omega. \tag{9.32}$$

From (9.7), (9.31) and (9.32) we obtain

$$\begin{aligned}
&C_0 \|\,\mathrm{grad}\, v_\varepsilon\|_{0,\Omega}^2 \\
&\leq (\mathcal{A}(u_1)\,\mathrm{grad}\, v_\varepsilon, \mathrm{grad}\, v_\varepsilon)_{0,\Omega} \\
&= (\mathcal{A}(u_1)\,\mathrm{grad}(u_2 - u_1), \mathrm{grad}\, v_\varepsilon)_{0,\Omega_0\setminus\Omega_\varepsilon} \\
&= (\mathcal{A}(u_1)\,\mathrm{grad}\, u_2 - \mathcal{A}(u_1)\,\mathrm{grad}\, u_1, \mathrm{grad}\, v_\varepsilon)_{0,\Omega} \\
&= (\mathcal{A}(u_1)\,\mathrm{grad}\, u_2 - \mathcal{A}(u_2)\,\mathrm{grad}\, u_2, \mathrm{grad}\, v_\varepsilon)_{0,\Omega} + \langle \alpha(u_1 - u_2), v_\varepsilon \rangle_{0,\partial\Omega} \\
&\leq ((\mathcal{A}(u_1) - \mathcal{A}(u_2))\,\mathrm{grad}\, u_2, \mathrm{grad}\, v_\varepsilon)_{0,\Omega}\,.
\end{aligned} \tag{9.33}$$

The last scalar product can be further estimated by (9.28), (9.29), the Cauchy–Schwarz inequality and (9.6) in the following way

$$\begin{aligned}
&((\mathcal{A}(u_1) - \mathcal{A}(u_2))\,\mathrm{grad}\, u_2, \mathrm{grad}\, v_\varepsilon)_{0,\Omega_0\setminus\Omega_\varepsilon} \\
&\leq \|(\mathcal{A}(u_1) - \mathcal{A}(u_2))\,\mathrm{grad}\, u_2\|_{0,\Omega_0\setminus\Omega_\varepsilon} \|\mathrm{grad}\, v_\varepsilon\|_{0,\Omega_0\setminus\Omega_\varepsilon} \\
&\leq \varepsilon C_L d^2 \,\|\mathrm{grad}\, u_2\|_{0,\Omega_0\setminus\Omega_\varepsilon} \|\mathrm{grad}\, v_\varepsilon\|_{0,\Omega_0\setminus\Omega_\varepsilon}\,.
\end{aligned} \tag{9.34}$$

Combining (9.33) and (9.34), we get

$$\|\operatorname{grad} v_\varepsilon\|_{0,\Omega} \le \varepsilon C \|\operatorname{grad} u_2\|_{0,\Omega_0\setminus\Omega_\varepsilon}. \tag{9.35}$$

According to (9.29) and (9.26),

$$0 \le v_\varepsilon \le (u_2 - u_1)^+ = u_2 - u_1 \le \frac{1}{\alpha_0}(\alpha u_2 - \alpha u_1) \quad \text{on } \Gamma_3 \cap \overline{\Omega}_0,$$

and thus (since $v_\varepsilon = 0$ on $\Gamma_3 \setminus \overline{\Omega}_0$)

$$v_\varepsilon^2 \le \frac{1}{\alpha_0}(\alpha u_2 - \alpha u_1)v_\varepsilon \quad \text{on } \Gamma_3. \tag{9.36}$$

Moreover, by (9.33),

$$-\langle \alpha u_1 - \alpha u_2, v_\varepsilon\rangle_{0,\partial\Omega} + C_0\|\operatorname{grad} v_\varepsilon\|_{0,\Omega}^2 \le ((\mathcal{A}(u_1) - \mathcal{A}(u_2))\operatorname{grad} u_2, \operatorname{grad} v_\varepsilon)_{0,\Omega}.$$

Consequently, Friedrichs' inequality (3.42), (9.36), (9.34) and (9.35) imply that

$$\begin{aligned}
\|v_\varepsilon\|_{0,\Omega}^2 &\le C_1\left(\|v_\varepsilon\|_{0,\Gamma_3}^2 + \|\operatorname{grad} v_\varepsilon\|_{0,\Omega}^2\right)\\
&\le C_2\left(\langle \alpha u_2 - \alpha u_1, v_\varepsilon\rangle_{0,\Gamma_3} + \|\operatorname{grad} v_\varepsilon\|_{0,\Omega}^2\right)\\
&\le C_3\left(\langle \alpha u_2 - \alpha u_1, v_\varepsilon\rangle_{0,\partial\Omega} + C_0\|\operatorname{grad} v_\varepsilon\|_{0,\Omega}^2\right)\\
&\le C_3\left((\mathcal{A}(u_1) - \mathcal{A}(u_2))\operatorname{grad} u_2, \operatorname{grad} v_\varepsilon\right)_{0,\Omega}\\
&\le \varepsilon C_4\|\operatorname{grad} u_2\|_{0,\Omega_0\setminus\Omega_\varepsilon}\|\operatorname{grad} v_\varepsilon\|_{0,\Omega_0\setminus\Omega_\varepsilon}\\
&\le \varepsilon^2 C_5\|\operatorname{grad} u_2\|_{0,\Omega_0\setminus\Omega_\varepsilon}^2.
\end{aligned}$$

From here, (9.29), the facts that u_2 is fixed and $\Omega_\varepsilon \subset \Omega_0$, we arrive at

$$\operatorname{meas}\Omega_\varepsilon = \varepsilon^{-2}\int_{\Omega_\varepsilon} \varepsilon^2\,dx \le \varepsilon^{-2}\|v_\varepsilon\|_{0,\Omega}^2 \le C_5\|\operatorname{grad} u_2\|_{0,\Omega_0\setminus\Omega_\varepsilon}^2 \to 0 \quad \text{as } \varepsilon \to 0,$$

where $C_5 > 0$ is independent of ε. This, however, contradicts (9.27) and (9.28), since $\operatorname{meas}\Omega_\varepsilon \to \operatorname{meas}\Omega_0$. Consequently, $\operatorname{meas}\Omega_0 = 0$ and $u_1 \ge u_2$ a.e. in Ω. Interchanging the role of u_1 and u_2, the uniqueness follows. □

Recall that sufficient conditions for the uniqueness of the Galerkin approximation u_h (defined by (9.16)) have been established in Theorem 9.5.

9.4. Convergence of finite element approximations

From (9.22) and the compactness of the imbedding operator $H^1(\Omega) \to L^2(\Omega)$ we can easily prove the convergence of the Galerkin approximations in the $\|\cdot\|_0$-norm. To prove even the (strong) convergence in the $\|\cdot\|_1$-norm, we shall, in addition, require that

$$V_h \subset W_4^1(\Omega), \quad \|v_h\|_{1,4} \le C(v) \quad \forall h, \tag{9.37}$$

where v_h satisfies (9.21) and $C(v)$ is a constant independent of h.

Remark 9.9. Using the standard interpolation theory one can verify (9.21) and (9.37) for many families of finite element spaces (see, e.g., [Ciarlet, p. 123] or Remark 4.35). The functions v_h can be defined, e.g., as the V_h-interpolant of v. Then we find that

$$\|v_h\|_{1,4} \leq \|v - v_h\|_{1,4} + \|v\|_{1,4} \leq C(v).$$

Theorem 9.10. *Let the assumptions of Theorem 9.6 be fulfilled and let (9.37) hold. Then the convergence (9.22) is strong, i.e.,*

$$\|u - u_{\hat{h}}\|_1 \to 0 \quad \text{as } \hat{h} \to 0. \tag{9.38}$$

Moreover, if there exists precisely one solution of the problem (9.12) then (9.38) holds for the whole sequence $\{u_h\}$.

Proof. Throughout the proof we use some ideas of [Douglas, Dupont] and [Nitsche], where the Dirichlet problem is investigated. Consider the sequence $\{u_{\hat{h}}\}$ from (9.22). For simplicity we again write only h instead of $\hat{h}$.

Since $u \in H^1(\Omega)$, there exists a sequence $\{w^m\}_{m=1}^{\infty} \subset C^\infty(\overline{\Omega})$ such that

$$\|u - w^m\|_1 \to 0 \quad \text{as } m \to \infty.$$

By the assumptions (9.21) and (9.37), there exist sequences $\{w_h^m\}_{h\to 0}, m = 1, 2, \ldots,$ such that $w_h^m \in V_h \cap W_4^1(\Omega)$,

$$\|w^m - w_h^m\|_1 \to 0 \quad \text{as } h \to 0, \tag{9.39}$$

$$\|w_h^m\|_{1,4} \leq C(w^m) \quad \forall h \, \forall m. \tag{9.40}$$

For brevity, let us set

$$v_h^m = u_h - w_h^m \in V_h.$$

Using (9.13) and the definitions (9.12), (9.16), we may write

$$\begin{aligned} C_0\|v_h^m\|_1^2 &\leq a(u_h; v_h^m, v_h^m) = a(u_h; u_h, v_h^m) - a(u_h; w_h^m, v_h^m) \\ &= a(u; u, v_h^m) - a(u_h; w_h^m, v_h^m) \\ &= a(u; u - w_h^m, v_h^m) + [a(u; w_h^m, v_h^m) - a(u_h; w_h^m, v_h^m)] \equiv E_1 + E_2. \end{aligned}$$

Making use of (9.14), the Cauchy–Schwarz and Hölder inequality (cf. Section 3.2), (9.40), the Lipschitz continuity and boundedness of $\mathcal{A}$, we may write

$$\begin{aligned} |E_1| &\leq C\|u - w_h^m\|_1\|v_h^m\|_1, \\ |E_2| &\leq |((\mathcal{A}(u) - \mathcal{A}(u_h)) \operatorname{grad} w_h^m, \operatorname{grad} v_h^m)_0| \\ &\leq \|(\mathcal{A}(u) - \mathcal{A}(u_h)) \operatorname{grad} w_h^m\|_0 \|\operatorname{grad} v_h^m\|_0 \\ &\leq d^2\|\mathcal{A}(u) - \mathcal{A}(u_h)\|_{0,4}\|\operatorname{grad} w_h^m\|_{0,4}\|v_h^m\|_1 \\ &\leq C\|u - u_h\|_{0,4}\|w_h^m\|_{1,4}\|v_h^m\|_1. \end{aligned}$$

Combining the above estimates, we arrive at

$$C_0'\|v_h^m\|_1 \le \|u - w_h^m\|_1 + \|u - u_h\|_{0,4}\|w_h^m\|_{1,4},$$

where $C_0' > 0$. From here and (9.40) we obtain

$$C_0'\|v_h^m\|_1 \le \|u - w^m\|_1 + \|w^m - w_h^m\|_1 + C(w^m)\|u - u_h\|_{0,4}.$$

Using (9.39), (9.22) and the compactness of the imbedding operator $H^1(\Omega) \to L^4(\Omega)$ for $d \in \{2,3\}$ (i.e., $\|v\|_{0,4} \le C\|v\|_{1,2}$ – see [Nečas, 1967]) we easily find that

$$\limsup_{h\to 0} \|v_h^m\|_1 \le C\|u - w^m\|_1 \quad \forall m \in \{1,2,\ldots\}, \tag{9.41}$$

where C is independent of m. Since $w_h^m - u_h = -v_h^m$, we may write

$$\|u - u_h\|_1 \le \|u - w^m\|_1 + \|w^m - w_h^m\|_1 + \|v_h^m\|_1.$$

Passing to the lim sup with $h \to 0$, we obtain by (9.41) and (9.39) that

$$\limsup_{h\to 0} \|u - u_h\|_1 \le (1 + C)\|u - w^m\|_1 \quad \forall m.$$

Passing to the limit with $m \to \infty$, we prove the strong convergence (9.38). □

If the weak solution is unique then, by Theorem 9.6, any cluster point of the sequence $\{u_h\}$ coincides with the weak solution. Consequently, the whole sequence tends to u in $H^1(\Omega)$.

Remark 9.11. We can prove the uniqueness of u_h only under very strong assumptions (see Theorem 9.5 or [Hlaváček, Křížek, Malý, p. 174]). On the other hand, from Theorem 9.8 the uniqueness of u follows under very weak assumptions. Unfortunately, we cannot prove the uniqueness of u_h in the same way as the uniqueness of u, since the test function (cf. (9.30))

$$v_{\varepsilon h} = \min(\varepsilon, (u_{1h} - u_{2h})^+)$$

does not belong to V_h, in general. Although we have no general theorem on the uniqueness of the Galerkin approximations (it is an open problem at present), we see that each u_h has to be near to u in the H^1-norm due to the convergence result (9.38).

If u is smooth enough, one can derive the rate of convergence as in [Douglas, Dupont, p. 691].

9.5. An application

The magnetic circuit (core, kernel) of large oil-immersed transformers (see Figure 9.1) consists of several right-angled polyhedral domains which are separated by cooling ducts. These can be seen in Figure 9.7 which shows the cross-section of the inner leg.

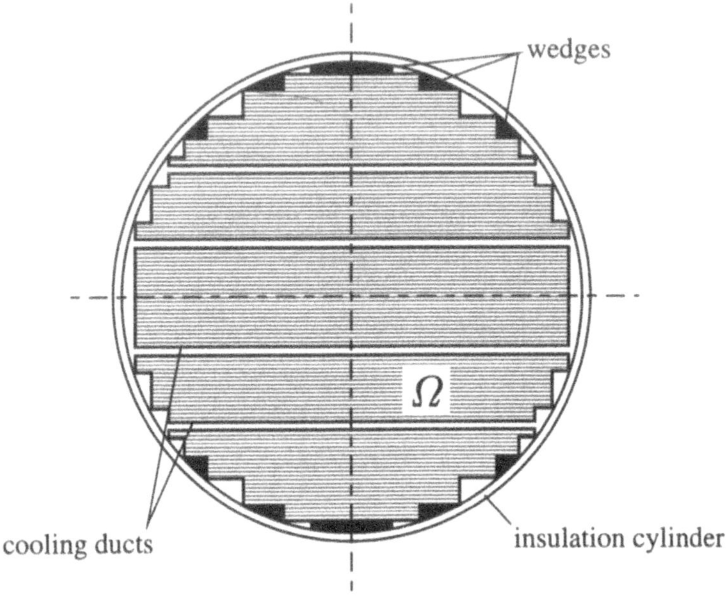

Figure 9.7.

It is convenient to calculate the temperature distribution in each of such domain individually. For simplicity, we shall only solve a two-dimensional case. Due to symmetry, we can deal with the right lower quarter alone which consists of three polygonal domains (there are no curved parts due to technological reasons).

For instance, consider the middle domain which we denote by Ω. On the left part of $\partial\Omega = \Gamma_2$, we prescribe the homogeneous Neumann symmetry boundary conditions (i.e., $\alpha = 0$ in (9.3)) and let $\alpha = 100$ [$\mathrm{Wm^{-1}K^{-1}}$] be the value of the heat transfer coefficient on the remaining part of the boundary. Let $g = \alpha u_o$ (cf. (9.3)), where $u_o = 55$ [°C] is the temperature of cooling oil. Suppose further that the density of heat sources $f = 10\,000$ [$\mathrm{Wm^{-2}}$] is constant in the whole Ω (heat sources are due to the alternating electromagnetic field). Finally, let the matrix $\mathcal{A}$ of heat conductivities (see (9.1) and (9.3)) be diagonal and independent of $x \in \Omega$. In Figures 9.2 and 9.3, we see the temperature dependence of the heat conductivity coefficients $a_{11}(u)$ and $a_{22}(u)$ across and along the lamination, respectively. Their domain of definition can be extended to $\mathbf{R}^1$ so that (9.5)–(9.7) remain valid.

The space V_h in (9.16) is generated by bilinear elements ($\dim V_h = 290$). We use Kačanov's method (9.17) and the centroid rule (see Section 9.6) to approximate the Galerkin solution u_h. For the constant initial temperature 90°C the corresponding temperature field after one iteration is displayed in Figure 9.8. We start from various initial temperatures and we always got the same "numerical" solution (see Figure 9.9) after 5 iterations of Kačanov's method. Increasing the dimension of V_h has not caused any essential quantitative changes. The black color in Figures 9.9 and 9.10 corresponds to parts with the lowest temperature. The maximum temperature was 128°C after one iteration and 133°C after five iterations always in the middle of the left part of $\partial\Omega$. Note that the knowledge of the temperature distribution is very important to avoid local overheating. If the temperature exceeds prescribed limits, the cooling oil starts to boil which may cause destruction of the whole transformer. Numerical results for a three-dimensional magnetic circuit are presented in [Preiningerová, Křížek, Kahoun].

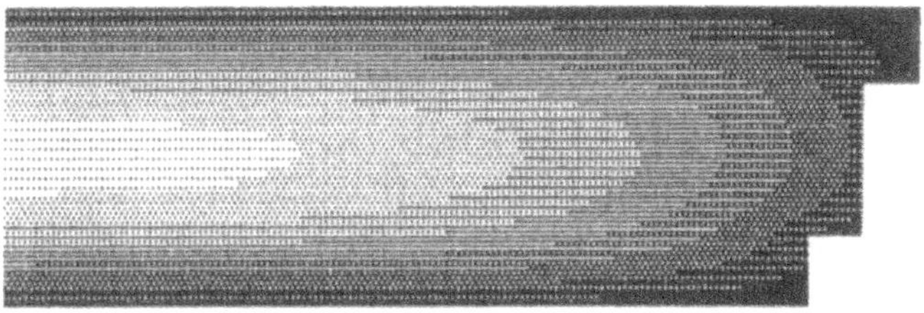

Figure 9.8.

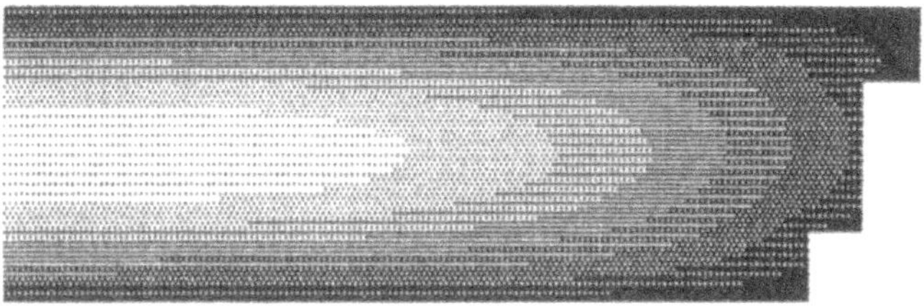

Figure 9.9.

Remark 9.12. Let Γ be that part of the boundary $\partial\Omega$ which is cooled by oil. Setting $v = 1$, we get by (9.8), Definition 9.1 and (9.9)

$$\int_\Gamma \alpha u\, ds = \int_{\partial\Omega} \alpha u\, ds = a(u; u, 1) = F(1) = (f, 1)_{0,\Omega} + \langle g, 1\rangle_{0,\partial\Omega}$$
$$= \int_\Omega f\, dx + \int_\Gamma \alpha u_o\, ds.$$

From here we observe that the average surface temperature rise ϑ_Γ on Γ does not depend upon the type of nonlinearity of the heat conduction coefficients,

$$\vartheta_\Gamma \equiv \frac{1}{\operatorname{meas}_1 \Gamma} \int_\Gamma (u - u_o)\, ds = \frac{1}{\alpha\big|_\Gamma \operatorname{meas}_1 \Gamma} \int_\Omega f\, dx,$$

provided $\alpha|_\Gamma$ is constant. Note that the total temperature flux on Γ is also independent of the heat conduction coefficients (which follows from Green's theorem). Finally, let us emphasize that it is not sufficient to calculate only the surface temperature, because the oil penetrates into the magnetic core.

9.6. Effect of numerical integration

In this section we confine our considerations only to the case of the Dirichlet boundary conditions for simplicity. We introduce a method for approximating nonhomogeneous Dirichlet boundary conditions. Moreover, we allow the use of a numerical integration. The convergence of approximate solutions in the H^1-norm without any regularity assumption upon the true solution will be again proved.

In practical calculations of the nonlinear problem (9.1)–(9.3) some numerical quadrature has to be used in general and the boundary conditions are approximated. We shall use a technique similar to that proposed in [Feistauer, Sobotíková], [Feistauer, Ženíšek, 1987, 1988], [Ženíšek, 1990b]. However, our assumptions upon the matrix $\mathcal{A}$ are not covered by the assumptions in these papers.

We shall be concerned with finite element approximations of a quasilinear elliptic problem (see [Feistauer, Křížek, Sobotíková])

$$-\operatorname{div}(\mathcal{A}(x,u)\operatorname{grad} u) = f(x) \quad \text{in } \Omega, \tag{9.42}$$

$$u = \overline{u} \quad \text{on } \partial\Omega, \tag{9.43}$$

where $\Omega \subset R^d$, $d \in \{1,2,...\}$, is a bounded domain with a Lipschitz continuous polyhedral boundary $\partial\Omega$ and $\mathcal{A} = (a_{ij})_{i,j=1}^d$ is again a uniformly positive definite matrix with respect to all variables, $f \in L^2(\Omega)$ and $\overline{u} \in H^1(\Omega)$.

To state a weak formulation of the problem (9.42)–(9.43) we set

$$V = H_0^1(\Omega).$$

Since quadrature formulae will be employed, we need stronger smoothness assumptions: Let $\mathcal{A} = (a_{ij})_{i,j=1}^d$, $a_{ij} = a_{ij}(x,\xi)$, $x \in \overline{\Omega}$, $\xi \in R^1$ and let a_{ij}, $\partial a_{ij}/\partial x_k$, $\partial a_{ij}/\partial \xi$ be continuous and bounded in $\overline{\Omega} \times R^1$ for all $i,j,k = 1,...,d$. Thus there exists a constant C such that

$$|a_{ij}(x,\xi)| \le C, \quad \left|\frac{\partial a_{ij}}{\partial x_k}(x,\xi)\right| \le C, \quad \left|\frac{\partial a_{ij}}{\partial \xi}(x,\xi)\right| \le C \quad \forall x \in \Omega \quad \forall \xi \in R^1. \tag{9.44}$$

The boundedness of the derivatives $\partial a_{ij}/\partial\xi$ obviously implies the Lipschitz-continuity of a_{ij} with respect to ξ, see (9.6).

Set

$$a(y;w,v) = (\mathcal{A}(\cdot\,,y)\operatorname{grad} w, \operatorname{grad} v)_0, \quad y,w,v \in H^1(\Omega), \tag{9.45}$$

$$F(v) = (f,v)_0, \quad v \in H^1(\Omega). \tag{9.46}$$

From the properties of the matrix $\mathcal{A}$ it follows that

$$|a(y;w,v)| \leq C\|w\|_1\|v\|_1 \quad \forall y,w,v \in H^1(\Omega). \tag{9.47}$$

Moreover, from the Lipschitz continuity (9.6) we have

$$\begin{aligned} &|a(y;w,v) - a(z;w,v)| \leq |((\mathcal{A}(\cdot\,,y) - \mathcal{A}(\cdot\,,z))\,\mathrm{grad}\,w, \mathrm{grad}\,v)_0| \\ &\leq \|\mathcal{A}(\cdot\,,y) - \mathcal{A}(\cdot\,,z)\|_0\|\,\mathrm{grad}\,w\|_0\|\,\mathrm{grad}\,v\|_{0,\infty} \\ &\leq d^2 C_L\|y-z\|_0\|w\|_1\|v\|_{1,\infty} \quad \forall y,z,w \in H^1(\Omega) \quad \forall v \in W^1_\infty(\Omega). \end{aligned} \tag{9.48}$$

Recall that a function $u \in H^1(\Omega)$ is said to be a *weak solution* of the problem (9.42)–(9.43) if

$$u - \overline{u} \in V, \tag{9.49}$$

$$a(u;u,v) = F(v) \quad \forall v \in V. \tag{9.50}$$

We denote by $\mathcal{T}_h$ a decomposition (triangulation) of $\overline{\Omega}$ into simplices in the usual sense. Set

$$X_h = \{v \in C(\overline{\Omega}) \mid v|_K \in P_1(K) \quad \forall K \in \mathcal{T}_h\}, \qquad V_h = V \cap X_h,$$

where $P_1(K)$ is the space of linear polynomials over K and h is the maximum diameter of all simplices from $\mathcal{T}_h$. For any $v_h \in X_h$ and any $K \in \mathcal{T}_h$ we further set

$$v_K = v_h|_K.$$

Recall that a family $\mathcal{F} = \{\mathcal{T}_h\}$, $h \in (0,h_0)$, of decompositions of $\overline{\Omega}$ is said to be ***strongly regular*** if there exists a constant $\varkappa > 0$ such that for any $\mathcal{T}_h \in \mathcal{F}$ and any simplex $K \in \mathcal{T}_h$ we have $\varkappa h \leq \rho_K$, where ρ_K is radius of a ball contained in K. For any strongly regular family of decompositions and $q \in [1,\infty]$ there exists a constant C such that the following inverse inequality holds, see [Ciarlet, Theorem 3.2.6]

$$|v_h|_{1,q} \leq \frac{C}{(h^d)^{\max(0,1/2-1/q)}}|v_h|_1 \quad \forall h \in (0,h_0) \quad \forall v_h \in X_h. \tag{9.51}$$

Throughout this section we assume that all decompositions $\mathcal{T}_h$ belong to a strongly regular family $\mathcal{F}$.

Concerning the function $\overline{u}$ representing the Dirichlet condition (9.43), we assume, moreover, that

$$\overline{u} \in W^1_p(\Omega),$$

where $p > d$ and $p \geq 2$. (Thus, $p > 2$ for $d \geq 2$ and $p \geq 2$ for $d = 1$.) Then in view of the Sobolev Imbedding Theorem ($W^k_p(\Omega) \subset C(\overline{\Omega})$ for $pk > d$) we have

$\overline{u} \in C(\overline{\Omega})$ and it makes sense to define the Lagrange interpolant $\pi_h \overline{u} \in X_h$ (i.e., $\pi_h \overline{u}(P) = \overline{u}(P)$ for every vertex P of simplices from $\mathcal{T}_h$). Let us set

$$\overline{u}_h = \pi_h \overline{u}.$$

Analogously to [Feistauer, Sobotíková, Proof of Lemma 3.1.3] we obtain that

$$\lim_{h \to 0} \|\overline{u} - \overline{u}_h\|_1 = 0, \tag{9.52}$$

and

$$\|\overline{u}_h\|_{1,q} \leq C(q), \quad h \in (0, h_0), \quad q \in [1, p]. \tag{9.53}$$

To approximate the Galerkin solution of the problem (9.49)–(9.50), we shall employ the following integration formula over an element $K \in \mathcal{T}_h$

$$\int_K g(x)\, dx \approx \operatorname{meas} K \sum_{k=1}^{M_K} \omega_{K,k} g(x_{K,k}), \tag{9.54}$$

where the weights $\omega_{K,k} \in R^1$ are such that

$$\omega_{K,k} > 0 \quad \text{and} \quad \sum_{k=1}^{M_K} \omega_{K,k} = 1, \tag{9.55}$$

and the nodes $x_{K,k} \in K$ for $k = 1, ..., M_K$. For instance if $M_K = 1$, $\omega_{K,1} = 1$ and $x_{K,1}$ is the centre of gravity of K, we obtain the well-known *centroid rule* which is exact for all linear polynomials and, moreover, for $d \leq 2$ we have

$$\left| \int_K g(x)\, dx - g(x_{K,1}) \operatorname{meas} K \right| \leq C h_K^2 |g|_{2,1,K} \quad \forall g \in W_1^2(K).$$

The proof of this estimate can be done along the lines of [Křížek, Neittaanmäki, 1990, p. 53], since the functions from $W_1^2(K)$ are continuous (see [Kufner, John, Fučík, p. 300]).

Definition 9.13. A function $u_h \in X_h$ is said to be an approximate solution of the problem (9.42)–(9.43) if

$$u_h - \overline{u}_h \in V_h, \tag{9.56}$$

$$a_h(u_h; u_h, v_h) = F_h(v_h) \quad \forall v_h \in V_h, \tag{9.57}$$

where

$$a_h(y_h; w_h, v_h) = \sum_{K \in \mathcal{T}_h} \operatorname{meas} K \sum_{k=1}^{M_K} \omega_{K,k} (\mathcal{A}(x_{K,k}, y_K(x_{K,k})) \operatorname{grad} w_K) \cdot \operatorname{grad} v_K \tag{9.58}$$

for $y_h, w_h, v_h \in X_h$ and

$$F_h(v_h) = \sum_{K\in\mathcal{T}_h} \operatorname{meas} K \sum_{k=1}^{M_K} \omega_{K,k} f(x_{K,k}) v_K(x_{K,k}), \quad v_h \in X_h. \tag{9.59}$$

Lemma 9.14. *There exist positive constants C_1 and C_2 independent of h such that*

$$a_h(y_h; v_h, v_h) \ge C_1 \|v_h\|_1^2 \quad \forall y_h \in X_h \quad \forall v_h \in V_h \tag{9.60}$$

and

$$|a_h(y_h; w_h, v_h)| \le C_2 \|w_h\|_1 \|v_h\|_1 \quad \forall y_h, w_h, v_h \in X_h. \tag{9.61}$$

Proof. By (9.58), (9.7), (9.55) and Friedrichs' inequality (3.43), we get

$$\begin{aligned} a_h(y_h, v_h, v_h) &= \sum_{K\in\mathcal{T}_h} \operatorname{meas} K \sum_{k=1}^{M_K} \omega_{K,k} (\mathcal{A}(x_{K,k}, y_K(x_{K,k})) \operatorname{grad} v_K) \cdot \operatorname{grad} v_K \\ &\ge \sum_{K\in\mathcal{T}_h} \operatorname{meas} K \sum_{k=1}^{M_K} \omega_{K,k} C_0 \|\operatorname{grad} v_K\|^2 \\ &= C_0 \sum_{K\in\mathcal{T}_h} \operatorname{meas} K \|\operatorname{grad} v_K\|^2 \\ &= C_0 |v_h|_1^2 \ge C_1 \|v_h\|_1^2 \end{aligned}$$

for any $v_h \in V_h$.

The relation (9.61) follows from (9.44) in a similar way. □

Lemma 9.15. *Let $f \in W^1_\infty(\Omega)$ and let (9.55) hold. Then there exists a constant C such that*

$$|F(v_h) - F_h(v_h)| \le Ch \|f\|_{1,\infty} \|v_h\|_1 \quad \forall v_h \in X_h. \tag{9.62}$$

Proof. By (9.55), the formula (9.54) is exact for any constant function on K. Then for any $q \in (d, \infty]$ and any $K \in \mathcal{T}_h$ we have by [Ciarlet, Theorem 4.1.5]

$$\begin{aligned} &\Big| \int_K f v_K \, dx - \operatorname{meas} K \sum_{k=1}^{M_K} \omega_{K,k} f(x_{K,k}) v_K(x_{K,k}) \Big| \le \\ &\le C h_K (\operatorname{meas} K)^{1/2 - 1/q} \|f\|_{1,q,K} \|v_K\|_{1,2,K}, \end{aligned} \tag{9.63}$$

where $h_K = \operatorname{diam} K$. Taking $q = \infty$, summing (9.63) over all simplices from $\mathcal{T}_h$ and using the Cauchy–Schwarz inequality

$$\sum_{K\in\mathcal{T}_h} h_K (\operatorname{meas} K)^{1/2} \|v_K\|_{1,2,K} \le \Big(\sum_{K\in\mathcal{T}_h} h_K^2 \operatorname{meas} K \Big)^{1/2} \Big(\sum_{K\in\mathcal{T}_h} \|v_K\|_{1,2,K}^2 \Big)^{1/2},$$

we get (9.62). □

Theorem 9.16. *Let the assumptions of Lemma 9.15 and the assumptions on a_{ij} be satisfied. Then, for any decomposition $\mathcal{T}_h$ of $\overline{\Omega}$ into simplices, there exists a solution u_h of the discrete problem (9.56)–(9.57).*

Proof. Let V_h with h fixed be given. From (9.62), (9.46) and the Cauchy–Schwarz inequality we see that F_h is a linear continuous functional on V_h, since

$$|F_h(v)| \leq |F_h(v) - F(v)| + |F(v)| \leq C\|v\|_1 \quad \forall v \in V_h. \tag{9.64}$$

In view of (9.56), we can write

$$u_h = \overline{u}_h + z_h, \quad z_h \in V_h. \tag{9.65}$$

Now let us consider a mapping S defined as follows: if $z \in V_h$ then $Sz \in V_h$ is the solution of the problem

$$a_h(\overline{u}_h + z; \overline{u}_h + Sz, v) = F_h(v) \quad \forall v \in V_h. \tag{9.66}$$

This identity can be rewritten in the form

$$a_h(\overline{u}_h + z; Sz, v) = G_h(v) := F_h(v) - a_h(\overline{u}_h + z; \overline{u}_h, v) \quad \forall v \in V_h. \tag{9.67}$$

For $z \in V_h$ fixed, G_h is obviously a continuous linear functional on V_h and, by (9.61), (9.64) and (9.53),

$$|G_h(v)| \leq C'\|v\|_1 + C''\|\overline{u}_h\|_1\|v\|_1 \leq C\|v\|_1, \quad v \in V_h. \tag{9.68}$$

The Lax–Milgram Lemma 3.3, (9.60) and (9.61) now imply that the mapping S is uniquely determined.

Setting $v = Sz$ in (9.67), we may write by (9.60) and (9.68)

$$C_1\|Sz\|_1^2 \leq a_h(\overline{u}_h + z; Sz, Sz) = G_h(Sz) \leq C\|Sz\|_1. \tag{9.69}$$

Consequently, the mapping S is bounded,

$$\|Sz\|_1 \leq C/C_1 \quad \forall z \in V_h. \tag{9.70}$$

Similarly to (9.20) we show now that the mapping S is Lipschitz continuous, i.e., there exists a constant $C > 0$ such that

$$\|Sy - Sz\|_1 \leq C\|y - z\|_1 \quad \forall y, z \in V_h. \tag{9.71}$$

So let $y, z \in V_h$ be arbitrary and let $v = Sy - Sz$. Hence, $v \in V_h$. Using (9.60), (9.66), (9.58), (9.55), (9.6), (9.53), (9.70), the Cauchy–Schwarz inequality and the equivalence of all norms in V_h, we obtain

$$\begin{aligned}
&C_1\|v\|_1^2 \le a_h(\overline{u}_h + y; v, v)\\
&= a_h(\overline{u}_h + y; \overline{u}_h + Sy, v) - a_h(\overline{u}_h + y; \overline{u}_h + Sz, v)\\
&= F_h(v) - a_h(\overline{u}_h + y; \overline{u}_h + Sz, v)\\
&= a_h(\overline{u}_h + z; \overline{u}_h + Sz, v) - a_h(\overline{u}_h + y; \overline{u}_h + Sz, v)\\
&= \sum_{K\in\mathcal{T}_h} \operatorname{meas} K \sum_{k=1}^{M_K} \omega_{K,k}([\mathcal{A}(x_{K,k}, (\overline{u}_h + z)(x_{K,k})) - \mathcal{A}(x_{K,k}, (\overline{u}_h + y)(x_{K,k}))]\\
&\operatorname{grad}(\overline{u}_h + Sz)|_K) \cdot \operatorname{grad} v|_K\\
&\le \sum_{K\in\mathcal{T}_h} \operatorname{meas} K\, d^2 C_L \|z - y\|_{0,\infty,K} \|\operatorname{grad}(\overline{u}_h + Sz)|_K\|\, \|\operatorname{grad} v|_K\|\\
&\le C\|z - y\|_{0,\infty}(|\overline{u}_h|_1 + |Sz|_1)|v|_1\\
&\le C'\|z - y\|_1\|v\|_1,
\end{aligned}$$

and thus (9.71) is valid.

The existence of a fixed point $z = Sz$ now follows from (9.70), the continuity of S and the Brouwer fixed-point Theorem 7.19. Setting $u_h = \overline{u}_h + z$, we obtain the approximate solution. □

Another possibility to prove the existence of the approximate solution is to adapt the approach used in [Feistauer, Ženíšek, 1988, Theorem 4.2.1] or [Feistauer, Sobotíková, Theorem 2.3.4].

Lemma 9.17. *Let (9.55) hold, let the assumptions on a_{ij} be satisfied and let $p > d$, $p \ge 2$. Then there exists a constant C independent of h such that*

$$|a(y_h; w_h, v_h) - a_h(y_h; w_h, v_h)| \le Ch|w_h|_1|v_h|_1 + Ch^{1-d/p}|y_h|_1|w_h|_{1,\alpha}|v_h|_{1,\beta} \quad (9.72)$$

for all $y_h, w_h, v_h \in X_h$, where either $\alpha = p$, $\beta = 2$ or $\alpha = 2$, $\beta = p$.

Proof. Let $y_h, w_h, v_h \in X_h$ be arbitrary. For any $K \in \mathcal{T}_h$ define the matrix $\Phi(K) = (\Phi_{ij}(K))_{i,j=1}^d$ by

$$\Phi(K) = \int_K \mathcal{A}(x, y_h(x))\, dx - \operatorname{meas} K \sum_{k=1}^{M_K} \omega_{K,k}\mathcal{A}(x_{K,k}, y_h(x_{K,k})).$$

Then we can write

$$|a(y_h; w_h, v_h) - a_h(y_h; w_h, v_h)| = \Big|\sum_{K\in\mathcal{T}_h} (\Phi(K) \operatorname{grad} w_K) \cdot \operatorname{grad} v_K\Big|. \quad (9.73)$$

Applying (9.63) for $v_K \equiv 1$, we have

$$|\Phi_{ij}(K)| \le Ch(\operatorname{meas} K)^{1-1/q}\|a_{ij}(\cdot, y_h)\|_{1,q,K}, \quad i,j = 1, ..., d, \quad (9.74)$$

whenever $q > d$. Using the properties (9.44), we come to

$$\begin{aligned}\|a_{ij}(\cdot\,, y_h)\|_{1,q,K} &\leq \Big(\int_K \Big(C^q + \Big(\Big|\frac{\partial a_{ij}}{\partial \xi}(x, y_K(x))\Big|\,\|\operatorname{grad} y_K\|\Big)^q\Big)\, dx\Big)^{1/q}\\ &\leq C'(\operatorname{meas} K)^{1/q}(1 + \|\operatorname{grad} y_K\|^q)^{1/q}\\ &\leq C'(\operatorname{meas} K)^{1/q}(1 + \|\operatorname{grad} y_K\|),\end{aligned}$$

where the last inequality is a special case of the Jensen inequality. From here, (9.73), (9.74), the Hölder inequality and taking r such that $1/p + 1/r = 1/2$, (for $p = 2$ we set $r = \infty$), we obtain

$$\begin{aligned}&|a(y_h; w_h, v_h) - a_h(y_h; w_h, v_h)|\\ &\leq Ch \sum_{K\in\mathcal{T}_h} \int_K (1 + \|\operatorname{grad} y_K\|)\|\operatorname{grad} w_K\|\|\operatorname{grad} v_K\|\, dx\\ &= Ch \int_\Omega (1 + \|\operatorname{grad} y_h\|)\|\operatorname{grad} w_h\|\|\operatorname{grad} v_h\|\, dx\\ &\leq Ch\Big(|w_h|_1|v_h|_1 + \Big(\int_\Omega \|\operatorname{grad} y_h\|^2\, dx\Big)^{1/2}\Big(\int_\Omega \|\operatorname{grad} w_h\|^2\|\operatorname{grad} v_h\|^2\, dx\Big)^{1/2}\Big)\\ &\leq Ch(|w_h|_1|v_h|_1 + |y_h|_1|w_h|_{1,p}|v_h|_{1,r})\\ &\leq Ch|w_h|_1|v_h|_1 + C'h^{1-d/p}|y_h|_1|w_h|_{1,p}|v_h|_1,\end{aligned}$$

where the last inequality follows from the inverse inequality (9.51). The case $\alpha = 2$ and $\beta = p$ is similar. □

Theorem 9.18. *Let the assumptions of Lemmas 9.15 and 9.17 be satisfied. Then*

$$u_h \rightharpoonup u \text{ (weakly) in } H^1(\Omega) \text{ as } h \to 0,$$

where u is the weak solution of the problem (9.42)–(9.43).

Proof. Let $q \in [1, \infty]$ be given. Then we can easily find that

$$\forall v \in V \cap C^\infty(\overline{\Omega}) \quad \exists \{v_h\}_{h\to 0} : \ v_h \in V_h,\ \|v - v_h\|_1 \to 0 \text{ as } h \to 0 \tag{9.75}$$

and

$$\|v_h\|_{1,q} \leq C(v) \quad \forall h, \tag{9.76}$$

where $C(v) > 0$ is a constant independent of h. The functions v_h can be defined, e.g., as the standard interpolate of v, i.e., $v_h = \pi_h v$. Then $\|v - v_h\|_{1,q} \leq Ch|v|_{2,q}$ for $h \to 0$ (see [Ciarlet, p. 123, p. 135]), and thus

$$\|v_h\|_{1,q} \leq \|v - v_h\|_{1,q} + \|v\|_{1,q} \leq C(v)$$

for sufficiently small h.

From (9.60), (9.57), (9.65), (9.64) and (9.61) we have

$$C_1\|z_h\|_1^2 \le a_h(u_h; z_h, z_h) = F_h(z_h) - a_h(u_h; \overline{u}_h, z_h) \le C\|z_h\|_1(1 + \|\overline{u}_h\|_1).$$

Hence,

$$\|z_h\|_1 \le C(1 + \|\overline{u}_h\|_1).$$

This and (9.53) imply that

$$\|u_h\|_1 = \|\overline{u}_h + z_h\|_1 \le C(1 + 2\|\overline{u}_h\|_1) \le C', \quad h \in (0, h_0). \tag{9.77}$$

Consequently, there exist an element $u \in H^1(\Omega)$ and a subsequence $\{u_{\hat{h}}\} \subset \{u_h\}$ such that $u_{\hat{h}} \rightharpoonup u$ as $\hat{h} \to 0$. For simplicity we write h instead of $\hat{h}$ in what follows. From the Rellich Theorem 3.9 on the compact imbedding of $H^1(\Omega)$ into $L^2(\Omega)$ we immediately see that

$$\|u - u_h\|_0 \to 0 \quad \text{as } h \to 0. \tag{9.78}$$

Let us show that u is a weak solution of (9.42)–(9.43). In view of (9.52) and (9.65), $z_h \rightharpoonup z := u - \overline{u}$ (weakly) in $H^1(\Omega)$. Since $z_h \in V$ and V is closed in $H^1(\Omega)$ and convex (which implies that V is weakly closed [Céa, 1978]), the weak limit z belongs to V, i.e., (9.49) holds. It remains to prove (9.50).

Consider an arbitrary $v \in V \cap C^\infty(\overline{\Omega})$ and $\{v_h\}$ satisfying (9.75)–(9.76). Then from the weak convergence of u_h to u in $H^1(\Omega)$, (9.57), (9.48), (9.47), (9.72), (9.62), (9.46), (9.75)–(9.78) and the inequality $p > d$ we obtain

$$\begin{aligned}
|a(u; u, v) - F(v)| &\le |a(u; u - u_h, v)| + |a(u; u_h, v) - a(u_h; u_h, v)| \\
&\quad + |a(u_h; u_h, v - v_h)| + |a(u_h; u_h, v_h) - a_h(u_h; u_h, v_h)| \\
&\quad + |F_h(v_h) - F(v_h)| + |F(v_h - v)| \\
&\le |a(u; u - u_h, v)| + d^2 C_L \|u - u_h\|_0 \|u_h\|_1 \|v\|_{1,\infty} \\
&\quad + C(\|u_h\|_1 \|v - v_h\|_1 + h\|u_h\|_1 \|v_h\|_1 \\
&\quad + h^{1-d/p}\|u_h\|_1^2 \|v_h\|_{1,p} + h\|v_h\|_1 + \|v - v_h\|_1) \to 0 \quad \text{as } h \to 0.
\end{aligned}$$

Hence,

$$a(u; u, v) = F(v) \quad \forall v \in V \cap C^\infty(\overline{\Omega}).$$

The density of $V \cap C^\infty(\overline{\Omega})$ in V, (9.46) and (9.47), imply that u is a weak solution of (9.42)–(9.43).

Similarly we can prove that any weak cluster (accumulation) point of the family $\{u_h\}$, $h \in (0, h_0)$, is a solution of (9.49)–(9.50). However, by Theorem 9.8 the weak solution u is unique, and thus the whole family $\{u_h\}$ weakly converges to u. □

Theorem 9.19. *Let the assumptions of Lemmas 9.15 and 9.17 be satisfied. Then*

$$u_h \to u \text{ (strongly) in } H^1(\Omega) \text{ as } h \to 0,$$

where u is the weak solution of the problem (9.42)–(9.43).

Proof. The solution u of the problem (9.49)–(9.50) can be written in the form $u = \overline{u} + z$ with $z \in V$. (We recall that $\overline{u} \in W_p^1(\Omega)$ with some $p > d$ and $p \geq 2$.) There exists a sequence $\{z^m\}_{m=1}^{\infty} \subset V \cap C^\infty(\overline{\Omega})$ such that

$$\|z - z^m\|_1 \to 0 \quad \text{as } m \to \infty.$$

Let m be arbitrary but fixed. By (9.75) and (9.76), there exists a family $\{z_h^m\}$, $h \in (0, h_0)$, such that $z_h^m \in V_h$,

$$\|z^m - z_h^m\|_1 \to 0 \quad \text{as } h \to 0 \tag{9.79}$$

and

$$\|z_h^m\|_{1,p} \leq C(z^m) \quad \forall h \in (0, h_0). \tag{9.80}$$

Let us set

$$v_h^m = z_h - z_h^m \in V_h,$$

where z_h is a function from the representation (9.65) of the approximate solution u_h.

Using (9.60), (9.57), (9.50), we can write

$$\begin{aligned} C_1\|v_h^m\|_1^2 &\leq a_h(u_h; v_h^m, v_h^m) \\ &= a_h(u_h; \overline{u}_h + z_h, v_h^m) - a_h(u_h; \overline{u}_h + z_h^m, v_h^m) \\ &= F_h(v_h^m) - a_h(u_h; \overline{u}_h + z_h^m, v_h^m) \\ &= F(v_h^m) - a_h(u_h; \overline{u}_h + z_h^m, v_h^m) + F_h(v_h^m) - F(v_h^m) \\ &= a(u; (\overline{u} + z) - (\overline{u}_h + z_h^m), v_h^m) \\ &\quad + [a(u; \overline{u}_h + z_h^m, v_h^m) - a(u_h; \overline{u}_h + z_h^m, v_h^m)] \\ &\quad + [a(u_h; \overline{u}_h + z_h^m, v_h^m) - a_h(u_h; \overline{u}_h + z_h^m, v_h^m)] \\ &\quad + [F_h(v_h^m) - F(v_h^m)] \\ &\equiv E_1 + E_2 + E_3 + E_4. \end{aligned} \tag{9.81}$$

By (9.47),

$$\begin{aligned} |E_1| &\leq C(\|\overline{u} - \overline{u}_h\|_1 + \|z - z_h^m\|_1)\|v_h^m\|_1 \\ &\leq C(\|\overline{u} - \overline{u}_h\|_1 + \|z - z^m\|_1 + \|z^m - z_h^m\|_1)\|v_h^m\|_1. \end{aligned} \tag{9.82}$$

Using the Cauchy–Schwarz inequality, the Hölder inequality, the Lipschitz-continuity and boundedness of $\mathcal{A}$, we obtain

$$\begin{aligned} |E_2| &\leq |((\mathcal{A}(\cdot, u) - \mathcal{A}(\cdot, u_h)) \operatorname{grad}(\overline{u}_h + z_h^m), \operatorname{grad} v_h^m)_0| \\ &\leq \|(\mathcal{A}(\cdot, u) - \mathcal{A}(\cdot, u_h)) \operatorname{grad}(\overline{u}_h + z_h^m)\|_0 \|\operatorname{grad} v_h^m\|_0 \\ &\leq d^2\|\mathcal{A}(\cdot, u) - \mathcal{A}(\cdot, u_h)\|_{0,q}\|\operatorname{grad}(\overline{u}_h + z_h^m)\|_{0,p}\|\operatorname{grad} v_h^m\|_0 \\ &\leq C\|u - u_h\|_{0,q}\|\overline{u}_h + z_h^m\|_{1,p}\|v_h^m\|_1, \end{aligned} \tag{9.83}$$

where $q \geq 2$ satisfies the relation $1/p + 1/q = 1/2$ (we set $q = \infty$ for $p = 2$). By virtue of Lemma 9.17,

$$|E_3| \leq C(h\|\overline{u}_h + z_h^m\|_1 + h^{1-d/p}\|u_h\|_1\|\overline{u}_h + z_h^m\|_{1,p})\|v_h^m\|_1. \tag{9.84}$$

Finally, by Lemma 9.15,

$$|E_4| \leq Ch\|f\|_{1,\infty}\|v_h^m\|_1. \tag{9.85}$$

In (9.82)–(9.85) the constant C is independent of h and m. Combining the above estimates (9.81)–(9.85), we arrive at

$$\begin{aligned} \|v_h^m\|_1 \leq C(&\|\overline{u} - \overline{u}_h\|_1 + \|z - z^m\|_1 + \|z^m - z_h^m\|_1 \\ &+ \|u - u_h\|_{0,q}(\|\overline{u}_h\|_{1,p} + \|z_h^m\|_{1,p}) \\ &+ h(\|\overline{u}_h\|_1 + \|z_h^m\|_1) + h^{1-d/p}\|u_h\|_1(\|\overline{u}_h\|_{1,p} + \|z_h^m\|_{1,p}) + h). \end{aligned} \tag{9.86}$$

Let $d \geq 2$. By the imbedding theorems (see, e.g., [Ciarlet, p. 114]) the identity operator $H^1(\Omega) \to L^q(\Omega)$ is compact if $1 \leq q < p^*$, where p^* satisfies $1/p^* = 1/2 - 1/d$. Since $1/q = 1/2 - 1/p$ and $p > d$, the bound

$$q < p^*$$

is satisfied. If $d = 1$ then the imbedding of $H^1(\Omega)$ into $L^\infty(\Omega)$ is also compact. Using these facts and Theorem 9.18, we find that

$$\|u - u_h\|_{0,q} \to 0 \quad \text{as } h \to 0.$$

Employing now (9.52), (9.53), (9.77), (9.79) and (9.80), we get from (9.86)

$$\limsup_{h\to 0} \|v_h^m\|_1 \leq C\|z - z^m\|_1 \quad \forall m \in \{1, 2, ...\}, \tag{9.87}$$

where C is independent of m. Further we have

$$\begin{aligned} \|u - u_h\|_1 &= \|(\overline{u} + z) - (\overline{u}_h + z_h)\|_1 \\ &\leq \|\overline{u} - \overline{u}_h\|_1 + \|z - z^m\|_1 + \|z^m - z_h^m\|_1 + \|z_h^m - z_h\|_1 \end{aligned}$$

and $z_h - z_h^m = v_h^m$. Passing to the lim sup with $h \to 0$, we obtain by (9.87), (9.52) and (9.79) that

$$\limsup_{h\to 0} \|u - u_h\|_1 \leq (1 + C)\|z - z^m\|_1 \quad \forall m.$$

Passing to the limit with $m \to \infty$, we get the strong convergence $u_h \to u$ in $H^1(\Omega)$. □

Remark 9.20. For efficient numerical quadrature formulae, see [Ciarlet], [Dunavant, 1985a,b], [Engels], [Křížek, Neittaanmäki, 1990], [Moan] and [Zienkiewicz].

9.7. Variational crimes

In practical implementations of the finite element method to boundary value problems, so-called *variational crimes* are usually committed: the given domain with a piecewise curved boundary is approximated by a polygonal (in $2d$) or polyhedral (in $3d$) domain, integrals are evaluated approximately with the use of numerical quadratures and boundary conditions are approximated. The theory of finite element variational crimes for linear problems was extensively studied, e.g., in [Ciarlet], [Ciarlet, Raviart], [Strang] and [Strang, Fix]. A detailed analysis of various nonlinear problems is given, e.g., in [Feistauer, Sobotíková], [Feistauer, Ženíšek, 1987, 1988], [Křížek, Lin], [Vanmaele, Van Keer] and [Ženíšek, 1990a,b].

In this section we deal with the convergence analysis of the finite element method in the case when an approximation of a curved boundary as well as numerical integration are used (see [Feistauer, Křížek, Sobotíková]). The main tool is Zlámal's concept [Zlámal, 1973] of an ideal interpolation which forces us to confine our considerations to two-dimensional case only (i.e., $d = 2$). The proof of the convergence of approximate solutions to the exact one is here more complex than that of Section 9.6, but it is similar. Therefore, the presentation will be a little bit more brief than before.

First we introduce assumptions on data. Let $\Omega \subset \mathbf{R}^2$ be a bounded domain with a Lipschitz continuous boundary which is piecewise of the class C^3. Let us consider families $\{\Omega_h\}$, $h \in (0, h_0)$, of polygonal approximations of Ω and $\{\mathcal{T}_h\}$, $h \in (0, h_0)$, of triangulations of $\overline{\Omega}_h$ (with $h_0 > 0$ sufficiently small). We suppose that triangulations $\mathcal{T}_h$ have usual properties [Ciarlet] that the corners of Ω_h are lying on $\partial\Omega$ and the family $\{\mathcal{T}_h\}$, $h \in (0, h_0)$, is strongly regular. Denoting by σ_h the set of all vertices of all triangles of $\mathcal{T}_h$, we assume that $\sigma_h \subset \overline{\Omega}$, $\sigma_h \cap \partial\Omega_h \subset \partial\Omega$ and the points of $\partial\Omega$, where the condition of its C^3-smoothness is not satisfied, are elements of σ_h.

Further, let $\tilde{\Omega} \subset \mathbf{R}^2$ be a bounded domain with a Lipschitz continuous boundary such that $\Omega \cup \Omega_h \subset \tilde{\Omega}$ for all $h \in (0, h_0)$ and let the coefficients $a_{ij} = a_{ij}(x, \xi)$, $x \in \tilde{\Omega}$, $\xi \in \mathbf{R}^1$, have the same properties as in Section 9.1, where Ω is replaced by $\tilde{\Omega}$. Let $f \in W^1_\infty(\tilde{\Omega})$ and $\overline{u} \in W^1_p(\tilde{\Omega})$ with $p > 2$.

For any $h \in (0, h_0)$ we set

$$
\begin{aligned}
X_h &= \{v_h \in C(\overline{\Omega}_h) \mid v_h|_K \in P_1(K) \quad \forall K \in \mathcal{T}_h\}, \\
V_h &= \{v_h \in X_h \mid v_h|_{\partial\Omega_h} = 0\}, \qquad\qquad (9.88)\\
\tilde{a}_h(y; w, v) &= (\mathcal{A}(y) \operatorname{grad} w, \operatorname{grad} v)_{0,\Omega_h}, \quad y, w, v \in H^1(\Omega_h), \\
\tilde{F}_h(v) &= (f, v)_{0,\Omega_h}, \quad v \in H^1(\Omega_h),
\end{aligned}
$$

and by a_h and F_h we denote the numerical integration approximations of $\tilde{a}_h$ and $\tilde{F}_h$ defined by (9.58) and (9.59), respectively. We again put

$$\overline{u}_h = \pi_h \overline{u} \in X_h.$$

Definition 9.21. *A function $u_h \in X_h$ is said to be an approximate solution of the problem (9.42)–(9.43) on the domain $\Omega \subset \mathbf{R}^2$ if (9.56)–(9.57) hold.*

By [Ženíšek, 1981], the seminorm $|\cdot|_{1,\Omega_h}$ (in $H^1(\Omega_h)$) is a norm on V_h uniformly equivalent to the norm $\|\cdot\|_{1,\Omega_h}$. This means that there exists a constant $C > 0$ such that

$$C\|v\|_{1,\Omega_h} \le |v|_{1,\Omega_h} \le \|v\|_{1,\Omega_h} \quad \forall v \in V_h \quad \forall h \in (0, h_0). \tag{9.89}$$

By [Ciarlet, Theorem 3.2.6], the inverse inequality (9.51), where Ω is replaced by Ω_h, holds.

Using (9.89), the inverse inequality and the same technique as in Section 9.6, we establish the following properties of the forms a_h and F_h and the error of numerical integration:

Lemma 9.22. *There exist constants $C_1, C_2, C > 0$ such that*

$$a_h(y_h; v_h, v_h) \ge C_1\|v_h\|^2_{1,\Omega_h} \qquad \forall y_h \in X_h \quad \forall v_h \in V_h, \tag{9.90}$$

$$|a_h(y_h; w_h, v_h)| \le C_2\|w_h\|_{1,\Omega_h}\|v_h\|_{1,\Omega_h} \quad \forall y_h, w_h, v_h \in X_h, \tag{9.91}$$

$$|F_h(v_h)| \le C\|v_h\|_{1,\Omega_h} \qquad \forall v_h \in X_h, \tag{9.92}$$

$$\begin{aligned} &|\tilde{a}_h(y_h; w_h, v_h) - a_h(y_h; w_h, v_h)| \\ &\le Ch|w_h|_{1,\Omega_h}|v_h|_{1,\Omega_h} + Ch^{1-2/p}|y_h|_{1,\Omega_h}|w_h|_{1,\alpha,\Omega_h}|v_h|_{1,\beta,\Omega_h} \end{aligned} \tag{9.93}$$

for all $y_h, w_h, v_h \in X_h$, where either $\alpha = p$, $\beta = 2$ or $\alpha = 2$, $\beta = p$,

$$|\tilde{F}_h(v_h) - F_h(v_h)| \le Ch\|f\|_{1,\infty,\tilde{\Omega}}\|v_h\|_{1,\Omega_h} \quad \forall v_h \in X_h. \tag{9.94}$$

Using (9.90)–(9.92) and the same approach as in Theorem 9.16, we obtain the solvability of the discrete problem:

Theorem 9.23. *For any $h \in (0, h_0)$ there exists an approximate solution $u_h \in X_h$ from Definition 9.21. Moreover, there exists a constant C such that*

$$\|u_h\|_{1,\Omega_h} \le C \quad \forall h \in (0, h_0). \tag{9.95}$$

In our further analysis we introduce some new notation and concepts. Let us put

$$\omega_h = \Omega \setminus \overline{\Omega}_h, \quad \tau_h = \Omega_h \setminus \overline{\Omega}.$$

In virtue of [Feistauer, 1987, Lemma 3.3.4], there exists a constant C such that

$$\text{meas}(\omega_h \cup \tau_h) \le Ch^2 \quad \forall h \in (0, h_0). \tag{9.96}$$

We assign to each element $K \in \mathcal{T}_h$ the so-called *ideal element* K^{id} which has the same vertices as K. Note that all sides of K are straight line segments. Sides of K^{id} are defined as follows. Let PQ be an arbitrary side of K. If the line segment PQ is not contained in $\partial\Omega_h$ then PQ is also one straight side of K^{id} . Otherwise, if $PQ \subset \partial\Omega_h$, the corresponding side of K^{id} is a part (possibly curved) of the boundary $\partial\Omega$ between the points P and Q. Note that $K = K^{\text{id}}$ if at most one vertex of K lies on $\partial\Omega$.

If we replace all triangles in $\mathcal{T}_h$ by their corresponding ideal elements, we obtain the *ideal triangulation* $\mathcal{T}_h^{\text{id}}$ of $\overline{\Omega}$ which is associated with $\mathcal{T}_h$. We assume, moreover, that all triangulations $\mathcal{T}_h$ are such that each ideal element has at most one curved side.

For $v_h \in X_h$ we define its *natural extension* $v_h^* \in C(\overline{\Omega} \cup \overline{\Omega}_h)$ as $v_h^* = v_h$ on $\overline{\Omega}_h$ and $v_h^*|_{K^{\text{id}}} = \ell|_{K^{\text{id}}}$, where ℓ is a linear polynomial such that

$$\ell|_K = v_h|_K.$$

Obviously, $v_h^* \in W_q^1(\Omega)$ for $q \in [1, \infty]$ (cf. [Ciarlet, Theorem 2.1.1]).

By [Feistauer, Ženíšek, 1987], for each $v_h \in V_h$ there exists $\hat{v}_h \in V$ (called the ideal interpolation of v_h^*) such that

$$\|v_h^* - \hat{v}_h\|_{1,\Omega} \le Ch\|v_h\|_{1,\Omega_h}, \tag{9.97}$$

where C is independent of h and v_h.

Furthermore, we use the following estimates established in a similar way as in [Feistauer, Ženíšek, 1987, Lemmas 3.3.12 and 3.3.11]:

$$\|v_h^*\|_{0,\omega_h \cup \tau_h} \le Ch\|v_h\|_{1,\Omega_h}, \tag{9.98}$$

$$\|v_h^*\|_{1,\omega_h \cup \tau_h} \le Ch^{1/2}\|v_h\|_{1,\Omega_h}, \tag{9.99}$$

$$|v_h^*|_{1,\alpha,\omega_h \cup \tau_h} \le Ch^{1/\alpha}|v_h|_{1,\alpha,\Omega_h}, \quad \forall v_h \in X_h \ \forall h \in (0, h_0) \ \forall \alpha \in [1, \infty] \tag{9.100}$$

$$\|v\|_{1,\omega_h \cup \tau_h} \le Ch\|v\|_{2,\mathbf{R}^2} \qquad \forall v \in H^2(\mathbf{R}^2) \ \forall h \in (0, h_0). \tag{9.101}$$

(Sobolev spaces over $\mathbf{R}^2$ are defined in [Nečas, 1967, p. 91].)

Using (9.100) and the general interpolation results of [Ciarlet], we find that

$$\|w - \pi_h w\|_{1,\Omega_h} \to 0 \quad \text{as } h \to 0, \tag{9.102}$$

$$\|\pi_h w\|_{1,q,\Omega_h} + |(\pi_h w)^*|_{1,q,\Omega} \le C(w, q) \tag{9.103}$$

for all $w \in H^2(\mathbf{R}^2)$, $h \in (0, h_0)$ and $q \in [1, \infty)$.

From (9.99) and (9.97) it follows that

$$\|v_h^*\|_{1,\Omega} + \|\hat{v}_h\|_{1,\Omega} \le C\|v_h\|_{1,\Omega_h} \quad \forall v_h \in V_h \ \forall h \in (0, h_0). \tag{9.104}$$

By virtue of the absolute continuity of the Lebesgue integral and (9.96), we have for any $w \in H^1(\mathbf{R}^2)$ that

$$\|w\|_{1,\omega_h \cup \tau_h} \to 0 \quad \text{as } h \to 0. \tag{9.105}$$

Analogously to [Feistauer, Sobotíková, 1990, Lemmas 3.1.3, 3.1.4, 3.1.6], we find by (9.99) and (9.100) that

$$\|\overline{u}_h - \overline{u}\|_{1,\Omega_h} + \|\overline{u}_h^* - \overline{u}\|_{1,\Omega} \to 0 \quad \text{as } h \to 0 \tag{9.106}$$

and

$$\begin{aligned} &\|\overline{u}_h\|_{1,\Omega_h} + \|\overline{u}_h^*\|_{1,\Omega} \le C, \\ &\|\overline{u}_h\|_{1,\alpha,\Omega_h} + |\overline{u}_h^*|_{1,\alpha,\Omega} \le C(\alpha) \quad \forall h \in (0, h_0)\ \forall \alpha \in [1, p]. \end{aligned} \tag{9.107}$$

If $G \subset \tilde{\Omega}$ is an open set, we put

$$\tilde{a}_G(y; w, v) = (\mathcal{A}(y) \operatorname{grad} w, \operatorname{grad} v)_{0,G}, \quad y, w, v \in H^1(G).$$

It obviously follows from the boundedness of $\mathcal{A}$ in $\tilde{\Omega} \times \mathbf{R}^1$ that

$$|\tilde{a}_G(y; w, v)| \le C\|w\|_{1,G}\|v\|_{1,G} \quad \forall y, w, v \in H^1(G). \tag{9.108}$$

Now, for each $h \in (0, h_0)$ we define a function $u_h' \in H^1(\Omega)$ associated with the approximate solution u_h (see Definition 9.21) as follows: if we express u_h in the form $u_h = \overline{u}_h + z_h$ with $z_h \in V_h$ (cf. (9.88)), then we set

$$u_h' = \overline{u}_h^* + \hat{z}_h. \tag{9.109}$$

Using now (9.95), (9.97), (9.107) and (9.109), we find that (cf. [Feistauer, Sobotíková, 1990, Lemma 3.1.6])

$$\|z_h\|_{1,\Omega_h} + \|u_h'\|_1 + \|u_h^*\|_1 + \|\hat{z}_h\|_1 + \|z_h^*\|_1 \le C \quad \forall h \in (0, h_0). \tag{9.110}$$

Finally, we can formulate the main convergence results.

Theorem 9.24. *If $u_h = \overline{u}_h + z_h$, $z_h \in V_h$, is a solution from Definition 9.21 and u_h' is the function associated with u_h by (9.109), then*

$$u_h' \rightharpoonup u \quad \textit{(weakly) in } H^1(\Omega) \quad \text{as } h \to 0,$$

where u is the weak solution of the problem (9.42)–(9.43).

Proof. According to (9.110), it is obviously sufficient to show that any weak cluster point of the family $\{u_h'\}$, $h \in (0, h_0)$, is the solution u of the problem (9.49)–(9.50) – cf. the proof of Theorem 9.18. So let us consider a subsequence of the family

$\{u'_h\}_{h\to 0}$ (denoted again by the same symbol) and assume the existence of some $u \in H^1(\Omega)$ such that

$$u'_h \rightharpoonup u \quad \text{(weakly) in } H^1(\Omega) \quad \text{as } h \to 0. \tag{9.111}$$

We want to prove that u is the solution of (9.49)–(9.50).

It is evident that u satisfies (9.49). Indeed, it follows from (9.109), (9.111), (9.106), the fact that $\hat{z}_h \in V$ and that $V = H_0^1(\Omega)$ is weakly closed, that $\hat{z}_h \rightharpoonup u - \overline{u} \in V$.

Now let us prove (9.50). To this end, we consider an arbitrary $v \in V \cap C^\infty(\overline{\Omega})$ and denote by $\tilde{v}$ the *Calderon extension* of v onto $\mathbf{R}^2$ in the H^2-topology. This means that $\tilde{v} \in H^2(\mathbf{R}^2)$, $\tilde{v}|_\Omega = v$ and $\|\tilde{v}\|_{2,2,\mathbf{R}^2} \le C\|v\|_{2,2,\Omega}$, where C is independent of v (see, e.g., [Nečas, 1967, p. 80]). If we put $v_h = \pi_h \tilde{v} \in V_h$ then, by (9.102), (9.103), (9.99) and (9.105),

$$\begin{aligned} &\|v - v_h^*\|_1 \to 0 &&\text{as } h \to 0,\\ &\|v_h\|_{1,q,\Omega_h} \le C(v,q), &&q \in [1,\infty). \end{aligned} \tag{9.112}$$

Taking into account that (cf. (9.109) and (9.97))

$$\|u_h^* - u'_h\|_1 = \|z_h^* - \hat{z}_h\|_1 \le ch\|z_h\|_{1,\Omega_h},$$

we can see from (9.110) and (9.111) that $u_h^* \rightharpoonup u$ in $H^1(\Omega)$. This and the compact imbedding $H^1(\Omega) \hookrightarrow L^2(\Omega)$ (see the Rellich Theorem 3.9) imply the strong convergence

$$u_h^* \to u \quad \text{in } L^2(\Omega). \tag{9.113}$$

We have

$$\begin{aligned} a(u;u,v) - F(v) = {} & a(u;u-u_h^*,v) + [a(u;u_h^*,v) - a(u_h^*;u_h^*,v)]\\ & + [a(u_h^*;u_h^*,v-v_h^*)] + [\tilde{a}_{\omega_h}(u_h^*;u_h^*,v_h^*) - \tilde{a}_{\tau_h}(u_h;u_h,v_h)]\\ & + [\tilde{a}_h(u_h;u_h,v_h) - a_h(u_h;u_h,v_h)] + [F_h(v_h) - \tilde{F}_h(v_h)]\\ & + [F(v_h^*) - F(v)] + \left[\int_{\tau_h} f v_h \,dx - \int_{\omega_h} f v_h^* \,dx\right]\\ \equiv {} & E_1 + \dots + E_8. \end{aligned}$$

The individual terms $E_1, \dots, E_8$ are estimated for $h \to 0$ as follows:

1. Since the mapping $y \in H^1(\Omega) \mapsto a(u,y,v) \in \mathbf{R}^1$ is a continuous linear functional, as follows from (9.47), and $u_h^* \rightharpoonup u$ in $H^1(\Omega)$ as $h \to 0$, we conclude that $E_1 \to 0$ as $h \to 0$.
2. By (9.48), (9.110), and (9.113),

$$|E_2| \le C\|u - u_h^*\|_0 \|u_h^*\|_1 \|v\|_{1,\infty} \le C\|u - u_h^*\|_0 \|v\|_{1,\infty} \to 0.$$

3. In virtue of (9.47), (9.110) and (9.112),

$$|E_3| \le C\|u_h^*\|_1\|v - v_h^*\|_1 \to 0.$$

4. It follows form (9.108), (9.99), (9.95) and (9.112) that

$$|E_4| \le C\|u_h^*\|_{1,\omega_h \cup \tau_h}\|v_h^*\|_{1,\omega_h \cup \tau_h} \le Ch\|u_h\|_{1,\Omega_h}\|v_h\|_{1,\Omega_h} \le C(v,2)Ch \to 0.$$

5. By (9.93), (9.95) and (9.112),

$$|E_5| \le Ch\|u_h\|_{1,\Omega_h}\|v_h\|_{1,\Omega_h} + Ch^{1-2/p}\|u_h\|_{1,\Omega_h}^2\|v_h\|_{1,p,\Omega_h} \to 0.$$

6. We have $|E_6| \to 0$ in virtue of (9.94) and (9.112).
7. Using the Cauchy-Schwarz inequality and (9.112), we obtain

$$|E_7| = |F(v) - F(v_h^*)| = |(f, v - v_h^*)_0| \le \|f\|_0\|v - v_h^*\|_0 \le C\|v - v_h^*\|_1 \to 0.$$

8. Obviously, by the Cauchy-Schwarz inequality, (9.98) and (9.112),

$$|E_8| \le \int_{\omega_h \cup \tau_h} |fv_h^*|\,dx \le \|f\|_{0,\tilde{\Omega}}\|v_h^*\|_{0,\omega_h \cup \tau_h} \le Ch\|v_h\|_{1,\Omega_h} \to 0.$$

Summarizing the above results, we see that u satisfies the identity

$$a(u; u, v) = F(v) \quad \forall v \in V \cap C^\infty(\overline{\Omega})$$

and, hence, by the density of $V \cap C^\infty(\overline{\Omega})$ in V and the continuity of a and F, the identity (9.50) follows. □

Finally, we will establish the strong convergence.

Theorem 9.25. *We have $u_h' \to u$ (strongly) in $H^1(\Omega)$ as $h \to 0$ and*

$$\lim_{h\to 0} \|\tilde{u} - u_h\|_{1,\Omega_h} = 0$$

where $\tilde{u} \in H^1(\mathbf{R}^2)$ is the standard Calderon extension of u onto $\mathbf{R}^2$ in the H^1-topology.

Proof. As above, we write $u = \overline{u} + z$, $z \in V$ and $u_h = \overline{u}_h + z_h$, $z_h \in V_h$. Similarly as in the proof of Theorem 9.19 let us consider a sequence $\{z^m\}_{m=1}^\infty \subset V \cap C^\infty(\Omega)$ such that

$$\|z - z^m\|_1 \to 0 \quad \text{as } m \to \infty. \tag{9.114}$$

Let $\tilde{z}^m \in H^2(\mathbf{R}^2)$ be the extension of z^m from Ω onto $\mathbf{R}^2$, i.e., $\tilde{z}^m|_\Omega = z^m$ and there exists a constant C such that $\|\tilde{z}^m\|_{2,\mathbf{R}^2} \le C\|z^m\|_{2,\Omega}$. Furthermore, let us set

$$z_h^m = \pi_h \tilde{z}^m.$$

Thus, $z_h^m \in V_h$ and

$$\begin{aligned} &\|\tilde{z}^m - z_h^m\|_{1,\Omega_h} \to 0 && \text{as } h \to 0, \\ &\|z_h^m\|_{1,q,\Omega_h} + |(z_h^m)^*|_{1,q,\Omega} \le C(z^m, q) && \forall h \in (0, h_0)\ \forall q \in [1, \infty), \end{aligned} \tag{9.115}$$

which follows from (9.102)–(9.103).

Let us define

$$v_h^m = z_h - z_h^m \,(\in V_h).$$

By (9.90), (9.57) and (9.50), we obtain

$$\begin{aligned} C_1\|v_h^m\|_{1,\Omega_h}^2 &\le a_h(u_h; v_h^m, v_h^m) \\ &= a_h(u_h; \overline{u}_h + z_h, v_h^m) - a_h(u_h; \overline{u}_h + z_h^m, v_h^m) \\ &= F_h(v_h^m) - a_h(u_h; \overline{u}_h + z_h^m, v_h^m) \\ &= [F_h(v_h^m) - \tilde{F}_h(v_h^m)] + \Big[\int_{\tau_h} f v_h^m \, dx - \int_{\omega_h} f(v_h^m)^* \, dx\Big] \\ &\quad + [F((v_h^m)^*) - F(\hat{v}_h^m)] + [a(u; u, \hat{v}_h^m) - a_h(u_h; \overline{u}_h + z_h^m, v_h^m)] \\ &\equiv G_1 + \dots + G_4. \end{aligned} \tag{9.116}$$

Further, we can write

$$\begin{aligned} G_4 &\equiv a(u; u, \hat{v}_h^m) - a_h(u_h; \overline{u}_h + z_h^m, v_h^m) \\ &= [a(u; \overline{u} + z - \overline{u}_h^* - (z_h^m)^*, \hat{v}_h^m)] \\ &\quad + [a(u; \overline{u}_h^* + (z_h^m)^*, \hat{v}_h^m) - a(u_h'; \overline{u}_h^* + (z_h^m)^*, \hat{v}_h^m)] \\ &\quad + [a(u_h'; \overline{u}_h^* + (z_h^m)^*, \hat{v}_h^m) - a(u_h'; \overline{u}_h^* + (z_h^m)^*, (v_h^m)^*)] \\ &\quad + [a(u_h'; \overline{u}_h^* + (z_h^m)^*, (v_h^m)^*) - a(u_h^*; \overline{u}_h^* + (z_h^m)^*, (v_h^m)^*)] \\ &\quad + [\tilde{a}_{\omega_h}(u_h^*; \overline{u}_h^* + (z_h^m)^*, (v_h^m)^*) - \tilde{a}_{\tau_h}(u_h; \overline{u}_h + z_h^m, v_h^m)] \\ &\quad + [\tilde{a}_h(u_h; \overline{u}_h + z_h^m, v_h^m) - a_h(u_h; \overline{u}_h + z_h^m, v_h^m)] \\ &\equiv H_1 + \dots + H_6. \end{aligned} \tag{9.117}$$

Let us estimate the individual terms G_1, G_2, G_3 and $H_1, \dots, H_6$. By (9.94),

$$|G_1| \le Ch\|f\|_{1,\infty,\tilde{\Omega}}\|v_h^m\|_{1,\Omega_h} \le C'h\|v_h^m\|_{1,\Omega_h}. \tag{9.118}$$

The Cauchy-Schwarz inequality and the estimate (9.98) imply that

$$|G_2| \le Ch\|v_h^m\|_{1,\Omega_h}. \tag{9.119}$$

Further, using the continuity of the functional F and (9.97), we get

$$|G_3| \le C\|(v_h^m)^* - \hat{v}_h^m\|_{1,\Omega} \le Ch\|v_h^m\|_{1,\Omega_h}. \tag{9.120}$$

In virtue of (9.47), the inequalities

$$\|z-(z_h^m)^*\|_1 \le \|z-z^m\|_1 + \|z^m-(z_h^m)^*\|_1,$$
$$\|z^m-(z_h^m)^*\|_1 \le C(\|\tilde{z}^m - z_h^m\|_{1,\Omega_h} + \|\tilde{z}^m\|_{1,\omega_h} + \|(z_h^m)^*\|_{1,\omega_h}),$$

(9.99), (9.115) with $q=2$ and (9.104), we have

$$|H_1| \le C(\|\overline{u}-\overline{u}_h^*\|_1 + \|z-z^m\|_1 + \|\tilde{z}^m - z_h^m\|_{1,\Omega_h} + \|\tilde{z}^m\|_{1,\omega_h} + C(z^m,2)h^{1/2})\|v_h^m\|_{1,\Omega_h}. \tag{9.121}$$

Similarly to (9.83), using (9.104), (9.107) and (9.115), we can show that

$$\begin{aligned}|H_2| &\le C\|u-u_h'\|_{0,q}|\overline{u}_h^* + (z_h^m)^*|_{1,p}\|\hat{v}_h^m\|_1 \\ &\le C'\|u-u_h'\|_{0,q}C(z^m,p)\|v_h^m\|_{1,\Omega_h}\end{aligned} \tag{9.122}$$

with $1/p+1/q=1/2$. The estimates (9.47), (9.107), (9.115) and (9.97) imply that

$$|H_3| \le C(z^m,2)Ch\|v_h^m\|_{1,\Omega_h}. \tag{9.123}$$

The term H_4 can be estimated in a similar way as H_2:

$$|H_4| \le C\|u_h'-u_h^*\|_{0,q}|\overline{u}_h^* + (z_h^m)^*|_{1,p}\|(v_h^m)^*\|_1.$$

Now, using the continuous imbedding $H^1(\Omega) \hookrightarrow L^q(\Omega)$ for $q<\infty$, (9.97), (9.110), (9.107), (9.115) and (9.104), we find that

$$|H_4| \le C(z^m,p)Ch\|v_h^m\|_{1,\Omega_h}. \tag{9.124}$$

In order to estimate H_5, we use (9.108), (9.99), (9.107) with $\alpha=2$ and (9.115) with $q=2$. Then

$$|H_5| \le Ch\|v_h^m\|_{1,\Omega_h}. \tag{9.125}$$

Finally, by (9.93) with $\alpha=p$, $\beta=2$, (9.95), (9.107) and (9.115)

$$\begin{aligned}|H_6| &\le C(h\|\overline{u}_h + z_h^m\|_{1,\Omega_h} + h^{1-2/p}\|u_h\|_{1,\Omega_h}\|\overline{u}_h + z_h^m\|_{1,p,\Omega_h})\|v_h^m\|_{1,\Omega_h} \\ &\le C'(h + C(z^m,p)h^{1-2/p})\|v_h^m\|_{1,\Omega_h}.\end{aligned} \tag{9.126}$$

Now taking into account (9.116) (9.126), we obtain the estimate

$$\begin{aligned}C_1\|v_h^m\|_{1,\Omega_h} \le C(h &+ \|\overline{u}-\overline{u}_h^*\|_1 + \|z-z^m\|_1 + C(z^m,2)h^{1/2} \\ &+ \|\tilde{z}^m - z_h^m\|_{1,\Omega_h} + \|\tilde{z}^m\|_{1,\omega_h} \\ &+ C(z^m,p)\|u-u_h'\|_{0,q} + C(z^m,p)h + C(z^m,p)h^{1-2/p}).\end{aligned}$$

Let us consider m fixed and pass to the limit for $h \to 0$. Using (9.115), (9.105) with $w = z^m$, (9.106), (9.111) and the compact imbedding $H^1(\Omega) \hookrightarrow L^q(\Omega)$ which imply that $u_h' \to u$ in $L^q(\Omega)$ as $h \to 0$, we conclude that

$$\limsup_{h\to 0} \|v_h^m\|_{1,\Omega_h} \leq C\|z - z^m\|_1, \quad m = 1, 2, \ldots,$$

where C is independent of m.

Now we proceed as in the proof of Theorem 9.19. Denoting by $\tilde{u} \in H^1(\mathbf{R}^2)$ the Calderon extension of the exact solution $u \in H^1(\Omega)$ and writing $\tilde{u} = \overline{u} + \tilde{z}$, we have

$$\begin{aligned} \|\tilde{u} - u_h\|_{1,\Omega_h} &= \|(\overline{u} + \tilde{z}) - (\overline{u}_h + z_h)\|_{1,\Omega_h} \\ &\leq \|\overline{u} - \overline{u}_h\|_{1,\Omega_h} + \|\tilde{z} - \tilde{z}^m\|_{1,\Omega_h} + \|\tilde{z}^m - z_h^m\|_{1,\Omega_h} + \|z_h^m - z_h\|_{1,\Omega_h}. \end{aligned}$$

Further,

$$\|\tilde{z} - \tilde{z}^m\|_{1,\Omega_h} \leq C(\|z - z^m\|_{1,\Omega} + \|\tilde{z}\|_{1,\omega_h \cup T_h} + \|\tilde{z}^m\|_{1,\omega_h \cup T_h}).$$

Taking into account that $z_h - z_h^m = v_h^m$, and passing to the limit for $h \to 0$, we obtain by (9.106), (9.115) and (9.105) that

$$\limsup_{h\to 0} \|\tilde{u} - u_h\|_{1,\Omega_h} \leq C\|z - z^m\|_{1,\Omega}, \quad m = 1, 2, \ldots .$$

Passing to the limit for $m \to \infty$ and using (9.114), we find that

$$\lim_{h\to 0} \|\tilde{u} - u_h\|_{1,\Omega_h} = 0.$$

The strong convergence $u_h' \to u$ in $H^1(\Omega)$ is a consequence of (9.97), (9.99), (9.105) and (9.110). □

Remark 9.26. For simplicity we confined our considerations to the case of the Dirichlet boundary conditions. The extension to mixed Dirichlet–Neumann conditions is similar to [Feistauer, Ženíšek, 1987, 1988] and [Feistauer, Sobotíková, 1990]. An analysis of a nonlinear Newton boundary condition used in [Hlaváček, Křížek, Malý, 1994] in the case of the approximation of the boundary and the use of the numerical integration in boundary integrals is a nontrivial and interesting unsolved problem. Also a generalization of the used technique to three-dimensional domains with piecewise curved boundaries seems to be an open problem until now.

10. Stationary semiconductor equations

10.1. Classical formulation

The behaviour of a semiconductor device is usually modelled by three coupled nonlinear partial differential equations of elliptic type. Such a system for the transport of mobile charge carriers was first introduced by Van Roosbroeck [Van Roosbroeck] in 1950. Nowadays there are many models which differ in their choice of unknowns, scales, various types of nonlinearities etc. (see, e.g., [Brezzi], [Gröger], [Markowich], [Markowich, Ringhofer, Schmeiser], [Mock, 1972], [Polak, den Heijer, Schilders, Markowich], [Pospíšek], [Pospíšek, Segeth, Šilhán], [Selberherr], [Sze], [Zlámal, 1986]).

Throughout this chapter we will examine the following equations (cf. (2.20)) in a domain $\Omega \in \mathcal{L}$, $\Omega \subset \mathbf{R}^d$, $d = 1, 2, 3$,

$$-\operatorname{div}(\varepsilon \operatorname{grad} u_0) = f + e_1(u_0, u_1) - e_2(u_0, u_2), \tag{10.1}$$

$$-\operatorname{div}(\mu_1 e_1(u_0, u_1) \operatorname{grad} u_1) = k(u_1, u_2)(1 - \mathrm{e}^{u_1+u_2}), \tag{10.2}$$

$$-\operatorname{div}(\mu_2 e_2(u_0, u_2) \operatorname{grad} u_2) = k(u_1, u_2)(1 - \mathrm{e}^{u_1+u_2}), \tag{10.3}$$

where $u = (u_0, u_1, u_2)$ is unknown,

$$e_1 = e_1(u_0, u_1) = \mathrm{e}^{u_1 - u_0}, \qquad e_2 = e_2(u_0, u_2) = \mathrm{e}^{u_2 + u_0}, \tag{10.4}$$

u_0 is the electrostatic potential, u_1 and u_2 are the electrochemical potentials of holes and electrons, $f \in L^\infty(\Omega)$ is the net density of charges of ionized impurities (the so-called doping), k is a continuous positive function, which represents the rate of generation of holes and electrons, $\varepsilon \in L^\infty(\Omega)$ is the dielectric permittivity, $\mu_1, \mu_2 \in L^\infty(\Omega)$ are the mobilities of holes and electrons. Moreover, let

$$\varepsilon \geq m_0, \qquad \mu_i \geq m_0, \quad i = 1, 2,$$

for some positive constant m_0.

As usual $(-\operatorname{grad} u_0)$ is the electric field corresponding to the potential u_0 occurring in (10.1). The highly nonlinear equations (10.2) and (10.3) have arisen from the continuity equations for holes and electrons. The equations (10.4) represent a special case of Boltzmann's statistics, see (2.19).

We will consider the mixed boundary conditions

$$u = \overline{u} \quad \text{on } \Gamma_1, \tag{10.5}$$

$$\alpha u_0 + \frac{\partial u_0}{\partial n} = \frac{\partial u_1}{\partial n} = \frac{\partial u_2}{\partial n} = 0 \quad \text{on } \Gamma_2, \tag{10.6}$$

where $\Gamma_1 \not\subseteq \emptyset$ and Γ_2 are disjoint and relatively open in $\partial\Omega$, $\overline{\Gamma}_1 \cup \overline{\Gamma}_2 = \partial\Omega$, n is the outward unit normal to Γ_2, $\overline{u} = (\overline{u}_0, \overline{u}_1, \overline{u}_2) \in (H^1(\Omega) \cap L^\infty(\Omega))^3$ is a given function, $\alpha \in L^\infty(\Gamma_2)$ and $\alpha \geq 0$. The part Γ_1 of the boundary corresponds to metal contacts, where voltages are prescribed.

Remark 10.1. Gröger [Gröger] discusses a model with nonlinear boundary conditions where the mobilities μ_i, $i = 1, 2$, of holes and electrons depend nonlinearly on $\operatorname{grad} u_0$ and $\operatorname{grad} u_i$ so that the associated operator is strongly monotone.

10.2. Existence of a weak solution

In this section we prove the existence of a weak solution of the problem (10.1)–(10.6) using a simple approach based on a fixed point theorem.

Define the space of test functions

$$V = \{z \in H^1(\Omega) \mid z = 0 \text{ on } \Gamma_1\}$$

and the forms

$$a(w_0, v_0) = \int_\Omega \varepsilon \operatorname{grad} w_0 \cdot \operatorname{grad} v_0 \, dx + \int_{\Gamma_2} \alpha w_0 v_0 \, ds, \tag{10.7}$$

$$\mathcal{A}(y; w, v) = \sum_{i=1}^{2} \int_\Omega \mu_i y_i \operatorname{grad} w_i \cdot \operatorname{grad} v_i \, dx, \tag{10.8}$$

$$\mathcal{F}(y, z; v) = \int_\Omega k(y)(1 - e^z)(v_1 + v_2) \, dx, \tag{10.9}$$

where $v_0, w_0 \in V$, $y = (y_1, y_2) \in (L^\infty(\Omega))^2$, $v = (v_1, v_2) \in V \times V$, $w = (w_1, w_2) \in V \times V$, $z \in L^\infty(\Omega)$ and "$\cdot$" is the standard scalar product in $\mathbf{R}^d$.

We shall examine the solvability of the following weak formulation of the problem (10.1)–(10.6): Find $u = (u_0, u_1, u_2)$ such that

$$\begin{aligned} & u - \overline{u} \in (V \cap L^\infty(\Omega))^3, \\ & a(u_0, v_0) = \int_\Omega (f + e_1 - e_2) v_0 \, dx \quad \forall v_0 \in V, \\ & \mathcal{A}(e; \hat{u}, v) = \mathcal{F}(e, u_1 + u_2; v) \qquad \forall v \in V \times V, \end{aligned} \tag{10.10}$$

where $e = (e_1, e_2)$ is defined in (10.4) and

$$\hat{u} = (u_1, u_2).$$

The solution u of (10.10) is called a *weak solution.* If u is sufficiently smooth, then it is easy to verify that u solves (10.1)–(10.6) if and only if it solves (10.10).

To prove that (10.10) has a solution, we set

$$K = \operatorname*{ess\,sup}_{i=1,2} \|\overline{u}_i\|_{0,\infty} \tag{10.11}$$

and choose

$$M \geq \|\overline{u}_0\|_{0,\infty} \tag{10.12}$$

so that

$$\begin{aligned} f + e^{K-M} - e^{M-K} &\leq 0, \\ f + e^{M-K} - e^{K-M} &\geq 0 \end{aligned} \tag{10.13}$$

almost everywhere.

Next, for $R > 0$, $x \in \Omega$ and $z \in L^2(\Omega)$ we define the function

$$(P_R z)(x) = \begin{cases} -R & \text{if } z(x) \leq -R, \\ z(x) & \text{if } -R \leq z(x) \leq R, \\ R & \text{if } R \leq z(x). \end{cases}$$

Using this cut-off function, we introduce a modified weak formulation of the problem (10.1)–(10.6): Find $u = (u_0, u_1, u_2)$ such that

$$\begin{aligned} u - \overline{u} &\in V \times V \times V, \\ a(u_0, v_0) &= \int_\Omega (f + E_1 - E_2) v_0 \, dx \quad \forall v_0 \in V, \\ \mathcal{A}(E; \hat{u}, v) &= \mathcal{F}(E, P_K u_1 + P_K u_2; v) \quad \forall v \in V \times V, \end{aligned} \tag{10.14}$$

where $E = (E_1, E_2)$ is such that

$$E_1 = E_1(u_0, u_1) = e^{P_K u_1 - P_M u_0}, \qquad E_2 = E_2(u_0, u_2) = e^{P_K u_2 + P_M u_0}. \tag{10.15}$$

Lemma 10.2. *Any solution u of (10.14) is also a solution of (10.10).*

Proof. Assume that u is a solution of (10.14). Taking suitable test functions (cf. (9.30)), we prove the lemma showing that $\|u_0\|_{0,\infty} \leq M$ and $\|u_i\|_{0,\infty} \leq K$ for $i = 1, 2$.

First, let $v_0 = (u_0 - M)^+$. Since $u_0 - \overline{u}_0 = 0$ on Γ_1, we see, by (10.12), that $u_0 - M \leq 0$ on Γ_1, and thus $v_0 = 0$ on Γ_1. As in the proof of Theorem 9.8, we find that $v_0 \in V$. Moreover, by [Gilbarg, Trudinger, p. 145], we have

$$\operatorname{grad} w^+ = \begin{cases} \operatorname{grad} w & \text{if } w > 0 \\ 0 & \text{if } w \leq 0 \end{cases} \qquad \forall w \in H^1(\Omega),$$

and thus

$$\operatorname{grad} u_0 \cdot \operatorname{grad} v_0 = \operatorname{grad}(u_0 - M)^+ \cdot \operatorname{grad} v_0 = \|\operatorname{grad} v_0\|^2.$$

From here, (10.14) and (10.15) we obtain

$$\int_\Omega \varepsilon \| \operatorname{grad} v_0 \|^2 \, dx + \int_{\Gamma_2} \alpha u_0 v_0 \, ds - \int_\Omega (f + \mathrm{e}^{P_K u_1 - P_M u_0} - \mathrm{e}^{P_K u_2 + P_M u_0}) v_0 \, dx = 0. \tag{10.16}$$

Since $v_0 = (u_0 - M)^+$, we see that if $u_0(x) \geq M$ (> 0) for some $x \in \overline{\Omega}$ then

$$\alpha(x) u_0(x) v_0(x) \geq 0$$

and

$$\mathrm{e}^{P_K u_2(x) + P_M u_0(x)} v_0(x) \geq \mathrm{e}^{-K+M} v_0(x),$$

otherwise, if $u_0(x) < M$ then $v_0(x) = 0$, and thus

$$\alpha(x) u_0(x) v_0(x) = 0$$

and

$$\mathrm{e}^{P_K u_2(x) + P_M u_0(x)} v_0(x) = \mathrm{e}^{-K+M} v_0(x) = 0.$$

Hence, $\alpha u_0 v_0 \geq 0$ a.e. on Γ_2, the boundary integral in (10.16) is nonnegative and, by (10.13),

$$(f + \mathrm{e}^{P_K u_1 - P_M u_0} - \mathrm{e}^{P_K u_2 + P_M u_0}) v_0 \leq (f + \mathrm{e}^{K-M} - \mathrm{e}^{-K+M}) v_0 \leq 0.$$

Therefore, from (10.16) we get

$$m_0 \int_\Omega \| \operatorname{grad} v_0 \|^2 \, dx \leq 0,$$

which implies that $\operatorname{grad} v_0 = 0$ and thus $v_0 = 0$ as $v_0 \in V$. In other words, $u_0 \leq M$. Similarly we obtain that $u_0 \geq -M$ by applying the test function $v_0 = (u_0 + M)^-$ and using (10.13).

Second, let $v_i = (u_i - K)^+$ for $i = 1, 2$. If $u_i(x) - \overline{u}_i(x) = 0$ then, by (10.11), $u_i(x) - K \leq 0$ and thus $v_i(x) = 0$. Hence, we again find that $v_i \in V$ and also $\operatorname{grad} u_i \cdot \operatorname{grad} v_i = \| \operatorname{grad} v_i \|^2$. According to (10.14), (10.8) and (10.9), we have

$$\sum_{i=1}^{2} \int_\Omega (\mu_i E_i \| \operatorname{grad} v_i \|^2 - k(E)(1 - \mathrm{e}^{P_K u_1 + P_K u_2}) v_i) \, dx = 0. \tag{10.17}$$

Since k is a positive function and $v_i = (u_i - K)^+$, we have

$$k(E)(1 - \mathrm{e}^{P_K u_1 + P_K u_2}) v_i \leq 0$$

and clearly

$$\mu_i E_i \geq m_0 \mathrm{e}^{-K-M} > 0.$$

Therefore, from (10.17) we get

$$\sum_{i=1}^{2} m_0 \mathrm{e}^{-K-M} \int_\Omega \| \operatorname{grad} v_i \|^2 \, dx \leq 0, \tag{10.18}$$

which again implies that $v_i = 0$ for $i = 1, 2$, i.e., $u_i \leq K$ for $i = 1, 2$. Similarly we obtain $u_i \geq -K$ by applying the test functions $v_i = (u_i + K)^-$ for $i = 1, 2$. □

Next we introduce a generalization of the Brouwer Fixed-Point Theorem 7.19. This generalization is sometimes called the Schauder Fixed-Point Theorem.

Theorem 10.3. *Let B be a nonempty convex and compact subset of a Banach space and let $S: B \to B$ be a continuous mapping. Then there exists $x \in B$ such that $Sx = x$.*

For the proof see [Tichonov].

Theorem 10.4. *There exists a solution of the problem (10.10).*

Proof. First we transform the problem of the existence of a solution u into a fixed point problem and then we apply Theorem 10.3 as in [Gröger].

Set $H = (L^2(\Omega))^3$. Let $\tilde{u} = (\tilde{u}_0, \tilde{u}_1, \tilde{u}_2) \in H$ be given. We define $\widetilde{E} = (\widetilde{E}_1, \widetilde{E}_2) \in (L^\infty(\Omega))^2$ by

$$\widetilde{E}_1 = \mathrm{e}^{P_K \tilde{u}_1 - P_M \tilde{u}_0}, \qquad \widetilde{E}_2 = \mathrm{e}^{P_K \tilde{u}_2 + P_M \tilde{u}_0}, \tag{10.19}$$

and determine $u_0 \in H^1(\Omega)$ as the unique solution of the linear problem

$$\begin{aligned} & u_0 - \overline{u}_0 \in V, \\ & a(u_0, v_0) = \int_\Omega (f + \widetilde{E}_1 - \widetilde{E}_2) v_0 \, dx \quad \forall v_0 \in V. \end{aligned} \tag{10.20}$$

Next we determine $(u_1, u_2) \in (H^1(\Omega))^2$ as the solution of another linear problem

$$\begin{aligned} u_i - \overline{u}_i &\in V & &\text{for } i = 1, 2, \\ \mathcal{A}(\widetilde{E}; u_1, u_2, v) &= \mathcal{F}(\widetilde{E}, P_K \tilde{u}_1 + P_K \tilde{u}_2; v) & &\forall v \in V \times V. \end{aligned} \tag{10.21}$$

Define now the operator $S: H \to H$ by

$$S\tilde{u} = u,$$

where $u = (u_0, u_1, u_2)$. If this u is a fixed point of S, then we easily find that u solves the problem (10.14).

Now we will verify the assumptions of Theorem 10.3. Using (10.20) and (10.21), we prove that $S: H \to H$ is continuous. First, we show that the mapping

$$\tilde{u} \in H \mapsto u_0 \in H^1(\Omega), \tag{10.22}$$

given by the problem (10.20), is Lipschitz continuous.

So let $\tilde{u} = (\tilde{u}_0, \tilde{u}_1, \tilde{u}_2)$ and $\tilde{u}' = (\tilde{u}'_0, \tilde{u}'_1, \tilde{u}'_2)$ from H be given. Denote by $u_0, u'_0 \in H^1(\Omega)$ their images. Let $\widetilde{E}'_1$ and $\widetilde{E}'_2$ correspond to $\tilde{u}'$ like in (10.19).

Then, by Friedrichs' inequality (3.42), (10.7) and (10.20), we have

$$\begin{aligned} C\|u_0 - u_0'\|_1^2 &\le \int_\Omega \varepsilon \|\operatorname{grad}(u_0 - u_0')\|^2 \, dx \\ &\le \int_\Omega \varepsilon \operatorname{grad}(u_0 - u_0') \cdot \operatorname{grad}(u_0 - u_0') \, dx \\ &\quad + \int_{\Gamma_2} \alpha(u_0 - u_0')(u_0 - u_0') \, ds \\ &= a(u_0, u_0 - u_0') - a(u_0', u_0 - u_0') \\ &= \int_\Omega (\tilde{E}_1 - \tilde{E}_2 - \tilde{E}_1' + \tilde{E}_2')(u_0 - u_0') \, dx \\ &\le (\|\tilde{E}_1 - \tilde{E}_1'\|_0 + \|\tilde{E}_2 - \tilde{E}_2'\|_0)\|u_0 - u_0'\|_1. \end{aligned} \tag{10.23}$$

Clearly,

$$\zeta, \xi \in [-M, M] \Longrightarrow |\mathrm{e}^\zeta - \mathrm{e}^\xi| \le \mathrm{e}^M |\zeta - \xi|. \tag{10.24}$$

Hence, according to (10.19) and (10.24), we come to

$$\begin{aligned} &\|\tilde{E}_1 - \tilde{E}_1'\|_0 = \|\mathrm{e}^{P_K \tilde{u}_1} \mathrm{e}^{-P_M \tilde{u}_0} - \mathrm{e}^{P_K \tilde{u}_1'} \mathrm{e}^{-P_M \tilde{u}_0'}\|_0 \\ &= \|\mathrm{e}^{P_K \tilde{u}_1}(\mathrm{e}^{-P_M \tilde{u}_0} - \mathrm{e}^{-P_M \tilde{u}_0'}) + (\mathrm{e}^{P_K \tilde{u}_1} - \mathrm{e}^{P_K \tilde{u}_1'})\mathrm{e}^{-P_M \tilde{u}_0'}\|_0 \\ &\le \mathrm{e}^K \|\mathrm{e}^{-P_M \tilde{u}_0} - \mathrm{e}^{-P_M \tilde{u}_0'}\|_0 + \mathrm{e}^M \|\mathrm{e}^{P_K \tilde{u}_1} - \mathrm{e}^{P_K \tilde{u}_1'}\|_0 \\ &\le \mathrm{e}^K \mathrm{e}^M (\| - P_M \tilde{u}_0 + P_M \tilde{u}_0'\|_0 + \|P_K \tilde{u}_1 - P_K \tilde{u}_1'\|_0) \\ &\le \mathrm{e}^{K+M} (\|\tilde{u}_0 - \tilde{u}_0'\|_0 + \|\tilde{u}_1 - \tilde{u}_1'\|_0), \end{aligned} \tag{10.25}$$

where the last inequality holds as P_K is a projection. Similarly, we can estimate the term $\|\tilde{E}_2 - \tilde{E}_2'\|_0$. Therefore, by (10.23) and (10.25), we find that

$$C\|u_0 - u_0'\|_1 \le 4\mathrm{e}^{K+M} \|\tilde{u} - \tilde{u}'\|_0,$$

which means that the mapping (10.22) is uniformly Lipschitz continuous.

In a similar way we can prove for the linear problem (10.21) that

$$\|u_i - u_i'\|_1 \le C\|\tilde{u} - \tilde{u}'\|_0, \quad i = 1, 2.$$

Hence, the operator $S\colon H \to H$ is continuous.

Further, by (10.20), we have

$$\begin{aligned} C\|u_0 - \overline{u}_0\|_1^2 &\le a(u_0 - \overline{u}_0, u_0 - \overline{u}_0) \\ &= \int_\Omega (f + \tilde{E}_1 - \tilde{E}_2)(u_0 - \overline{u}_0) \, dx - a(\overline{u}_0, u_0 - \overline{u}_0) \\ &\le C'(\|f\|_{0,\infty} + 2\mathrm{e}^{K+M} + \|\overline{u}_0\|_1)\|u_0 - \overline{u}_0\|_1, \end{aligned}$$

and similarly, by (10.8) and (10.21),

$$Ce^{-K-M}\sum_{i=1}^{2}\|u_i-\overline{u}_i\|_1^2 \le \mathcal{A}(\tilde{E};u_1-\overline{u}_1,u_2-\overline{u}_2,u_1-\overline{u}_1,u_2-\overline{u}_2)$$
$$=\mathcal{F}(\tilde{E},P_K\tilde{u}_1+P_K\tilde{u}_2;u_1-\overline{u}_1,u_2-\overline{u}_2)-\mathcal{A}(\tilde{E};\overline{u}_1,\overline{u}_2,u_1-\overline{u}_1,u_2-\overline{u}_2)$$
$$\le C_1(1+e^{2K})\sum_{i=1}^{2}\|u_i-\overline{u}_i\|_1+C_2\sum_{i=1}^{2}e^{K+M}\|\overline{u}_i\|_1\|u_i-\overline{u}_i\|_1,$$

where C_1 depends upon the maximum of k over the square $[0,e^{K+M}]\times[0,e^{K+M}]$. From here we see that $\|u\|_1$ is bounded independently of $\tilde{u}$. Thus S maps the whole space H into a closed ball in $(H^1(\Omega))^3$. This ball is evidently nonempty and convex. From Rellich Theorem 3.9 it is compact in $H=(L^2(\Omega))^3$. By Theorem 10.3, the mapping S has a fixed point, which is a solution of problem (10.14). According to Lemma 10.2, this solution solves problem (10.10) as well. □

10.3. Uniqueness and nonuniqueness

We again investigate problem (10.10) under all assumptions of Section 10.1. The following theorem guarantees the uniqueness of the weak solution provided the forces driving the flow vanish on the boundary Γ_1 (see [Gröger]). The function $\overline{u}_0$, which determines the boundary conditions for the electrostatic potential u_0 on Γ_1, can be arbitrary.

Theorem 10.5. *Let*

$$\operatorname{grad}\overline{u}_1=\operatorname{grad}\overline{u}_2=0 \quad \text{and} \quad \overline{u}_1+\overline{u}_2=0. \tag{10.26}$$

Then the problem (10.10) has a unique solution $u=(u_0,u_1,u_2)$. This solution satisfies the relations $u_i=\overline{u}_i$ for $i=1,2$.

Proof. Let u be a weak solution of (10.10) (cf. Theorem 10.4) and let $v_i=u_i-\overline{u}_i$ for $i=1,2$. Then from (10.8), (10.9) and (10.26) we have

$$0=\mathcal{A}(e;u_1,u_2,u_1-\overline{u}_1,u_2-\overline{u}_2)-\mathcal{F}(e;u_1+u_2,u_1-\overline{u}_1,u_2-\overline{u}_2)$$
$$=\int_\Omega\Big(\sum_{i=1}^{2}\mu_ie_i\|\operatorname{grad}u_i\|^2-k(e)\big(1-e^{u_1+u_2}\big)(u_1+u_2)\Big)\,dx.$$

Both parts of the integrand are nonnegative, since $(1-e^{\xi})\xi\le 0$ for any $\xi\in\mathbf{R}^1$. Hence, $\operatorname{grad}u_i=0$ for $i=1,2$, and thus we obtain $u_i=\overline{u}_i$ for $i=1,2$.

Finally, we prove that there exists at most one component u_0 which satisfies the nonlinear problem: Find $u_0\in H^1(\Omega)$ such that

$$\begin{aligned} &u_0-\overline{u}_0\in V\cap L^\infty(\Omega),\\ &a(u_0,v_0)=\int_\Omega\big(f+e^{\overline{u}_1-u_0}-e^{\overline{u}_2+u_0}\big)v_0\,dx \quad \forall v_0\in V. \end{aligned} \tag{10.27}$$

Assume that two solutions u_0 and u_0' satisfy (10.27). For any $y \in \mathbf{R}^1$ the functions $\xi \mapsto -e^{y-\xi}$ and $\xi \mapsto e^{y+\xi}$ are increasing, and thus (compare (7.10))

$$(-e^{y-\xi_1} + e^{y-\xi_2})(\xi_1 - \xi_2) \geq 0 \quad \forall \xi_1, \xi_2 \in \mathbf{R}^1,$$
$$(e^{y+\xi_1} - e^{y+\xi_2})(\xi_1 - \xi_2) \geq 0 \quad \forall \xi_1, \xi_2 \in \mathbf{R}^1.$$

From here, (10.27) with $v_0 = u_0 - u_0' \in V$ and the ellipticity of the bilinear form $a(\cdot\,,\cdot)$, we get

$$\begin{aligned} 0 &= a(u_0 - u_0', u_0 - u_0') \\ &+ \int_\Omega (-e^{\overline{u}_1 - u_0} + e^{\overline{u}_1 - u_0'} + e^{\overline{u}_2 + u_0} - e^{\overline{u}_2 + u_0'})(u_0 - u_0')\, dx \\ &\geq a(u_0 - u_0', u_0 - u_0') \geq C\|u_0 - u_0'\|_1^2. \end{aligned} \tag{10.28}$$

Hence, $u_0 = u_0'$. □

Other uniqueness theorems can be found, e.g., in [Gajewski], [Markowich], [Mock, 1983], [Polak, den Heijer, Schilders, Markowich]. However, let us point out that the semiconductor equations have more solutions, in general (see [Markowich, p. 43] or [Mock, 1982]), and some semiconductor devices (e.g., thyristor) admit several different stationary states for certain fixed boundary conditions.

10.4. Finite element approximation

Let $\mathcal{T}_h$ be a triangulation of a polyhedral domain $\Omega \in \mathcal{L}$ in $\mathbf{R}^d$, $d = 1, 2, 3$, and let

$$X_h = \{v_h \in H^1(\Omega) \mid v_h|_K \in P_K \quad \forall K \in \mathcal{T}_h\},$$

where P_K is a space of polynomials over K so that $P_K \supset P_1(K)$. Assume that $\mathcal{T}_h$ is consistent with Γ_1 and Γ_2.

According to Remark 3.13 (Part 3), the function $\overline{u} \in (H^1(\Omega) \cap L^\infty(\Omega))^3$ need not be continuous. Therefore, we will assume, in addition, that $\overline{u} \in (C(\overline{\Omega}))^3$ and set

$$\overline{u}_h = \pi_h \overline{u},$$

where $\pi_h \overline{u}$ is the $(X_h)^3$-interpolant of $\overline{u}$.

Introducing the space of test functions

$$V_h = V \cap X_h,$$

we define a finite element approximation of problem (10.10) as follows: Find $u_h = (u_{0h}, u_{1h}, u_{2h})$ such that

$$\begin{aligned} &u_h - \overline{u}_h \in V_h \times V_h \times V_h, \\ &a(u_{0h}, v_{0h}) = \int_\Omega (f + e_{1h} - e_{2h}) v_{0h}\, dx \quad \forall v_{0h} \in V_h, \\ &\mathcal{A}(e_h; \hat{u}_h, v_h) = \mathcal{F}(e_h, u_{1h} + u_{2h}; v_h) \qquad \forall v_h \in V_h \times V_h, \end{aligned} \tag{10.29}$$

where $\hat{u}_h = (u_{1h}, u_{2h})$ and

$$e_h = (e_{1h}, e_{2h}) = (\mathrm{e}^{u_{1h}-u_{0h}}, \mathrm{e}^{u_{2h}+u_{0h}}).$$

Remark 10.6. Unfortunately, we cannot prove the existence of u_h in the same way as the existence of u, since the test functions (compare the proof of Lemma 10.2)

$$v_{0h} = (u_{0h} - M)^+, \qquad v_{ih} = (u_{ih} - K)^+, \quad i = 1, 2,$$

do not belong to V_h, in general.

To construct a finite element solution satisfying (10.29), one could proceed in the following way. Let $H_h = (X_h)^3$ and let an initial approximation $\tilde{u}_h = (\tilde{u}_{0h}, \tilde{u}_{1h}, \tilde{u}_{2h}) \in H_h$ be given. Put $\tilde{e}_h = (\tilde{e}_{1h}, \tilde{e}_{2h})$, where

$$\tilde{e}_{1h} = \mathrm{e}^{\tilde{u}_{1h}-\tilde{u}_{0h}}, \qquad \tilde{e}_{2h} = \mathrm{e}^{\tilde{u}_{2h}+\tilde{u}_{0h}},$$

and determine $u_{0h} \in X_h$ as the unique solution of the linear problem

$$\begin{aligned} &u_{0h} - \overline{u}_{0h} \in V_h, \\ &a(u_{0h}, v_{0h}) = \int_\Omega (f + \tilde{e}_{1h} - \tilde{e}_{2h}) v_{0h}\, dx \quad \forall v_{0h} \in V_h. \end{aligned} \tag{10.30}$$

Next define $(u_{1h}, u_{2h}) \in X_h \times X_h$ as the unique solution of another linear problem

$$\begin{aligned} u_{ih} - \overline{u}_{ih} &\in V_h \qquad\qquad\qquad\qquad \text{for } i = 1, 2, \\ \mathcal{A}(\tilde{e}_h; u_{1h}, u_{2h}, v_h) &= \mathcal{F}(\tilde{e}_h, \tilde{u}_{1h} + \tilde{u}_{2h}; v_h) \quad \forall v_h \in V_h \times V_h. \end{aligned} \tag{10.31}$$

Now, define the operator $S_h\colon H_h \to H_h$ by

$$S_h \tilde{u}_h = u_h,$$

where $u_h = (u_{0h}, u_{1h}, u_{2h})$ is determined by (10.30) and (10.31). If S_h has a fixed point then, of course, this point is a solution of (10.29).

The existence of a finite element solution of a problem similar to (10.29) is proved in [Zlámal, 1986] provided an acute type condition is fulfilled. This condition requires that there are no obtuse angles in the triangulation, and implies the discrete maximum principle (see also [Křížek, Lin]). Some convergence analysis is done in [Markowich, Section 5.3].

Theorem 10.7. *Let* $\operatorname{grad} \overline{u}_{1h} = \operatorname{grad} \overline{u}_{2h} = 0$ *and* $\overline{u}_{1h} + \overline{u}_{2h} = 0$. *Then the problem (10.29) has at most one solution* $u_h = (u_{0h}, u_{1h}, u_{2h})$ *and this solution, if it exists, satisfies the relations* $u_{ih} = \overline{u}_{ih}$ *for* $i = 1, 2$.

Proof. Assume that u_h solves (10.29). Analogously to the proof of Theorem 10.5 we find that

$$\begin{aligned} 0 &= \mathcal{A}(e_h; u_{1h}, u_{2h}, u_{1h} - \overline{u}_{1h}, u_{2h} - \overline{u}_{2h}) - \mathcal{F}(e_h; u_{1h} + u_{2h}, u_{1h} - \overline{u}_{1h}, u_{2h} - \overline{u}_{2h}) \\ &= \int_\Omega \Big(\sum_{i=1}^2 \mu_i e_{ih} \|\operatorname{grad} u_{ih}\|^2 - k(e_h)(1 - \mathrm{e}^{u_{1h}+u_{2h}})(u_{1h} + u_{2h}) \Big)\, dx. \end{aligned}$$

Since both the parts of the integrand are nonnegative, we have $\operatorname{grad} u_{ih} = 0$ for $i = 1, 2$. Therefore, $u_{ih} = \overline{u}_{ih}$ for $i = 1, 2$.

Consider the problem: Find $u_{0h} \in X_h$ such that $u_{0h} - \overline{u}_{0h} \in V_h$ and

$$a(u_{0h}, v_{0h}) = \int_\Omega (f + \mathrm{e}^{\overline{u}_{1h} - u_{0h}} - \mathrm{e}^{\overline{u}_{2h} + u_{0h}}) v_{0h} \, dx \quad \forall v_{0h} \in V_h.$$

Similarly to (10.28) we can prove that there exists at most one u_{0h}. □

Remark 10.8. The uniqueness of a discrete solution of another but similar problem is proved in [Zlámal, 1986].

Example 10.9. Let us consider the system (10.1)–(10.3) in the case $d = 1$, $\Omega = (-r, r)$ with $r = 0.001$, $\varepsilon = 10^{-7}$, $k = 0$ and μ_i constant. Then we have

$$\begin{aligned} -\varepsilon u_0'' &= f + e_1 - e_2, \\ (e_1 u_1')' &= 0, \\ (e_2 u_2')' &= 0, \end{aligned} \tag{10.32}$$

where e_1 and e_2 are given by (10.4). Suppose that $f(x) = -3 \operatorname{sign}(x)/2$ and let us prescribe the following Dirichlet boundary conditions

$$\begin{aligned} u_0(-r) &= 1, & u_0(r) &= -1, \\ u_1(-r) &= 0.5, & u_1(r) &= 0, \\ u_2(-r) &= 0, & u_2(r) &= 0.5. \end{aligned} \tag{10.33}$$

The boundary value problem (10.32)–(10.33) serves as a simple model of a one-dimensional PN-diode of Figure 10.1 (a similar example is treated in [Polak, den Heijer, Schilders, Markowich, p. 833]). Figure 10.2 illustrates the corresponding electrostatic potential u_0 and the electrochemical potentials u_2 and u_3 computed by the method of Remark 10.6. Due to the interior layer of u_0, the construction of a triangulation near the P–N junction should be done with a particular care.

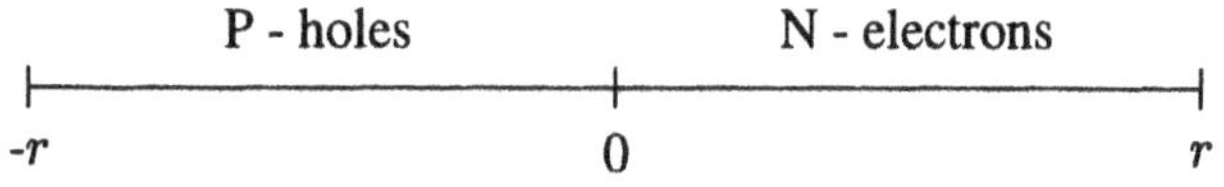

Figure 10.1.

Remark 10.10. The matrix form of the problem (10.29) requires us to evaluate the integrals

$$\int_K \mathrm{e}^{p(x)} \, dx. \tag{10.34}$$

If $d = 2$ and p is a linear function over a triangle $K \in \mathcal{T}_h$, we can employ a useful formula from [Pospíšek]. Denote by p_1, p_2 and p_3, the values of p at the three

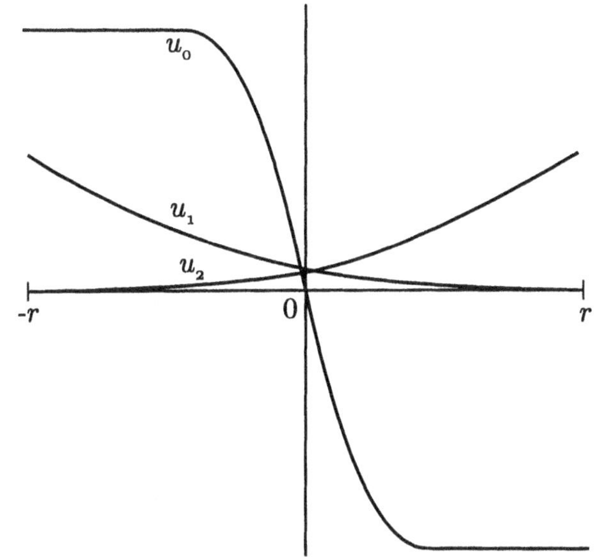

Figure 10.2.

vertices of K. If, for instance, $p_1 \neq p_2 \neq p_3 \neq p_1$, then after some calculation we have

$$\int_K e^{p(x)}\,dx = \frac{2\,\mathrm{meas}\,K}{p_2 - p_1}\left(\frac{e^{p_3} - e^{p_2}}{p_3 - p_2} - \frac{e^{p_3} - e^{p_1}}{p_3 - p_1}\right).$$

Hence, no numerical integration is necessary to evaluate the integral (10.34).

Example 10.11. Let Ω be a square domain. The position of the sets Γ_1 and Γ_2 is marked in Figure 10.3. Since the permittivity ε occurring in (10.1) is a very small number ($\varepsilon \approx 10^{-10}$), the elliptic character of the equation (10.1) is almost negligible. This makes numerical implementation of problem (10.1)–(10.6) difficult. Figure 10.4 illustrates the triangulation used here which respects the P–N junction. In Figure 10.5, we see contour lines of the electrostatic potential u_0 of a two-dimensional model problem computed by M. Pospíšek.

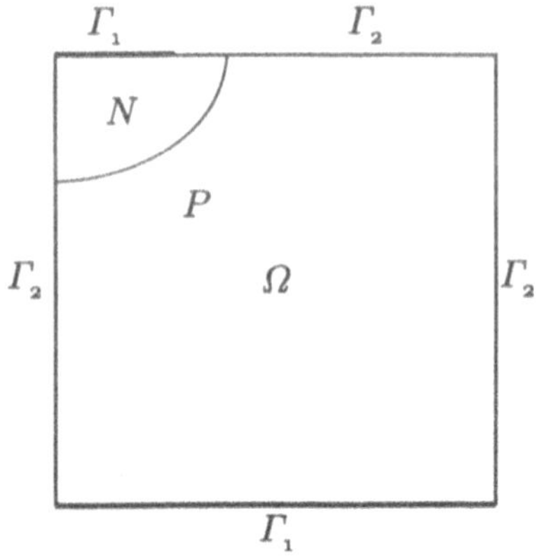

Figure 10.3.

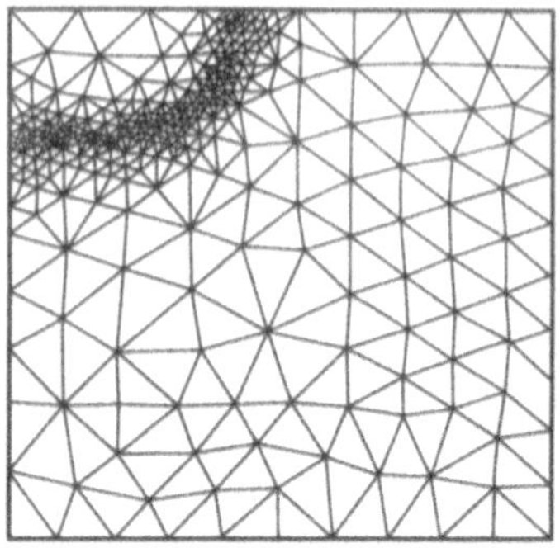

Figure 10.4.

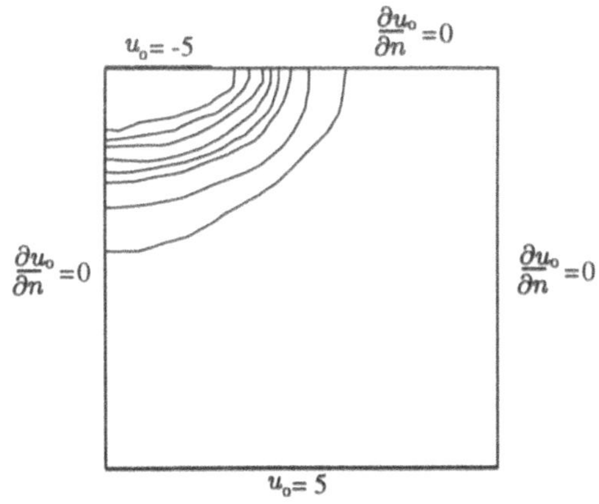

Figure 10.5.

11. Nonstationary heat conduction in a stator

11.1. Classical and weak formulations

In this chapter we describe a method for computing the $3d$ nonstationary temperature field in the lamination pack of a stator of a synchronous or asynchronous (induction) motor with a centrifugal, meander or chamber ventilation (see [Křížek, Preiningerová]). The stator of a motor has quite a complicated geometrical form. Moreover, it consists of anisotropic materials which have very different heat conductivities, e.g., 332.8 [W/mK] for copper wires and 0.2 [W/mK] for their insulations. This causes big jumps in coefficients of the appropriate heat conduction equation, and is the main source of numerical difficulties in practical calculations.

The stator of a large motor is usually divided by radial ventilation channels into several packs (see Figure 11.1).

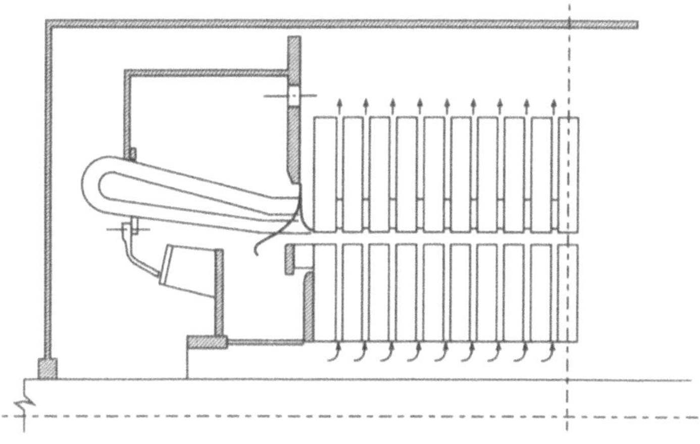

Figure 11.1.

A cross section of such a pack is illustrated in Figure 11.2.

Due to the "symmetry", we can restrict our considerations to a smaller part of the pack which is sketched in Figure 11.3. Denote this domain by Ω.

A nonstationary heat conduction problem in Ω is described by a partial differential equation of a parabolic type with initial-boundary conditions. The heat conduction equation is given by (see (2.10))

$$c\rho\frac{\partial u}{\partial t} - \operatorname{div}(\mathcal{A}\operatorname{grad} u) = f \quad \text{in } \Omega \times I, \tag{11.1}$$

where $t \in I = (0, T)$ is the time variable, $T > 0$, $\rho = \rho(x) > 0$ is the specific mass, $c = c(x) > 0$ is the specific heat capacity, $u = u(x, t)$ is the temperature,

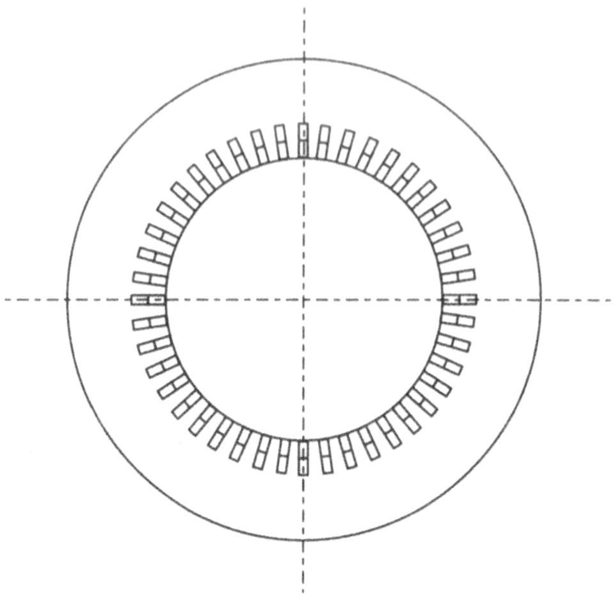

Figure 11.2.

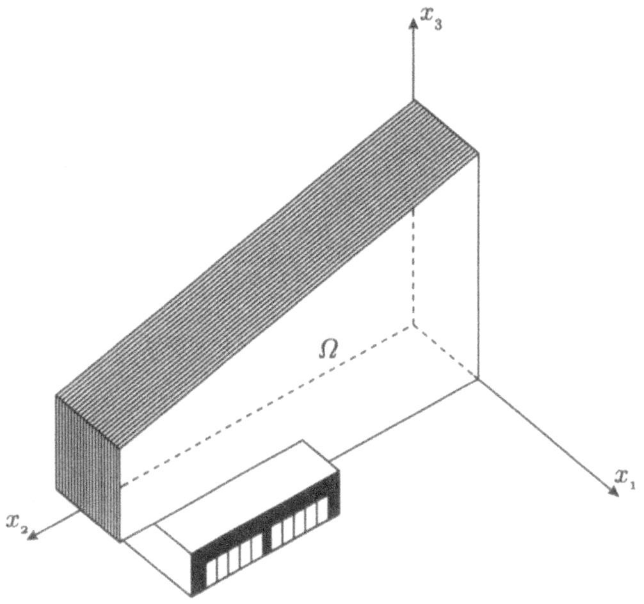

Figure 11.3.

$\mathcal{A} = (a_{ij}(x))_{i,j=1}^3$ is a symmetric and uniformly positive definite matrix of heat conductivities and $f = f(x,t)$ is the density of volume heat sources.

On the boundary $\partial\Omega$, we consider the Newton boundary condition which represents a heat transfer from Ω to its neighbourhood,

$$\alpha u + n^T \mathcal{A} \operatorname{grad} u = g \quad \text{on } \partial\Omega \times I. \tag{11.2}$$

Here $\alpha = \alpha(s) \geq 0$ ($s \in \partial\Omega$) is the heat transfer coefficient, n is the outward unit normal to Ω, the cogradient $\mathcal{A} \operatorname{grad} u$ represents the heat flux, $n^T \mathcal{A} \operatorname{grad} u$ is called the normal component of the heat flux, and $g = g(s,t)$ is the density of surface heat sources. The coefficient α is equal to zero at those parts of $\partial\Omega$, which correspond to a plane of symmetry. Otherwise it is positive.

Finally, we consider the initial condition

$$u(x,0) = u^0(x), \quad x \in \Omega, \tag{11.3}$$

where u^0 represents the temperature in Ω in time $t = 0$.

To introduce a weak formulation of the problem (11.1)–(11.3), we suppose that $c \in L^\infty(\Omega)$, $\rho \in L^\infty(\Omega)$, $c\rho \geq C > 0$, $a_{ij} \in L^\infty(\Omega)$, $\alpha \in L^\infty(\partial\Omega)$, $u^0 \in H^1(\Omega) \cap C(\overline{\Omega})$, $f \in C(\overline{I}, L^2(\Omega))$ and $g \in C(\overline{I}, L^2(\partial\Omega))$, where the symbol $C(\overline{I}, L^2(\cdot))$ stands for the space of all continuous mappings from $\overline{I}$ into $L^2(\cdot)$. Let

$$V = H^1(\Omega)$$

be the space of test functions. If the classical solution $u = u(x,t)$ of (11.1)–(11.3) is sufficiently smooth, then for $u = u(\cdot\,,t)$, $t \in I$ fixed, we get (cf. Section 3.4 or 9.1) that

$$(c\rho v, u'(t))_0 + a(v,u) = F(t;v) \quad \forall v \in V, \tag{11.4}$$

where $u' = \partial u/\partial t$ is the time derivative,

$$a(v,w) = (\mathcal{A} \operatorname{grad} v, \operatorname{grad} w)_{0,\Omega} + \langle \alpha v, w \rangle_{0,\partial\Omega}$$

and

$$F(t;v) = (f(\cdot\,,t), v)_{0,\Omega} + \langle g(\cdot\,,t), v \rangle_{0,\partial\Omega}.$$

A weak formulation of the classical problem (11.1)–(11.3) consists of finding $u \in H^1(I,V)$ (a weak solution) which satisfies (11.4) for a.e. fixed $t \in I$ and (11.3) for a.e. $x \in \Omega$. Here $H^1(I,V)$ denotes the space of all the mappings $t \in I \mapsto v(t) \in V$ such that the function $t \mapsto \|v(t)\|_{1,\Omega}$ is from $H^1(I)$. For the existence and uniqueness of the weak solution u we refer, e.g., to [Lions, Magenes, Chapter 3], [Rektorys, 1982, Chap. 11].

11.2. Finite element analysis

We divide $\overline{\Omega}$ into trihedral and quadrihedral prismatic elements (see Figure 11.4), which makes it possible to respect in detail the geometry of the stator, thermophysical properties of all materials used, and the spatial distribution of the

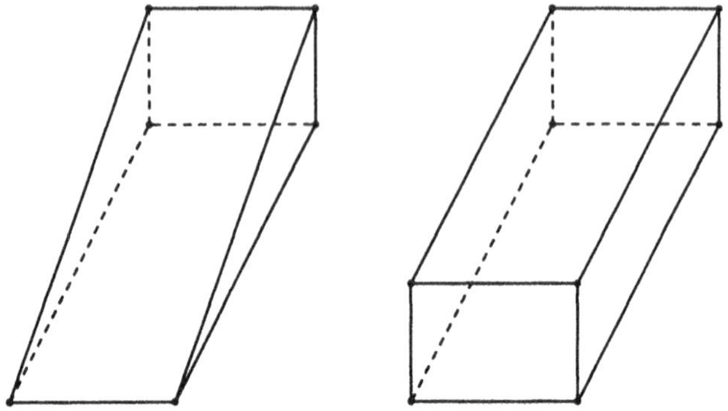

Figure 11.4.

density of heat losses and flowpaths of the cooling air. In particular, the functions c, ρ, a_{ij} are assumed to be constant on each element.

The finite element space consists of all continuous piecewise polynomial functions from V, which have the form

$$c_0 + c_1x_1 + c_2x_2 + c_3x_3 + c_4x_1x_2 + c_5x_1x_3$$

on trihedral prismatic elements and

$$c_0 + c_1x_1 + c_2x_2 + c_3x_3 + c_4x_1x_2 + c_5x_1x_3 + c_6x_2x_3 + c_7x_1x_2x_3$$

on quadrihedral prismatic elements. Of course, the finite element analysis described next can also be done for other elements. So until the end of this section we assume that $d \in \{1, 2, ...\}$, $\mathcal{T}_h$ is a triangulation of a polyhedron (polygon) $\overline{\Omega} \subset \mathbf{R}^d$ and let

$$V_h = \{v_h \in V \mid v_h|_K \in P_K \quad \forall K \in \mathcal{T}_h\},$$

where $P_1(K) \subset P_K$ for all $K \in \mathcal{T}_h$.

The semidiscrete Galerkin approximation of equation (11.4) consists of finding $u_h = u_h(\cdot, t)$ which belongs to V_h for every $t \in I$ and satisfies

$$(c\rho v_h, u_h'(t))_0 + a(v_h, u_h) = F(t; v_h) \quad \forall v_h \in V_h \tag{11.5}$$

for a.e. $t \in I$. This means that we have performed a discretization only in the space variables while the time variable is still continuous. The initial condition (11.3) can be approximated for instance as follows

$$(c\rho u_h(0), v_h)_0 = (c\rho u^0, v_h)_0 \quad \forall v_h \in V_h. \tag{11.6}$$

We shall consider only the Lagrange elements. Let $z_1, ..., z_N$ be the nodal points of the elements of the space discretization V_h, i.e., $N = \dim V_h$. Then the functions $v^i \in V_h$ such that

$$v^i(z_j) = \delta_{ij} \quad \text{for } i, j = 1, ..., N$$

are basis functions in V_h. Looking for u_h in the form

$$u_h(x,t) = \sum_{i=1}^{N} z_i(t)v^i(x), \quad x \in \Omega,\ t \in I,$$

we get by (11.5) a system of ordinary differential equations of the first order

$$\sum_{j=1}^{N}(c\rho v^i, v^j)_0 z_j'(t) + \sum_{j=1}^{N} a(v^i, v^j)z_j(t) = F(t; v^j), \quad i = 1, ..., N, \tag{11.7}$$

for unknown functions $z_1, ..., z_N$. The corresponding initial conditions for $z_j = z_j(t)$ can be determined by solving the system

$$\sum_{j=1}^{n}(c\rho v^i, v^j)_0 z_j(0) = (c\rho v^i, u^0)_0, \quad i = 1, ..., N, \tag{11.8}$$

where u^0 is the initial condition from the classical formulation (see (11.3)). Denoting $z(t) = (z_1(t), ..., z_N(t))^T$ the vector of unknowns, $M = ((c\rho v^i, v^j)_0)_{i,j=1}^N$ the mass matrix, $A = (a(v^i, v^j))_{i,j=1}^N$ the stiffness matrix, $\mathcal{F}(t) = (F(t; v^1), ..., F(t; v^N))^T$ the right-hand side and $z^0 = ((c\rho v^1, u^0)_0, ..., (c\rho v^N, u^0)_0)^T$ the initial condition, the problem (11.7)–(11.8) may equivalently be written in the matrix form

$$Mz'(t) + Az(t) = \mathcal{F}(t), \quad t \in I, \tag{11.9}$$

$$Mz(0) = z^0. \tag{11.10}$$

Remark 11.1. It is easy to see that the mass matrix M is symmetric and positive definite. Hence, by the Choleski factorization (see, e.g., [Křížek, Neittaanmäki, 1990, p. 89]) there exists a lower triangular matrix L such that

$$M = LL^T.$$

Introducing a new variable $y(t) = L^T z(t)$, the system (11.9)–(11.10) can be equivalently rewritten as the following initial problem for ordinary differential equations

$$y'(t) + By(t) = q(t), \quad t \in I, \tag{11.11}$$

$$y(0) = y^0, \tag{11.12}$$

where $B = L^{-1}AL^{-T}$ is a symmetric positive definite matrix, $L^{-T} = (L^{-1})^T$, $q = L^{-1}\mathcal{F}$, and $y^0 = L^{-1}z^0$. According to [Coddington, Levinson, Section III.4], the unique solution of (11.11)–(11.12) is explicitly given by

$$y(t) = \mathrm{e}^{-Bt}y^0 + \int_0^t \mathrm{e}^{-B(t-\tau)}q(\tau)\,d\tau, \quad t \in I, \tag{11.13}$$

since the system (11.11)–(11.12) is linear and the matrix B is constant (independent of time). Note that e^{-Bt} is defined through the Taylor expansion

$$\mathrm{e}^{-Bt} = I - tB + \tfrac{1}{2}t^2B^2 - \tfrac{1}{6}t^3B^3 + \cdots,$$

where I is the identity matrix.

Since the problem (11.5)–(11.6) is equivalent to the problem (11.11)–(11.12), we have the following corollary:

Corollary 11.2. *The problem (11.5)–(11.6) has a unique solution.*

Lemma 11.3. *Let $\{\mathcal{T}_h\}$ be a strongly regular family of triangulations. Then there exist positive constants C_1 and C_2 such that for any function*

$$v_h = \sum_{i=1}^{N} \eta_i v^i \in V_h$$

we have

$$C_1 h^d \|\eta\|^2 \leq \|v_h\|_0^2 \leq C_2 h^d \|\eta\|^2 \quad \text{for } \eta = (\eta_1, ..., \eta_N).$$

Proof. Let A_i, $i = 1, ..., r$, be nodal points of some element $K \in \mathcal{T}_h$. Then

$$C_1 h_K^d \sum_{i=1}^{r} |v_h(A_i)|^2 \leq \|v_h\|_{0,K}^2 \leq C_2 h_K^d \sum_{i=1}^{r} |v_h(A_i)|^2,$$

where $h_K = \operatorname{diam} K$. The desired estimate now follows directly by summation over all $K \in \mathcal{T}_h$. □

Remark 11.4. We can solve (11.9)–(11.10) or (11.11)–(11.12) using standard numerical methods for stiff initial value problems, since the eigenvalues of B satisfy $\lambda_{\min}(B) \ll \lambda_{\max}(B)$ for h small. Next we prove that the condition number of B is equal to $\kappa(B) = \mathcal{O}(h^{-2})$ provided the family $\{\mathcal{T}_h\}$ of triangulations is strongly regular. To see this we will first investigate the behaviour of $\kappa(A)$ and $\kappa(M)$ for $h \to 0$.

For the spectral matrix norm, which is induced by the Euclidean vector norm $\|\cdot\|$, we shall use the same symbol $\|\cdot\|$. Recall that

$$\|A\| = \max_{\eta \neq 0} \frac{\|A\eta\|}{\|\eta\|} = \max_{\eta \neq 0} \frac{\eta^T A \eta}{\|\eta\|^2} = \lambda_{\max}(A),$$

since A is symmetric and positive definite.

The strong regularity of $\{\mathcal{T}_h\}$ implies that (see [Ciarlet, p. 142])

$$\|v_h\|_1 \leq C h^{-1} \|v_h\|_0 \quad \forall v_h \in V_h.$$

This inequality is called the ***inverse inequality*** for finite element spaces. By Trace Theorem 3.8 and the inverse inequality, we obtain

$$\begin{aligned} a(v_h, v_h) &= (\mathcal{A} \operatorname{grad} v_h, \operatorname{grad} v_h)_{0,\Omega} + \langle \alpha v_h, v_h \rangle_{0,\partial\Omega} \\ &\leq C_1 |v_h|_{1,\Omega}^2 + C_2 \|v_h\|_{0,\partial\Omega}^2 \leq C' \|v_h\|_1^2 \leq C h^{-2} \|v_h\|_0^2 \qquad \forall v_h \in V_h. \end{aligned}$$

From here we get for

$$v_h = \sum_{i=1}^{N} \eta_i v^i$$

the following relation for the stiffness matrix A,

$$\frac{\eta^T A\eta}{\|\eta\|^2} = \frac{a(v_h, v_h)}{\|\eta\|^2} \le Ch^{-2}\frac{\|v_h\|_0^2}{\|\eta\|^2} \le C_2 h^{d-2} \quad \forall \eta = (\eta_1, ..., \eta_N)^T \in \mathbf{R}^N \setminus \{0\},$$

where the last inequality follows from Lemma 11.3.

Moreover, from the V-ellipticity of $a(\cdot\,,\cdot)$ and Lemma 11.3,

$$\frac{\eta^T A\eta}{\|\eta\|^2} = \frac{a(v_h, v_h)}{\|\eta\|^2} \ge C\frac{\|v_h\|_0^2}{\|\eta\|^2} \ge C_1 h^d \quad \forall \eta \ne 0.$$

Therefore,

$$C_1 h^d \le \lambda_{\min}(A) \le \|A\| = \lambda_{\max}(A) \le C_2 h^{d-2}$$

for h sufficiently small. This implies that A is ill-conditioned,

$$\kappa(A) = \frac{\lambda_{\max}(A)}{\lambda_{\min}(A)} = \mathcal{O}(h^{-2}).$$

Notice that the right-hand side is independent of the dimension d.

For the mass matrix M, we have by Lemma 11.3,

$$\frac{\eta^T M\eta}{\|\eta\|^2} = \frac{(c\rho v_h, v_h)_0}{\|\eta\|^2} \le C\frac{\|v_h\|_0^2}{\|\eta\|^2} \le C_2 h^d \quad \forall \eta \ne 0,$$

and similarly we find that

$$\frac{\eta^T M\eta}{\|\eta\|^2} \ge C_1 h^d \quad \forall \eta \ne 0.$$

Hence, there again exist positive constants C_1 and C_2 such that

$$C_1 h^d \le \lambda_{\min}(M) \le \|M\| = \lambda_{\max}(M) \le C_2 h^d$$

and thus

$$\kappa(M) = \frac{\lambda_{\max}(M)}{\lambda_{\min}(M)} = \mathcal{O}(1).$$

For the inverse matrix we have

$$\|M^{-1}\| = \left(\min_{\eta\ne 0}\frac{\eta^T M\eta}{\|\eta\|^2}\right)^{-1} = \mathcal{O}(h^{-d}).$$

Since the similarity transformation LBL^{-1} does not change the eigenvalues of the matrix B, we obtain the desired relation

$$\begin{aligned}\|B\| &= \|LBL^{-1}\| = \|LL^{-1}AL^{-T}L^{-1}\| \\ &\le \|A\|\,\|L^{-T}L^{-1}\| = \|A\|\|M^{-1}\| = \mathcal{O}(h^{-2}).\end{aligned}$$

Before we prove convergence of the approximate solutions u_h, we introduce one lemma.

Lemma 11.5. *Let $\mathcal{I}$ be an arbitrary index set and let B_t, $t \in \mathcal{I}$, be a set of symmetric positive semidefinite matrices. Then there exists a constant C such that*

$$\|B_t e^{-B_t}\| \leq C$$

holds for any $t \in \mathcal{I}$.

Proof. The eigenvalues of the symmetric matrix $B_t e^{-B_t}$ are of the form $\lambda_t e^{-\lambda_t}$, where λ_t is an eigenvalue of the matrix B_t. Since B_t is positive semidefinite, we have $\lambda_t \in [0, \infty)$. The real function $\lambda e^{-\lambda}$ of the real variable $\lambda \in [0, \infty)$, however, attains its maximum at $\lambda = 1$. This yields the statement of the lemma. □

The next theorem establishes the rate of convergence of the Galerkin solution u_h given by (11.5)–(11.6) to the weak solution u. Its proof has arisen by generalization of some tricks of [Johnson] and [Vitásek] (see also [Thomée, 1984], [Wheeler]).

Theorem 11.6. *Let $\{\mathcal{T}_h\}$ be a strongly regular family of triangulations of $\overline{\Omega} \subset \mathbf{R}^d$, $d = 1, 2, 3$, and let $u(t) \in H^2(\Omega)$ for any $t \in I$. Let the weak solution z of the problem*

$$a(z, v) = (\theta, v)_0 \quad \forall v \in V$$

belong to $H^2(\Omega)$ for any $\theta \in L^2(\Omega)$ and let

$$|z|_2 \leq \overline{C}\|\theta\|_0.$$

Then there exists a constant $C > 0$ such that

$$\max_{t \in I} \|u(t) - u_h(t)\|_0 \leq C h^2 \left(1 + \left|\ln \frac{T}{h^2}\right|\right) \max_{t \in I} |u(t)|_2.$$

Proof. We introduce the following dual problem for the unknown function $\varphi_h : (0, t) \to V_h$, $t \in I$,

$$-(c\rho v_h, \varphi_h'(\tau))_0 + a(v_h, \varphi_h(\tau)) = 0 \qquad \forall v_h \in V_h,\ \tau \in (0, t), \quad (11.14)$$

$$\varphi_h(t) = e_h(t), \quad (11.15)$$

where $e_h(\tau) = u_h(\tau) - \tilde{u}_h(\tau)$. The so-called elliptic projection $\tilde{u}_h(\tau) \in V_h$ of the true solution $u(\tau)$ satisfies the orthogonality relation

$$a(v_h, u(\tau) - \tilde{u}_h(\tau)) = 0 \quad \forall v_h \in V_h,\ \tau \in I. \quad (11.16)$$

For simplicity we set

$$\theta(\tau) = u(\tau) - \tilde{u}_h(\tau).$$

Employing (11.4), (11.5), (11.6) and (11.3), we find that

$$(c\rho\varphi_h, u_h')_0 + a(\varphi_h, u_h) = (c\rho\varphi_h, u')_0 + a(\varphi_h, u),$$

$$(c\rho u_h(0), \varphi_h(0))_0 = (c\rho u(0), \varphi_h(0))_0.$$

From here and the definitions of e_h and θ we get

$$(c\rho\varphi_h, e_h')_0 + a(\varphi_h, e_h) = (c\rho\varphi_h, \theta')_0 + a(\varphi_h, \theta), \tag{11.17}$$
$$(c\rho e_h(0), \varphi_h(0))_0 = (c\rho\theta(0), \varphi_h(0))_0. \tag{11.18}$$

Put $v_h = e_h(\tau)$ in (11.14) and (11.16). Integrating by parts in $(0,t)$, using the symmetry of $a(\cdot\,,\cdot)$, (11.14), (11.15), (11.17) and (11.18), and then again integrating by parts and using (11.16), we find that

$$\begin{aligned}(c\rho e_h(t), e_h(t))_0 &= \int_0^t [-(c\rho e_h(\tau), \varphi_h'(\tau))_0 + a(e_h(\tau), \varphi_h(\tau))]\, d\tau \\ &\quad + (c\rho e_h(t), \varphi_h(t))_0 \\ &= \int_0^t [(c\rho\varphi_h, e_h')_0 + a(\varphi_h, e_h)]\, d\tau + (c\rho e_h(0), \varphi_h(0))_0 \\ &= \int_0^t [(c\rho\varphi_h, \theta')_0 + a(\varphi_h, \theta)]\, d\tau + (c\rho\theta(0), \varphi_h(0))_0 \\ &= -\int_0^t (c\rho\theta, \varphi_h')_0\, d\tau + (c\rho\theta(t), \varphi_h(t))_0.\end{aligned} \tag{11.19}$$

Next, we derive some bounds for φ_h and its time derivative (cf. (11.28) and (11.29)). Analogously to Remark 11.1, problem (11.14)–(11.15) is equivalent to the system of homogeneous ordinary differential equations

$$\begin{aligned}-y'(\tau) + By(\tau) &= 0, \quad \tau \in (0,t), \\ y(t) &= y^0,\end{aligned} \tag{11.20}$$

where $B = L^{-1}AL^{-T}$ and $M = LL^T$. The solution of (11.20) is given by (cf. (11.13))

$$y(\tau) = \mathrm{e}^{-(t-\tau)B} y(t). \tag{11.21}$$

The matrix $\mathrm{e}^{-(t-\tau)B}$ is symmetric and its eigenvalues are equal to $\mathrm{e}^{-(t-\tau)\lambda_i}$, $i = 1, ..., N$, where λ_i are the eigenvalues of B. Since B is symmetric and positive definite, we have $\lambda_i > 0$, and thus all eigenvalues of the matrix $\mathrm{e}^{-(t-\tau)B}$ are not greater than one for $\tau \in \langle 0, t\rangle$. Consequently

$$\|\mathrm{e}^{-(t-\tau)B}\| \le 1,$$

and by (11.21) we arrive at

$$\|y(\tau)\| \le \|y(t)\| \quad \forall \tau \in (0,t). \tag{11.22}$$

Differentiating (11.21) with respect to τ, we obtain

$$y'(\tau) = B\mathrm{e}^{-(t-\tau)B} y(t) = \frac{1}{t-\tau}(t-\tau)B\mathrm{e}^{-(t-\tau)B} y(t).$$

From here and Lemma 11.5 we see that

$$\|y'(\tau)\| \le \frac{C}{t-\tau}\|y(t)\|. \tag{11.23}$$

Let us write

$$\int_0^t \|y'(\tau)\|\,d\tau = \int_0^{t-h^2} \|y'(\tau)\|\,d\tau + \int_{t-h^2}^t \|y'(\tau)\|\,d\tau. \tag{11.24}$$

The first term on the right-hand side can be estimated with the help of (11.23),

$$\int_0^{t-h^2} \|y'(\tau)\|\,d\tau \le C\|y(t)\|\left|\int_0^{t-h^2} \frac{d\tau}{t-\tau}\right| = C\|y(t)\|\left|\ln\frac{t}{h^2}\right|. \tag{11.25}$$

To estimate the second term on the right-hand side of (11.24) we use (11.20) and (11.22),

$$\int_{t-h^2}^t \|y'(\tau)\|\,d\tau = \int_{t-h^2}^t \|By(\tau)\|\,d\tau \le h^2\|B\|\|y(t)\| \le C\|y(t)\|, \tag{11.26}$$

where the last inequality follows from the fact that $\|B\| = \mathcal{O}(h^{-2})$, see Remark 11.4. Substituting (11.25) and (11.26) into (11.24), we get

$$\int_0^t \|y'(\tau)\|\,d\tau \le C\|y(t)\|\left(1+\left|\ln\frac{t}{h^2}\right|\right). \tag{11.27}$$

From Lemma 11.3 and (11.22) we have

$$\begin{aligned} C_1h^d\|y(t)\|^2 &\le \|\varphi_h(t)\|_0^2 \le C_2h^d\|y(t)\|^2, \\ C_1h^d\|y'(\tau)\|^2 &\le \|\varphi_h'(\tau)\|_0^2 \le C_2h^d\|y'(\tau)\|^2. \end{aligned} \tag{11.28}$$

Using (11.28) and a derivation like that used to prove (11.27), we obtain

$$\int_0^t \|\varphi_h'(\tau)\|_0\,d\tau \le C\|\varphi_h(t)\|_0\left(1+\left|\ln\frac{t}{h^2}\right|\right).$$

Therefore, by (11.15),

$$\int_0^t \|\varphi_h'(\tau)\|_0\,d\tau \le C\|e_h(t)\|_0\left(1+\left|\ln\frac{t}{h^2}\right|\right). \tag{11.29}$$

Now, we are almost finished. From (11.19) we find by the Cauchy–Schwarz inequality that

$$\begin{aligned} \|e_h(t)\|_0^2 &\le C(c\rho e_h(t), e_h(t))_0 \\ &\le C_1\|\theta(t)\|_0\|e_h(t)\|_0 + C_2\int_0^t \|\theta(\tau)\|_0\|\varphi_h'(\tau)\|_0\,d\tau. \end{aligned}$$

Hence, by (11.29),

$$\|e_h(t)\|_0^2 \le C_1\|e_h(t)\|_0\left(1 + C\left(1 + \left|\ln\frac{t}{h^2}\right|\right)\right)\max_{\tau\in[0,t]}\|\theta(\tau)\|_0$$

and thus

$$\|e_h(t)\|_0 \le C\left(1 + \left|\ln\frac{t}{h^2}\right|\right)\max_{\tau\in[0,t]}\|\theta(\tau)\|_0. \tag{11.30}$$

Writing the discretization error $u - u_h$ in the form

$$u - u_h = u - \tilde{u}_h + \tilde{u}_h - u_h = \theta - e_h$$

and using (11.30), we obtain

$$\|u(t) - u_h(t)\|_0 \le C\left(1 + \left|\ln\frac{t}{h^2}\right|\right)\max_{\tau\in[0,t]}\|\theta(\tau)\|_0. \tag{11.31}$$

Now, it is sufficient to estimate the norm of θ. Define $z \in V$ by the equation

$$a(z, v) = (\theta, v)_0 \quad \forall v \in V. \tag{11.32}$$

Then, by the assumptions of the theorem, we have $z \in H^2(\Omega)$ and

$$|z|_2 \le C\|\theta\|_0. \tag{11.33}$$

Let $\pi_h z \in V_h$ be the V_h-interpolant of z. It is well-defined due to the Sobolev Imbedding Theorem 3.11 for $d \le 3$. Using the standard approximation properties of π_h (see [Ciarlet, p. 124]) and (11.33), we get

$$\|z - \pi_h z\|_1 \le Ch|z|_2 \le C'h\|\theta\|_0. \tag{11.34}$$

Furthermore, since $\theta = u - \tilde{u}_h$, we have by (11.16) that

$$\|\theta\|_1 = \|u - \tilde{u}_h\|_1 \le Ch|u|_2. \tag{11.35}$$

Hence, $\theta \in V = H^1(\Omega)$ may be used as a test function in (11.32), and by (11.16) we see that

$$a(\pi_h z, \theta) = 0. \tag{11.36}$$

Employing (11.32), (11.36), the V-ellipticity of $a(\cdot\,,\cdot)$, (11.34) and (11.35), we obtain

$$\begin{aligned} \|\theta\|_0^2 &= (\theta, \theta)_0 = a(z, \theta) = a(z - \pi_h z, \theta) \\ &\le C_1\|z - \pi_h z\|_1\|\theta\|_1 \le C_2 h\|\theta\|_0 h|u|_2. \end{aligned} \tag{11.37}$$

This inequality and (11.31) lead to the assertion of the theorem. □

Remark 11.7. The H^2-regularity assumptions of Theorem 11.6 are not satisfied in the stator problem, since the associated thermophysical coefficients are discontinuous (piecewise constant). Using fractional Sobolev spaces, we can prove the convergence of u_h to u under less regularity assumptions. So let $\varepsilon \in (0,1)$ and let $u(t) \in H^{1+\varepsilon}(\Omega)$ for any $t \in I$. Suppose that the weak solution z of (11.32) satisfies the inequality (cf. (11.33))

$$|z|_{1+\varepsilon} \leq C\|\theta\|_0. \tag{11.38}$$

Then by the approximation properties of π_h (see [Feistauer, 1989]) we have

$$\|z - \pi_h z\|_1 \leq Ch^\varepsilon |z|_{1+\varepsilon} \leq C' h^\varepsilon \|\theta\|_0$$

(cf. (11.34)), and thus (11.37) may be rewritten as

$$\|\theta\|_0^2 \leq C_1 \|z - \pi_h z\|_1 \|\theta\|_1 \leq C_2 h^\varepsilon \|\theta\|_0 h^\varepsilon |u|_{1+\varepsilon}.$$

From here and (11.31) we finally get

$$\max_{t\in I} \|u(t) - u_h(t)\|_0 \leq Ch^{2\varepsilon}\left(1 + \left|\ln \frac{T}{h^2}\right|\right) \max_{t\in I} |u(t)|_{1+\varepsilon},$$

provided $\{\mathcal{T}_h\}$ is strongly regular.

The $H^{1+\varepsilon}$-regularity of the true solution from (11.38) is investigated, e.g., in [Babuška, 1970], [Babuška, Aziz, 1972, Chapter 8], [Nečas, 1967].

Remark 11.8. Convergence of the finite element method after time discretization of the problem (11.5)–(11.6) is studied in [Ciarlet, Lions], [Johnson] and [Thomée, 1984], see also [Čermák, Zlámal] and [Larsson, Thomée, Zhang] for a nonlinear evolution problem of parabolic type.

Remark 11.9. We may also discretize the problem (11.1)–(11.3) first only in time. This approach is based on the Rothe method (see, e.g., [Rektorys, 1982] for details).

11.3. Numerical example

If the difference between temperatures in one millimetre of the wire insulation is greater than some prescribed bound (e.g., 30 [° C]), then the heat flux becomes so large that the insulation starts to burn up. That is why a detailed knowledge of the temperature distribution is very useful when designing electrical machines.

We applied the method proposed in the previous section to calculate the temperature field in the stator of a quadrupole three-phase asynchronous motor with power output 4000 kW and voltage 6000 V. This motor has a centrifugal ventilation, 96 slots and 12 radial ventilation channels 0.01 m wide. The thickness of

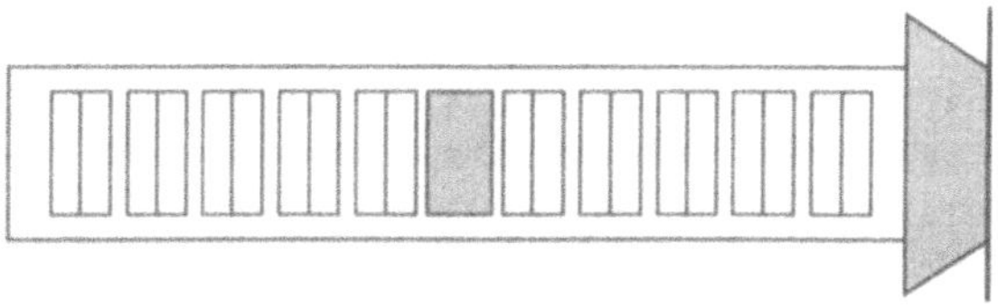

Figure 11.5.

each pack is 0.046 m. The cross section of one slot is sketched in Figure 11.5. It contains 10 composite copper wires covered by insulation.

The heat transfer coefficient α is equal to 0 at four planes of symmetry (two of them contain the axis of revolution and the other two are perpendicular to this axis - see Figures 11.1–11.3). In ventilation channels, the distribution of values of the coefficient α is quite complicated; it belongs to the interval from 44 to 179 [W/m^{2}K]. Thermophysical properties of copper wires (Cu), iron sheets (Fe) and insulation (In) are the following: the specific heat c_{Cu}=386 [J/kgK], c_{Fe}=451 [J/kgK], c_{In}=856 [J/kgK], the specific mass ρ_{Cu}=8890 [kg/m^3], ρ_{Fe}=7850 [kg/m^3], ρ_{In}=2300 [kg/m^3], the heat conductivities a_{Cu}=332.8 [W/mK], a_{Fe}=40 [W/mK] along stator sheets and a'_{Fe}=1.1 [W/mK] across sheets, a_{In}=0.5 [W/mK] along insulation layers and a'_{In}=0.2 [W/mK] across these layers. The density of heat sources f (and g) in stator sheets and copper wires is calculated by special subroutines, which use some information about alternating electric currents and magnetic fields. We assume that $f = 0$ in the wire insulation. The initial condition u^0 is chosen constant over the whole Ω and equal to room temperature 20° C.

To discretize in time, we apply the well-known *Crank–Nicolson scheme.* Denote by Δt a fixed time step (e.g., Δt=5 [s]). Let f^r be the value of $\mathcal{F}$ at the rth time step. We approximate (11.9) symmetrically with respect to $(r+\frac{1}{2})\Delta t$. Then we obtain

$$M\frac{z^{r+1}-z^r}{\Delta t}+A\frac{z^{r+1}+z^r}{2}=\frac{f^{r+1}+f^r}{2},$$

where the vector z^r stands for the approximate values of z on the rth time level. We may therefore compute z^{r+1} from the recurrence equation

$$(M+\tfrac{1}{2}\Delta tA)z^{r+1}=(M-\tfrac{1}{2}\Delta tA)z^r+\tfrac{1}{2}\Delta t(f^{r+1}+f^r), \quad r=0,1,...,$$
$$z^0=(z_1(0),...,z_N(0))^T.$$

At the start of the computation we compute once the factorization

$$LL^T=M+\tfrac{1}{2}\Delta tA,$$

and then we can easily find the solution z^r on each time level.

Starting from room temperature, it is necessary to compute until about $t =$ 20 minutes to achieve a numerical steady-state temperature distribution in the considered stator.

The highest difference between temperatures in one millimetre of insulation is 7.6 [° C] in the plane of symmetry x_2x_3 (see Figure 11.3). This can be computed from the solution of the stationary problem, i.e., for $t \to \infty$. In this case the parabolic equation (11.1) becomes an elliptic one, and the system of ordinary differential equations (11.9) has to be replaced by a system of algebraic equations $Az = \mathcal{F}$. This system can be solved by several methods. It is found that classical iterative methods (like Jacobi or Gauss–Seidel) converge extremely slowly, since the condition number of A is too large. One of the best method seems to be the preconditioned conjugate gradient method (see Chapter 5). However, this method is quite sensitive to roundoff errors since one constructs sequences of vectors which have to be mutually orthogonal with respect to a certain scalar product. Moreover, when solving problems with highly variable coefficients this method loses its good convergence properties. In such a case it is difficult to obtain higher accuracy of the solution in some subdomains of special interest, because the conjugate gradient method (and its variants) converge "globally" (see Theorem 5.19). We get quick convergence of the preconditioned conjugate gradient method at those parts of the stator which correspond to iron and copper. But convergence in the insulation is very slow, because the associated entries of the stiffness matrix were very small. To overcome these complications, after the application of the preconditioned conjugate gradient method, we perform several iterations of the SOR method which converges "locally" and has good smoothing properties. Furthermore, note that the SOR method is not so sensitive to roundoff errors. This is because we can always consider the last iterate as the new initial vector.

In Figure 11.6 we see the stationary temperature field on the upper part of the domain from Figure 11.3.

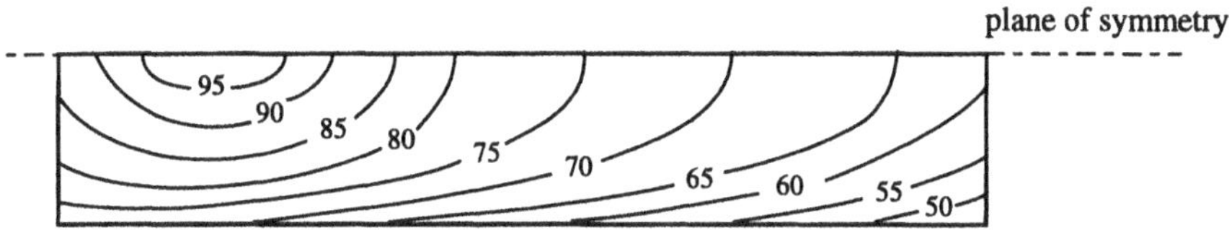

Figure 11.6.

12. The time-harmonic Maxwell equations

12.1. Problem with inhomogeneous conductivities

In this chapter we shall see that the solution of the time-harmonic Maxwell equations with real coefficients can be transformed to time independent partial differential equations with complex coefficients. Then we introduce a finite element approximation proposed in [Křížek, Neittaanmäki, 1989]. A similar technique is analyzed in [Křížek, Neittaanmäki, 1984b], [Monk, 1992a] (for fully time dependent problems see, e.g., [Monk 1992b,c]).

Consider the Maxwell equations from Section 2.2,

$$\begin{aligned} \frac{\partial \mathcal{D}}{\partial t} &= \operatorname{rot} \mathcal{H} - \mathcal{J}, \\ \frac{\partial \mathcal{B}}{\partial t} &= -\operatorname{rot} \mathcal{E}, \end{aligned} \tag{12.1}$$

in a three-dimensional domain $\Omega \in \mathcal{L}$, where $\mathcal{D} = \varepsilon \mathcal{E}$ is electric flux density, $\mathcal{B} = \mu \mathcal{H}$ is magnetic flux density, $\mathcal{E}$ is electric field, $\mathcal{H}$ is magnetic field, ε is permittivity, μ is permeability, $\mathcal{J} = \sigma \mathcal{E}$ is electric current and σ is a real 3×3 matrix of electric conductivities. We assume $\mathcal{E}$ and $\mathcal{H}$ to be time-harmonic with an angular frequency $\omega > 0$, i.e.,

$$\begin{aligned} \mathcal{E}(x_1, x_2, x_3, t) &= \operatorname{Re}(E(x_1, x_2, x_3) e^{i\omega t}), \\ \mathcal{H}(x_1, x_2, x_3, t) &= \operatorname{Re}(H(x_1, x_2, x_3) e^{i\omega t}), \end{aligned} \tag{12.2}$$

where E and H are complex-valued vector functions independent of time and $\operatorname{Re}(v_1, v_2, v_3) = (\operatorname{Re} v_1, \operatorname{Re} v_2, \operatorname{Re} v_3)$. Moreover, we shall assume that $\partial \mathcal{D} / \partial t = 0$ and that μ is a real positive constant in the whole domain. For instance, the permeability of copper or aluminium wires, insulation, oil, air, etc. is almost equal to vacuum permeability. On the other hand, the conductivity σ may vary considerably. Under the above assumptions, the relations (12.1) and (12.2) enable us to deal with the following time-harmonic system (where time does not occur)

$$\begin{aligned} \operatorname{rot} H &= \sigma E \quad && \text{in } \Omega, \\ \operatorname{rot} E &= -i\omega\mu H \quad && \text{in } \Omega. \end{aligned} \tag{12.3}$$

From here we see that H can be directly computed from a knowledge of E,

$$H = -\frac{\operatorname{rot} E}{i\omega\mu},$$

and for E we get the equation

$$\operatorname{rot} \operatorname{rot} E + i\omega\mu\sigma E = 0 \quad \text{in } \Omega. \tag{12.4}$$

Note that H formally satisfies the same equation at those parts of Ω, where σ is constant,

$$\operatorname{rot}\operatorname{rot} H + \mathrm{i}\omega\mu\sigma H = 0.$$

Further, we prescribe the boundary condition

$$n \wedge E = n \wedge \tilde{E} \quad \text{on } \partial\Omega, \tag{12.5}$$

where $n = (n_1, n_2, n_3)$ is the outward unit normal to $\partial\Omega$ and $\tilde{E}$ is a given vector function which is for convenience defined over the whole of $\overline{\Omega}$. Since div rot $\equiv 0$ (compare (3.109)), the first equation of (12.3) yields $\operatorname{div}\sigma E = 0$ in Ω ($\operatorname{div}\sigma E$ will always mean $\operatorname{div}(\sigma E)$). So we shall look for a complex-valued vector function $u = E - \tilde{E}$ such that

$$\begin{aligned} \operatorname{rot}\operatorname{rot} u + \mathrm{i}\omega\mu\sigma u &= -\operatorname{rot}\operatorname{rot}\tilde{E} - \mathrm{i}\omega\mu\sigma\tilde{E} && \text{in } \Omega, \\ \operatorname{div}\sigma u &= -\operatorname{div}\sigma\tilde{E} && \text{in } \Omega, \\ n \wedge u &= 0 && \text{on } \partial\Omega. \end{aligned} \tag{12.6}$$

Here we assume that $u, \tilde{E} \in (C^2(\overline{\Omega}))^3$, and $\sigma \in (C^1(\overline{\Omega}))^{3\times 3}$ is a real symmetric and uniformly positive matrix function, i.e., (12.6) is satisfied in the classical sense.

For a weak formulation of problem (12.6), it suffices to assume that $\sigma \in (L^\infty(\Omega))^{3\times 3}$. We introduce the space

$$\begin{aligned} H(\operatorname{div}\sigma) &= \{v \in (L^2(\Omega))^3 \mid \exists F \in L^2(\Omega) : (\sigma v, \operatorname{grad} z)_0 = -(F, z)_0 \quad \forall z \in C_0^\infty(\Omega)\} \\ &= \{v \in (L^2(\Omega))^3 \mid \operatorname{div}\sigma v \in L^2(\Omega)\}, \end{aligned}$$

where $(\cdot,\cdot)_0$ is the usual L^2-scalar product in complex space, that is

$$(w, z)_0 = \sum_{j=1}^{r} \int_\Omega (\operatorname{Re} w_j + \mathrm{i}\operatorname{Im} w_j)(\operatorname{Re} z_j - \mathrm{i}\operatorname{Im} z_j)\, dx$$

for $w = (w_1, ..., w_r)$, $z = (z_1, ..., z_r) \in (L^2(\Omega))^r$. The function F is called the divergence of σv in the sense of distributions. If σ is the identity matrix then obviously $H(\operatorname{div}) = H(\operatorname{div}\sigma)$. Moreover, we set

$$H(\operatorname{div}^0\sigma) = \{v \in H(\operatorname{div}) \mid \operatorname{div}\sigma v = 0 \text{ in } \Omega\}.$$

Let us introduce the space of test functions

$$V = H(\operatorname{div}\sigma) \cap H_0(\operatorname{rot})$$

equipped with the norm

$$|\!|\!| v |\!|\!|_\Omega = (\|v\|_0^2 + \|\operatorname{div}\sigma v\|_0^2 + \|\operatorname{rot} v\|_0^2)^{1/2}, \quad v \in V.$$

For the definition of the space $H_0(\operatorname{rot})$ see Section 3.8.

Lemma 12.1. *Let $\Omega \in \mathcal{L}$ be a three-dimensional domain. Then*

$$(\operatorname{rot} s, v)_0 = (s, \operatorname{rot} v)_0 \quad \forall s \in H(\operatorname{rot})\ \forall v \in H_0(\operatorname{rot}). \tag{12.7}$$

Proof. Let $s \in H(\operatorname{rot})$ and $v \in H_0(\operatorname{rot})$ be given. By [Girault, Raviart, p. 32], the set $(C_0^\infty(\Omega))^3$ is dense in $H_0(\operatorname{rot})$ with respect to $H(\operatorname{rot})$-norm. Hence, there exists a sequence $\{v_j\}_{j=1}^\infty \subset (C_0^\infty(\Omega))^3$ such that

$$\|v - v_j\|_0 + \|\operatorname{rot}(v - v_j)\|_0 \to 0 \quad \text{as } j \to \infty.$$

This and the Cauchy–Schwarz inequality imply that

$$(\operatorname{rot} s, v_j)_0 \to (\operatorname{rot} s, v)_0$$

and

$$(s, \operatorname{rot} v_j)_0 \to (s, \operatorname{rot} v)_0$$

for $j \to \infty$. From here and Green's formula (see (3.106))

$$(\operatorname{rot} s, v_j)_0 - (s, \operatorname{rot} v_j)_0 = \langle n \wedge s, v_j \rangle_{\partial\Omega} = 0, \quad j = 1, 2, \dots,$$

the assertion of the lemma follows. □

Suppose now that $u \in V$ (with an appropriate smoothness) satisfies (12.6). Using (12.7) for $s = \operatorname{rot} u$ and then for $s = \operatorname{rot} \tilde{E}$, we find that for any $v \in V$

$$\begin{aligned}(\operatorname{rot} v, \operatorname{rot} u)_0 + (v, \mathrm{i}\omega\mu\sigma u)_0 &= -(\operatorname{rot} v, \operatorname{rot} \tilde{E})_0 - (v, \mathrm{i}\omega\mu\sigma\tilde{E})_0,\\ (\operatorname{div} \sigma v, \operatorname{div} \sigma u)_0 &= -(\operatorname{div} \sigma v, \operatorname{div} \sigma\tilde{E})_0.\end{aligned}$$

Consequently,

$$a(v, u) = F(v) \quad \forall v \in V, \tag{12.8}$$

where a is the sesquilinear form

$$a(v, w) = (\operatorname{div} \sigma v, \operatorname{div} \sigma w)_0 + (\operatorname{rot} v, \operatorname{rot} w)_0 - \mathrm{i}\omega\mu(v, \sigma w)_0, \quad v, w \in V, \tag{12.9}$$

and F is the linear form defined by

$$F(v) = -(\operatorname{div} \sigma v, \operatorname{div} \sigma\tilde{E})_0 - (\operatorname{rot} v, \operatorname{rot} \tilde{E})_0 + \mathrm{i}\omega\mu(v, \sigma\tilde{E})_0, \quad v \in V. \tag{12.10}$$

In Remark 12.3 we further show that if u is sufficiently smooth and satisfies (12.8) then u satisfies also (12.6). Hence, we are justified to call (12.8) the weak (least squares) formulation of problem (12.6).

Theorem 12.2. *Let $\Omega \in \mathcal{L}$ be a three-dimensional domain with a connected boundary and let $\tilde{E} \in H(\operatorname{div}\sigma) \cap H(\operatorname{rot})$. Then there exists a unique $u \in V$ such that (12.8) holds.*

Proof. Let $v \in V$ be arbitrary and let $z \in H_0^1(\Omega)$ be the weak solution of the Dirichlet problem

$$\begin{aligned} -\operatorname{div}(\sigma \operatorname{grad} z) &= -\operatorname{div}\sigma v \quad \text{in } \Omega,\\ z &= 0 \quad\quad\quad\quad \text{on } \partial\Omega. \end{aligned} \tag{12.11}$$

Then obviously

$$\|z\|_1 \le C_1 \|\operatorname{div}\sigma v\|_0. \tag{12.12}$$

From Green's formula (3.98) and (3.107) we see that $\operatorname{grad} z \in H(\operatorname{div}\sigma) \cap H_0(\operatorname{rot}^0)$, i.e., by (12.11) we have

$$w = v - \operatorname{grad} z \in H(\operatorname{div}^0 \sigma) \cap H_0(\operatorname{rot}). \tag{12.13}$$

Since $\sigma w \in H(\operatorname{div}^0)$, by Lemma 3.36 there exists a unique stream function $s \in H_0(\operatorname{div}^0) \cap H(\operatorname{rot}) \cap (H_0(\operatorname{div}^0) \cap H(\operatorname{rot}^0))^\perp$, such that

$$\sigma w = \operatorname{rot} s. \tag{12.14}$$

From here, the symmetry and uniform positive definiteness of σ, (12.7) and Lemma 3.36, we obtain

$$\begin{aligned} \|w\|_0^2 &= \|\sigma^{-1}\operatorname{rot} s\|_0^2 \le C\|\sigma^{-1/2}\operatorname{rot} s\|_0^2\\ &= C(\sigma^{-1/2}\operatorname{rot} s, \sigma^{-1/2}\operatorname{rot} s)_0 = C(\operatorname{rot} s, w)_0\\ &= C(s, \operatorname{rot} w)_0 \le C\|s\|_0\|\operatorname{rot} w\|_0 \le C'\|\operatorname{rot} s\|_0\|\operatorname{rot} w\|_0\\ &\le C_2\|\sigma^{-1}\operatorname{rot} s\|_0\|w\|_0 = C_2\|w\|_0\|\operatorname{rot} w\|_0, \end{aligned}$$

where $\sigma^{-1/2}$ denotes the square root of the real positive definite matrix σ^{-1}. This and (12.13), (12.12) and (3.107) yield

$$\begin{aligned} \|v\|_0 \le \|\operatorname{grad} z\|_0 + \|w\|_0 &\le C_1\|\operatorname{div}\sigma v\|_0 + C_2\|\operatorname{rot} w\|_0\\ &\le C(\|\operatorname{div}\sigma v\|_0 + \|\operatorname{rot} v\|_0). \end{aligned} \tag{12.15}$$

It is easy to show that V is a Hilbert space with the scalar product

$$(v, w)_\Omega = (\operatorname{div}\sigma v, \operatorname{div}\sigma w)_0 + (\operatorname{rot} v, \operatorname{rot} w)_0 + (v, w)_0, \quad v, w \in V,$$

for which $|\!|\!| v |\!|\!|_\Omega^2 = (v, v)_\Omega, v \in V$. Employing (12.15) to (12.9), we arrive at

$$\begin{aligned} |a(v,v)| &\ge \operatorname{Re} a(v,v) = \|\operatorname{div}\sigma v\|_0^2 + \|\operatorname{rot} v\|_0^2\\ &\ge \frac{1}{4C^2}\|v\|_0^2 + \frac{1}{2}\|\operatorname{div}\sigma v\|_0^2 + \frac{1}{2}\|\operatorname{rot} v\|_0^2 \ge c|\!|\!| v |\!|\!|_\Omega^2 \quad (c > 0) \end{aligned} \tag{12.16}$$

for any $v \in V$, i.e., the sesquilinear form $a(\cdot,\cdot)$ is V-elliptic. We immediately see that $a(\cdot,\cdot)$ is continuous, $|a(v,w)| \le C|||v|||_\Omega\,|||w|||_\Omega$ for all $v, w \in V$, and that the linear form $b(\cdot)$ is also continuous when $\tilde{E} \in H(\operatorname{div}\sigma) \cap H(\operatorname{rot})$. Thus the rest of the proof follows from the Lax–Milgram Lemma 3.3. □

Remark 12.3. Let $u \in V$ satisfy (12.8) and let the smoothness assumptions upon u, $\tilde{E}$ and σ, which are stated just after (12.6), hold. Then we show that u satisfies all equations of (12.6).

That the boundary condition $n \wedge u = 0$ on $\partial\Omega$ holds is obvious, since $u \in V \subset H_0(\operatorname{rot};\Omega)$.

Next we show that

$$\operatorname{div}\sigma u = -\operatorname{div}\sigma\tilde{E} \quad \text{in } \Omega. \tag{12.17}$$

So let $f \in L^2(\Omega)$ be arbitrary and let us consider the Dirichlet problem

$$\begin{aligned} -\operatorname{div}\sigma\operatorname{grad} z - \mathrm{i}\omega\mu z &= -f \quad \text{in } \Omega,\\ z &= 0 \quad \text{on } \partial\Omega. \end{aligned} \tag{12.18}$$

By the uniform positive definiteness of σ and Friedrichs' inequality (3.43), the associated sesquilinear form is $H_0^1(\Omega)$-elliptic as

$$\begin{aligned} |(\operatorname{grad} z, \sigma\operatorname{grad} z)_0 - \mathrm{i}\omega\mu(z,z)_0| &\ge (\operatorname{grad} z, \sigma\operatorname{grad} z)_0\\ &\ge c(\operatorname{grad} z, \operatorname{grad} z)_0 \ge C\|z\|_1^2 \quad \forall z \in H_0^1(\Omega), \end{aligned}$$

where $C > 0$ is independent of z. Hence, there exists a weak solution $z \in H_0^1(\Omega)$ of problem (12.18). Setting $v = \operatorname{grad} z$, we find by (3.108) and (12.18) that $v \in V$. Consequently, from (12.8), (12.9), (12.10), (3.107), (3.98) and (12.18) we get

$$\begin{aligned} 0 = a(v,u) - F(v) &= (\operatorname{div}\sigma v, \operatorname{div}\sigma(u+\tilde{E}))_0 - \mathrm{i}\omega\mu(\operatorname{grad} z, \sigma(u+\tilde{E}))_0\\ &= (\operatorname{div}\sigma\operatorname{grad} z, \operatorname{div}\sigma(u+\tilde{E}))_0 + \mathrm{i}\omega\mu(z, \operatorname{div}\sigma(u+\tilde{E}))_0\\ &= (f, \operatorname{div}\sigma(u+\tilde{E}))_0. \end{aligned}$$

Thus (12.17) is valid.

Further, let $v \in (C_0^\infty(\Omega))^3$ be arbitrary. We see that $\operatorname{div}\sigma v \in C(\overline{\Omega}) \subset L^2(\Omega)$ and thus $v \in V$. Due to (12.7), (12.9), (12.8), (12.17) and (12.10) we obtain

$$\begin{aligned} (v, \operatorname{rot}\operatorname{rot} u + \mathrm{i}\omega\mu\sigma u)_0 &= (\operatorname{rot} v, \operatorname{rot} u)_0 - \mathrm{i}\omega\mu(v, \sigma u)_0\\ &= a(v,u) - (\operatorname{div}\sigma v, \operatorname{div}\sigma u)_0 = F(v) + (\operatorname{div}\sigma v, \operatorname{div}\sigma\tilde{E})_0\\ &= -(\operatorname{rot} v, \operatorname{rot}\tilde{E})_0 + \mathrm{i}\omega\mu(v, \sigma\tilde{E})_0 = -(v, \operatorname{rot}\operatorname{rot}\tilde{E} + \mathrm{i}\omega\mu\sigma\tilde{E})_0. \end{aligned}$$

Since $(C_0^\infty(\Omega))^3$ is dense in $(L^2(\Omega))^3$, the first equation of (12.6) holds.

Remark 12.4. Let $\sigma = 0$ in some subdomain $\Omega_0 \ne \emptyset$ of Ω, i.e., σ is not uniformly positive definite. Then the problem (12.6) is not uniquely solvable. Putting $E_0 = \operatorname{grad}\varphi$ for some $\varphi \in \{z \in C_0^\infty(\Omega) \mid z = 0 \text{ in } \Omega \setminus \Omega_0\}$, $\varphi \ne 0$, we see that

$$\begin{aligned} \operatorname{rot}\operatorname{rot} E_0 + \mathrm{i}\omega\mu\sigma E_0 &= 0 \quad \text{in } \Omega,\\ \operatorname{div}\sigma E_0 &= 0 \quad \text{in } \Omega,\\ n \wedge E_0 &= 0 \quad \text{on } \partial\Omega. \end{aligned}$$

This means that $u + E_0$ is a solution of (12.6) as well.

Finally note that Theorem 12.2 remains valid even when $\omega = 0$.

12.2. FE-approximation of the three-dimensional problem

Let $\Omega \in \mathcal{L}$ be a polyhedral domain and let $\mathcal{T}_h$ be a decomposition of $\overline{\Omega}$ into elements (tetrahedra, prism, parallelepipeds, etc.) in the usual sense (cf. Section 4.1), i.e., any face of any $K \in \mathcal{T}_h$ is either a subset of $\partial\Omega$, or a face of another element $K' \in \mathcal{T}_h$. Let

$$W_h = \{v_h \in (L^2(\Omega))^3 \mid v_h|_K \in (P_K)^3 \quad \forall K \in \mathcal{T}_h\},$$

where P_K is a space of complex-valued polynomials such that $P_K \supset P_1(K)$. A discrete analogue of the problem (12.8) consists of finding $u_h \in V_h$ such that

$$a(v_h, u_h) = F(v_h) \quad \forall v_h \in V_h, \tag{12.19}$$

where

$$V_h = V \cap W_h.$$

From the V-ellipticity of $a(\cdot,\cdot)$ we see that a unique solution u_h exists provided $V_h \neq \emptyset$. First we establish some properties of the space V_h.

Definition 12.5. *Let $v_h \in W_h$. Then the tangential components of v_h are said to be continuous at element interfaces if for any two adjacent elements $K_1, K_2 \in \mathcal{T}_h$ we have*

$$n \wedge v_h|_{K_1} = n \wedge v_h|_{K_2} \quad \text{on } S = K_1 \cap K_2, \tag{12.20}$$

where n is a normal to S.

Lemma 12.6. *If $v_h \in V_h$ then the tangential components of v_h are continuous at element interfaces.*

Proof. Let $K_1, K_2 \in \mathcal{T}_h$ be adjacent elements and let $v_h \in V_h \subset H(\mathrm{rot})$ be arbitrary. Using Green's formula (3.106) for $z \in (C_0^\infty(K_1 \cup K_2))^3$, we get

$$\int_{K_j} (z \operatorname{rot} v_h - v_h \operatorname{rot} z)\, dx = \int_S (n_{K_j} \wedge v_h|_{K_j}) z \, ds, \quad j = 1, 2, \tag{12.21}$$

where n_{K_j} denotes the outward unit normal to ∂K_j. Summing (12.21) over $j = 1, 2$, we find by (3.106) that

$$0 = \int_S (n_{K_1} \wedge v_h|_{K_1} - n_{K_1} \wedge v_h|_{K_2}) z \, ds$$

which holds particularly for all $z \in (C_0^\infty(S))^3$. Hence, (12.20) is valid. □

For the remainder of this section we shall assume that the first derivatives of σ exist and that they are continuous over each element $K \in \mathcal{T}_h$, i.e., for any $v_h \in W_h$ the divergence of σv_h exists on each $K \in \mathcal{T}_h$ in the classical sense.

Definition 12.7. *Let $v_h \in W_h$. Then the normal component of σv_h is said to be continuous at element interfaces if for any two adjacent elements $K_1, K_2 \in \mathcal{T}_h$ we have*

$$n \cdot (\sigma v_h)|_{K_1} = n \cdot (\sigma v_h)|_{K_2} \quad \text{on } S = K_1 \cap K_2, \tag{12.22}$$

where n is a normal to S.

Lemma 12.8. *If $v_h \in V_h$ then the normal component of σv_h is continuous at element interfaces.*

The proof is analogous to that of Lemma 12.6 (now Green's formula (3.98) has to be used).

Note that when σ is, e.g., a diagonal matrix and $\sigma_{11} = \sigma_{22} = \sigma_{33} \in C^1(\overline{\Omega})$, then any $v_h \in V_h$ is continuous on the whole $\overline{\Omega}$ due to Lemmas 12.6 and 12.8.

Lemma 12.9. *Let (12.20) and (12.22) be satisfied for some $v_h \in W_h$ and let $n \wedge v_h = 0$ on $\partial\Omega$. Then $v_h \in V_h$.*

Proof. We show that $v_h \in V = H_0(\text{rot}) \cap H(\text{div}\,\sigma)$. Let $G \in (L^2(\Omega))^3$ be defined through the relation

$$G|_K = \text{rot}\, v_h|_K, \quad K \in \mathcal{T}_h.$$

Denoting by n_K the outward unit normal to ∂K, we obtain by Green's formula (3.106)

$$\begin{aligned}
\int_\Omega v_h \,\text{rot}\, z \,dx &= \sum_{K\in\mathcal{T}_h} \int_K v_h \,\text{rot}\, z\,dx \\
&= \sum_{K\in\mathcal{T}_h} \left(\int_K z \,\text{rot}\, v_h \,dx - \int_{\partial K} (n_K \wedge v_h) z\,ds \right) \\
&= \int_\Omega Gz\,dx \qquad \forall z \in (C_0^\infty(\Omega))^3,
\end{aligned}$$

where the sum of the boundary integrals vanish since $n_{K_1} + n_{K_2} = 0$ on $S = K_1 \cap K_2$ for adjacent elements $K_1, K_2 \in \mathcal{T}_h$. Hence, $G = \text{rot}\, v_h$ and $v_h \in H_0(\text{rot})$ because of the assumption that $n \wedge v_h = 0$ on $\partial\Omega$.

Defining $F \in L^2(\Omega)$ by

$$F|_K = \text{div}\,\sigma v_h|_K, \quad K \in \mathcal{T}_h,$$

we get by (3.98)

$$\begin{aligned}
\int_\Omega (\sigma v_h)\,\text{grad}\, z\,dx &= \sum_{K\in\mathcal{T}_h} \int_K (\sigma v_h)\,\text{grad}\, z\,dx \\
&= \sum_{K\in\mathcal{T}_h} \left(-\int_K z \,\text{div}\,\sigma v_h\,dx + \int_{\partial K} n_K \cdot (\sigma v_h) z\,ds \right) \\
&= -\int_\Omega Fz\,dx \qquad \forall z \in C_0^\infty(\Omega),
\end{aligned}$$

i.e., $v_h \in H(\operatorname{div} \sigma)$. □

Remark 12.10. Let us consider problem (12.19) in the case of linear tetrahedral elements, i.e.,

$$P_K = P_1(K) \quad \forall K \in \mathcal{T}_h. \tag{12.23}$$

We will briefly outline the proof that under certain assumptions the rate of convergence takes the form

$$|\!|\!| u - u_h |\!|\!|_\Omega = O(h) \quad \text{as } h \to 0. \tag{12.24}$$

We suppose that σ is piecewise constant; more precisely, let there exist mutually disjoint polyhedral domains Ω_j, $j = 1, ..., r$, such that

$$\overline{\Omega} = \bigcup_{j=1}^{r} \overline{\Omega}_j$$

and

$$\sigma|_{\Omega_j} \in (P_0(\overline{\Omega}_j))^{3\times 3} \quad \forall j.$$

Then evidently

$$\| \operatorname{div} \sigma v \|_{0,\Omega_j} \le C \| v \|_{1,\Omega_j} \quad \forall v \in (H^1(\Omega_j))^3,$$

and thus

$$|\!|\!| \cdot |\!|\!|_{\Omega_j} \le C' \| \cdot \|_{1,\Omega_j}, \tag{12.25}$$

where $C, C' > 0$ do not depend on j.

We shall consider only such decompositions $\mathcal{T}_h$ of $\overline{\Omega}$ into tetrahedra for which

$$\partial\Omega_j \cap K^0 = \emptyset \quad \forall K \in \mathcal{T}_h, \tag{12.26}$$

where K^0 denotes the interior of K. Further, let $\{\mathcal{T}_h\}_{h\to 0}$ be a regular family of decompositions of $\overline{\Omega}$ into tetrahedra.

Moreover, we will require the following piecewise regularity of $u \in V$:

$$u|_{\Omega_j} \in (H^2(\Omega_j))^3, \quad j = 1, ..., r. \tag{12.27}$$

Note that u need not be continuous across $\partial\Omega_j$, since σ may possess jumps and the normal component of σu has to be continuous at $\partial\Omega_j$. Assuming (12.27), it is well-known that

$$\| u - \pi_h^j u \|_{1,\Omega_j} \le Ch \| u \|_{2,\Omega_j}, \quad j = 1, ..., r, \tag{12.28}$$

where $\pi_h^j u \in (H^1(\Omega_j))^3$ is the standard linear interpolant of $u|_{\Omega_j}$.

Let $v_h \in W_h$ be defined as

$$v_h = \pi_h^j u \quad \text{on } \Omega_j, \ j = 1, ..., r. \tag{12.29}$$

We show now that $v_h \in V_h$. Since $\pi_h^j u = u$ at the vertices of each tetrahedron $K \subset \overline{\Omega}_j$, $j \in \{1, ..., r\}$, the condition $n \wedge u = 0$ on any face $S \subset \partial\Omega$ and relation (12.23) imply $n \wedge \pi_h^j u = 0$ on S, so $n \wedge v_h = 0$ on $\partial\Omega$. According to the Sobolev imbedding $H^2(K) \subset C(K)$, (12.27) and (12.26), u is continuous on every $K \in \mathcal{T}_h$. Thus analogously to the proof of Lemma 12.6 we obtain that

$$v \wedge u|_{K_1} = v \wedge u|_{K_2} \quad \text{on } S = K_1 \cap K_2$$

for any two adjacent tetrahedra $K_1, K_2 \in \mathcal{T}_h$. This equality is valid particularly at the vertices of S and thus from (12.29) it is easy to see that (12.20) holds. The constraint (12.22) may be derived in a similar way, since $\sigma|_K$ is constant for every $K \in \mathcal{T}_h$. Therefore, $v_h \in V_h$ by Lemma 12.9.

Next we slightly modify the proof of Céa's lemma (see [Ciarlet, p. 104]) to the complex case. As $u_h - v_h \in V_h$, we get from (12.8) and (12.19) that

$$a(u_h - v_h, u - u_h) = 0. \tag{12.30}$$

Now using the V-ellipticity condition (12.16), (12.30) and the continuity of $a(\cdot\,,\cdot)$, we find that

$$\begin{aligned} c|||u - u_h|||_\Omega^2 &\le |a(u - u_h, u - u_h)| \\ &\le |a(u - v_h, u - u_h)| \le C|||u - v_h|||_\Omega |||u - u_h|||_\Omega. \end{aligned} \tag{12.31}$$

Hence, by (12.31), (12.29), (12.25) and (12.28) we obtain

$$\begin{aligned} |||u - u_h|||_\Omega^2 \le \tilde{C}|||u - v_h|||_\Omega^2 &= \tilde{C}\sum_{j=1}^r |||u - \pi_h^j u|||_{\Omega_j}^2 \\ &\le \overline{C}\sum_{j=1}^r \|u - \pi_h^j u\|_{1,\Omega_j}^2 = O(h^2), \end{aligned}$$

whence (12.24) as required.

12.3. Magnetic potential of transformer shieldings

Windings of large transformers are surrounded by radial magnetic shieldings (see [Doležel]). Then the magnetic field produced by one winding does not interfere with magnetic fields of the other windings. Magnetic shieldings are made either from iron or aluminium. In the case of iron shielding, the magnetic potential A_φ in the transformer window Ω is described by the equation (compare (6.11))

$$-\frac{\partial}{\partial r}\left(\frac{1}{\mu r}\frac{\partial(rA_\varphi)}{\partial r}\right) - \frac{\partial}{\partial z}\left(\frac{1}{\mu r}\frac{\partial(rA_\varphi)}{\partial z}\right) = J_\varphi \quad \text{in } \Omega, \tag{12.32}$$

where $J_\varphi \in L^2(\Omega)$ is the current density, and $\Omega = (R_1, R_2) \times (0, Z)$, $0 < R_1 < R_2$, $Z > 0$. The difference between this equation and equation (6.11) is that the permeability μ is not constant in (12.32). In (12.32) the function μ is equal to the vacuum permeability μ_0 outside the iron shielding, but μ depends on the first derivatives of A_φ inside the shielding. This case can be, in fact, investigated along the lines of Section 7.2.

As a further complication let us deal with dynamic aluminium transformer shieldings. The corresponding magnetic potential A_φ is then described by the partial differential equation with complex coefficients

$$-\frac{\partial}{\partial r}\left(\frac{1}{r}\frac{\partial(rA_\varphi)}{\partial r}\right) - \frac{\partial}{\partial z}\left(\frac{1}{r}\frac{\partial(rA_\varphi)}{\partial z}\right) - \mathrm{i}\omega\gamma A_\varphi = \mu_0 J_\varphi, \tag{12.33}$$

where ω is the circular frequency of the alternating electric current, γ is zero outside the shielding, and γ is equal to the magnetic conductivity of aluminium inside the shielding. The additional term $-\mathrm{i}\omega\gamma A_\varphi$ arises as a consequence of the induction of eddy currents (compare (6.11) with (12.33)). Moreover, the function A_φ in (6.11) is real, whereas that in (12.33) is complex. We consider the boundary conditions of the Neumann type

$$\frac{\partial A_\varphi}{\partial r} + \frac{A_\varphi}{r} = 0 \quad \text{on } \Gamma_1, \tag{12.34}$$

$$\frac{\partial A_\varphi}{\partial z} = 0 \quad \text{on } \Gamma_2, \tag{12.35}$$

where Γ_2 is the upper and lower part of the boundary $\partial\Omega$ and $\Gamma_1 = \partial\Omega \setminus \overline{\Gamma}_2$ (see Figure 12.1). The components of the magnetic field are given by

$$H_r = \frac{1}{\mu_0 r}\frac{\partial(rA_\varphi)}{\partial z}, \quad H_z = \frac{1}{\mu_0 r}\frac{\partial(rA_\varphi)}{\partial r}. \tag{12.36}$$

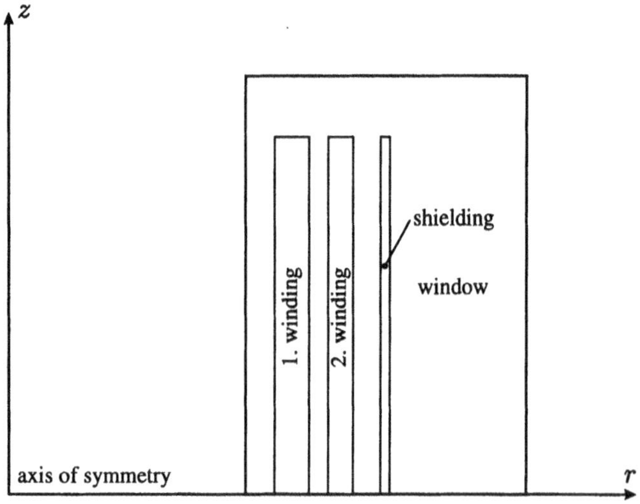

Figure 12.1.

Introducing a modified magnetic potential

$$u = \frac{1}{\mu_0} r A_\varphi, \tag{12.37}$$

problem (12.33)–(12.35) can be rewritten as follows

$$-\frac{\partial}{\partial r}\left(\frac{1}{r}\frac{\partial u}{\partial r}\right) - \frac{\partial}{\partial z}\left(\frac{1}{r}\frac{\partial u}{\partial z}\right) - \mathrm{i}\omega\gamma\frac{u}{r} = J_\varphi \quad \text{in } \Omega, \tag{12.38}$$

$$\frac{\partial u}{\partial n} = 0 \quad \text{on } \partial\Omega. \tag{12.39}$$

Note that

$$H_r = \frac{1}{r}\frac{\partial u}{\partial z}, \quad H_z = \frac{1}{r}\frac{\partial u}{\partial r},$$

which follows from (12.36) and (12.37).

A weak formulation of problem (12.38)–(12.39) consists of finding $u \in V = H^1(\Omega)$ such that

$$a(v, u) = F(v) \quad \forall v \in V, \tag{12.40}$$

where

$$a(v, w) = \int_\Omega \left(\frac{\partial v}{\partial r}\frac{\partial w^C}{\partial r} + \frac{\partial v}{\partial z}\frac{\partial w^C}{\partial z} - \mathrm{i}\omega\gamma\mu_0 v w^C\right)\frac{1}{r}\,dr\,dz, \tag{12.41}$$

$$F(v) = \int_\Omega J_\varphi v\,dr\,dz, \tag{12.42}$$

and where w^C stands for the conjugate number to w.

Theorem 12.11. *Problem (12.40) has one and only one solution.*

Proof. The form $F(\cdot)$ is obviously linear and continuous, and the form $a(\cdot,\cdot)$ is sesquilinear. Since the distance R_1 of the window Ω from the axis z is positive, there is no singularity in the weighted coefficient $\frac{1}{r}$. Consequently,

$$\begin{aligned} |a(v,w)| &\leq \frac{1}{R_1}\left(\left\|\frac{\partial v}{\partial r}\right\|_0\left\|\frac{\partial w}{\partial r}\right\|_0 + \left\|\frac{\partial v}{\partial z}\right\|_0\left\|\frac{\partial w}{\partial z}\right\|_0 + C\|\gamma\|_{0,\infty}\|v\|_0\|w\|_0\right) \\ &\leq C'\|v\|_1\|w\|_1 \qquad \forall v, w \in V, \end{aligned}$$

which means that $a(\cdot,\cdot)$ is continuous.

To prove the V-ellipticity of $a(\cdot,\cdot)$, we furthermore denote by Ω_0 the subdomain of Ω which corresponds to the shielding. Recall that γ is equal to zero in $\Omega \setminus \Omega_0$. Hence,

$$a(v,v) = \int_\Omega \left(\frac{\partial v}{\partial r}\frac{\partial v^C}{\partial r} + \frac{\partial v}{\partial z}\frac{\partial v^C}{\partial z}\right)\frac{1}{r}\,dr\,dz - \int_{\Omega_0} \mathrm{i}\omega\gamma\mu_0 v v^C \frac{1}{r}\,dr\,dz,$$

where the first integral is real but the second one is imaginary. Using the trivial fact that

$$|x - iy| \geq \frac{1}{2}|x| + \frac{1}{2}|y| \quad \forall x, y \in \mathbf{R}^1,$$

we obtain

$$\begin{aligned} |a(v,v)| &\geq \frac{1}{2}\int_\Omega \left(\frac{\partial v}{\partial r}\frac{\partial v^C}{\partial r} + \frac{\partial v}{\partial z}\frac{\partial v^C}{\partial z}\right)\frac{1}{r}\,dr\,dz + \frac{1}{2}\omega\mu_0 \int_{\Omega_0} \gamma v v^C \frac{1}{r}\,dr\,dz \\ &\geq \frac{1}{2R_2}|v|^2_{1,\Omega} + \frac{C}{2R_2}\|v\|^2_{0,\Omega_0} \geq C'\|v\|^2_{1,\Omega} \qquad \forall v \in V, \end{aligned}$$

where the last inequality follows from (3.42). The existence of a unique solution $u \in V$ is now guaranteed by the Lax–Milgram Lemma 3.3. □

Remark 12.12. The proof of Céa's Lemma 4.11 can be again easily modified to finite element spaces of complex-valued functions (cf. (12.31)). Similarly to [Křížek, Neittaanmäki, 1990, p. 161] we can prove that $u \in H^2(\Omega)$, even though γ has jumps. Consequently, the rate of convergence of the FE-solution u_h is at least linear in the H^1-norm, where u_h is defined by the equation

$$a(v_h, u_h) = F(v_h) \tag{12.43}$$

for all v_h from some finite element subspace of V.

In Figure 12.2, we see the real part of the computed modified magnetic potential in the upper half of a window of a real-life transformer. The system of complex algebraic equations corresponding to (12.43) has been solved by the preconditioned biconjugate gradient method of Section 5.3. The method converged very fast. All eigenvalues of the associated stiffness matrix are accumulated near the positive part of the real axis.

Problems analogous to (12.38)–(12.39) are investigated, e.g., in [Červ, Segeth], [Haslinger, Neittaanmäki, 1984], [Křížek, Neittaanmäki, 1989], [Monk, 1992a], [Neittaanmäki, Saranen, 1981a,b,c], [Saranen, 1981], [Vanmaele, Morton, Süli, Borzì].

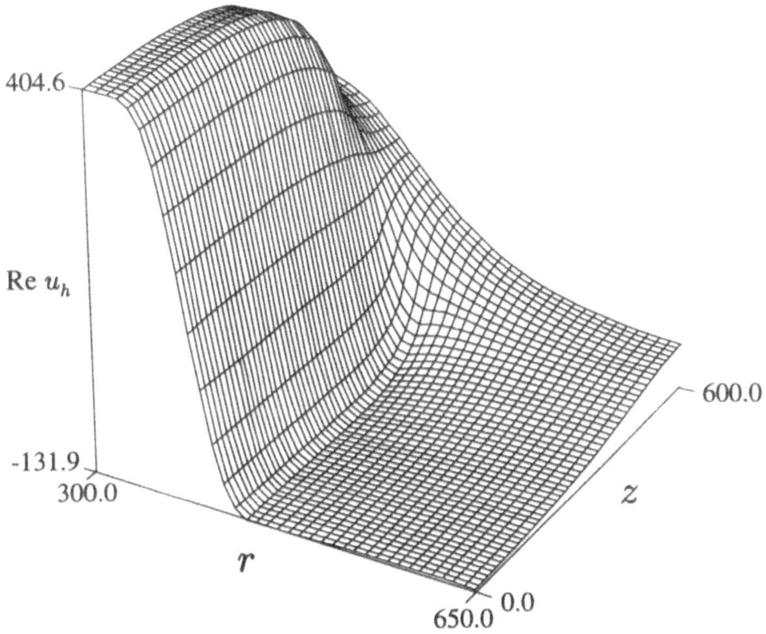

Figure 12.2.

13. Approximation of the Maxwell equations in anisotropic inhomogeneous media

13.1. The initial-boundary value problem and its semidiscrete approximation

Let $\Omega \in \mathcal{L}$ be in $\mathbb{R}^2$. We consider the initial-boundary value problem

$$
\begin{aligned}
&\operatorname{rot} E(x,t) + \mu(x)\frac{\partial}{\partial t}H(x,t) = J(x,t), && \\
& && (x,t) \in \Omega \times (0,T], \\
&\operatorname{curl} H(x,t) - \varepsilon(x)\frac{\partial}{\partial t}E(x,t) = K(x,t), && \\
&n \wedge E(x,t) = 0, && (x,t) \in \partial\Omega \times (0,T], \\
&(E(x,0), H(x,0)) = (E_0(x), H_0(x)), && x \in \overline{\Omega}.
\end{aligned}
\tag{13.1}
$$

In the two-dimensional case E is a vector field $E = (E_1, E_2)$ (electric field) and H is a scalar (magnetic field). The notation for curl are given formally by $\operatorname{rot} E = \partial_1 E_2 - \partial_2 E_1$, $\operatorname{curl} H = (\partial_2 H, -\partial_1 H)$ and correspondingly

$$n \wedge E = n_1 E_2 - n_2 E_1.$$

In equation (13.1), $\varepsilon = \varepsilon(x)$ is a smooth positive definite matrix on $\overline{\Omega}$ and $\mu = \mu(x)$ is a smooth positive function on $\overline{\Omega}$.

Let us assume that a pair (E, H) satisfies (13.1). Then, as usual, we shall derive from (13.1) an equation for E. To explain the method let us suppose that the fields are sufficiently smooth for the following differentiations to be carried out (we shall not need this in practice). From (13.1) we obtain

$$\varepsilon\frac{\partial^2}{\partial t^2}E + \operatorname{curl}(\mu^{-1}\operatorname{rot} E) = \operatorname{curl}(\mu^{-1}J) - \frac{\partial}{\partial t}K \tag{13.2}$$

and

$$\operatorname{div} \varepsilon E = -\int_0^t \operatorname{div} K(x,\tau)\, d\tau + \operatorname{div}(\varepsilon(x)E(x,0)). \tag{13.3}$$

One can easily check that the term $\operatorname{curl}(\mu^{-1}\operatorname{rot} E)$ in (13.2) containing the spatial derivatives of the electric field is not elliptic. We shall make an elliptization of this term by adding the term $-\varepsilon \operatorname{grad} \operatorname{div} \varepsilon E$. (By $\operatorname{div} \varepsilon E$ we mean $\operatorname{div}(\varepsilon E)$.)

Equations (13.2) and (13.3) imply

$$\varepsilon\frac{\partial^2}{\partial t^2}E + \operatorname{curl}(\mu^{-1}\operatorname{rot} E) - \varepsilon \operatorname{grad} \operatorname{div} \varepsilon E = F, \tag{13.4}$$

where

$$F(x,t) = \operatorname{curl}(\mu^{-1}(x)J(x,t)) - \frac{\partial}{\partial t}K(x,t) + \varepsilon(x)\left[\int_0^1 \operatorname{grad}\operatorname{div} K(x,\tau)\,d\tau - \operatorname{grad}\underline{\operatorname{div}(\varepsilon(x)E(x,0))}\right]. \tag{13.5}$$

Hence, F is known if $\operatorname{div}\varepsilon E$ is given for $t=0$.

From now on we assume that (13.4), with the additional condition $\operatorname{div}\varepsilon E\big|_{\partial\Omega} = 0$, is satisfied in the classical sense.

Furthermore, we recall the following spaces appropriate for $\mathbf{R}^2$ (cf. Sections 3.6 and 3.8),

$$\begin{aligned} H(\mathrm{rot}) &= \{E \in (L^2(\Omega))^2 \mid \operatorname{rot} E \in L^2(\Omega)\},\\ H_0(\mathrm{rot}) &= \{E \in H(\mathrm{rot}) \mid (\operatorname{rot} E, \varphi)_0 = (E, \operatorname{curl}\varphi)_0 \quad \forall \varphi \in H^1(\Omega)\},\\ H(\operatorname{div}\varepsilon) &= \{E \in (L^2(\Omega))^2 \mid \operatorname{div}\varepsilon E \in L^2(\Omega)\},\\ H(\operatorname{div}^0\varepsilon) &= \{E \in H(\operatorname{div}\varepsilon) \mid \operatorname{div}\varepsilon E = 0\}. \end{aligned}$$

The derivatives appearing above are defined distributionally in the standard way. If ε is the identity matrix we simply write $H(\mathrm{div})$ and $H(\mathrm{div}^0)$ instead of $H(\operatorname{div}\varepsilon)$ and $H(\operatorname{div}^0\varepsilon)$, respectively.

We now define the bilinear form $a(\cdot\,,\cdot)$ by

$$a(\psi,\phi) = (\alpha \operatorname{rot}\psi, \operatorname{rot}\phi)_0 + (\operatorname{div}\varepsilon\psi, \operatorname{div}\varepsilon\phi)_0 \tag{13.6}$$

for $\psi,\phi \in H(\mathrm{rot}) \cap H(\operatorname{div}\varepsilon)$, and we define the space V by

$$V = H_0(\mathrm{rot}) \cap H(\operatorname{div}\varepsilon).$$

If $F(\cdot\,,t) \in (L^2(\Omega))^2$, the weak form of (13.4) with the boundary conditions $n \wedge E(\cdot\,,t) = 0$, $\operatorname{div}\varepsilon E(\cdot\,,t)\Big|_{\partial\Omega} = 0$ is given by

$$\left(\varepsilon \frac{\partial^2}{\partial t^2}E(\cdot\,,t),\phi\right)_0 + a(E(\cdot\,,t),\phi) = (F(\cdot\,,t),\phi)_0 \tag{13.7}$$

for every $\phi \in V$ and $t \in (0,T]$.

In addition to (13.7) the initial data of $E(\cdot\,,t)$ at the time $t=0$ are assumed to be known and satisfy

$$(E(\cdot\,,0),\phi)_0 = (E_0,\phi)_0 \quad \forall\phi \in (L^2(\Omega))^2, \tag{13.8}$$

$$\left(\frac{\partial}{\partial t}E(\cdot\,,0),\phi\right)_0 = (E_1,\phi)_0 \quad \forall\phi \in (L^2(\Omega))^2. \tag{13.9}$$

We remark that equation (13.7) contains the condition $\operatorname{div}\varepsilon E(\cdot\,,t) \in H_0^1(\Omega)$ as a natural boundary condition (cf. [Saranen, 1980, Lemma 4.1]).

We suppose that the domain Ω is bounded and simply connected and that the boundary $\partial\Omega$ of Ω is smooth. This guarantees that the bilinear form a defined in (13.6) is coercive over V. In order to verify the coerciveness we need the following lemma.

Lemma 13.1. *Let $\Omega \subset \mathbf{R}^2$ be a bounded simply connected domain with a smooth boundary. Then there exists a constant $C > 0$ such that*

$$\|\phi\|_{1,\Omega}^2 \le Ca(\phi, \phi) \tag{13.10}$$

for every $\phi \in V$.

Proof. See [Křížek, Neittaanmäki, 1984a].

Using Lemma 13.1 and the techniques of [Saranen, 1980, Satz 2.2] the following result can be proved.

Lemma 13.2. *Under the assumptions of Lemma 13.1 the inequality*

$$\|\phi\|_{1,\Omega}^2 \le C\big(a(\phi, \phi) + \|n \wedge \phi\|_{\frac{1}{2},\partial\Omega}^2\big) \tag{13.11}$$

holds for all $\phi \in (H^1(\Omega))^2$.

In order to obtain a semidiscrete approximation for E we use a piecewise linear trial space. Let $\mathcal{F} = \{\mathcal{T}_h\}$, be a regular family of triangulations of $\overline{\Omega}$ with $\overline{\Omega} = \cup_{K \in \mathcal{T}_h} K$. We use the following trial spaces

$$S^h = \big\{\phi \in (C(\overline{\Omega}))^2 \mid \phi_i\big|_K \in P_i(K),\ i = 1, 2,\ n \wedge \phi = 0 \text{ at the node points of } \partial\Omega\big\}.$$

The semidiscrete approximation E^h of the solution E of (13.7)–(13.9) is given by solving the initial value problem

$$\left(\varepsilon \frac{\partial^2}{\partial t^2} E^h(\cdot\,, t), \phi\right)_0 + a(E^h(\cdot\,, t), \phi) = (F(\cdot\,, t), \phi)_0 \tag{13.12}$$

for all $\phi \in S^h$, $0 < t \le T$, with the initial conditions

$$(E^h(\cdot\,, 0), \phi)_0 = (P^h E(\cdot\,, 0), \phi)_0 \qquad \forall \phi \in (L^2(\Omega))^2, \tag{13.13}$$

$$\left(\frac{\partial}{\partial t} E^h(\cdot\,, 0), \phi\right)_0 = \left(\frac{\partial}{\partial t} P^h E(\cdot\,, 0), \phi\right)_0 \quad \forall \phi \in (L^2(\Omega))^2. \tag{13.14}$$

Here $P^h E(\cdot\,, t)$ denotes the projection defined by

$$a_0(P^h E(\cdot\,, t) - E(\cdot\,, t), \phi) = 0 \tag{13.15}$$

for all $\phi \in S^h$, $0 < t \le T$, with

$$a_0(\psi, \phi) = a(\psi, \phi) + (\psi, \phi)_0.$$

Let us note that $P^h E(\cdot\,, t)$ can be computed at the time $t = 0$, since $E(\cdot\,, 0)$ is known.

In order to prove that a is uniformly coercive over S^h we need the following lemma (see [Neittaanmäki, Picard, 1980]).

Lemma 13.3. *There exist two constants c_1 and c_2 independent of h and ϕ such that*

$$\|n \wedge \phi\|_{0,\partial\Omega} + h^{1/2}\|n \wedge \phi\|_{1/2,\partial\Omega} \le ch^2\|\phi\|_{1,\Omega} + ch^{3/2}\|\phi\|_{1,\Omega_0}$$

for all $\phi \in S^h$, where $\Omega_0 = \cup\{K \in \mathcal{T}_h \mid K \cap \partial\Omega \neq \emptyset\}$.

From Lemmas 13.2 and 13.3 we immediately have:

Lemma 13.4. *Under the assumptions of Lemma 13.2, there exist constants C_1 and h_1 such that*

$$\|\phi\|_{1,\Omega}^2 \le C_1 a(\phi, \phi)$$

for all $\phi \in S^h$ and $0 < h \le h_1$.

According to Lemma 13.4 one can conclude that for the system (13.12)–(13.14) there exists a unique twice differentiable solution

$$E^h(\cdot\,, t) \colon I \to S^h$$

for all $h \in (0, h_1)$, where $I = [0, T]$.

In [Neittaanmäki, Saranen, 1981] the following error estimate for the approximation is proved.

Theorem 13.5. *Let $\Omega \subset \mathbf{R}^2$ be a simply connected domain with a smooth boundary. Let E be the solution of (13.7) with initial conditions (13.8)–(13.9). Suppose that E, $\frac{\partial}{\partial t}E$ and $\frac{\partial^2}{\partial t^2}E$ belong to $L^\infty(I, (H^2(\Omega))^2)$. Let E^h be the semidiscrete approximation of E defined by (3.12)–(3.14). Then there exists a constant $C > 0$ independent of E and h such that*

$$\|E - E^h\|^2_{L^\infty(I,(H^1(\Omega))^2)} + \left\|\frac{\partial}{\partial t}(E - E^h)\right\|^2_{L^\infty(I,(L^2(\Omega))^2)} \le Ch^2 \sum_{k=0}^{2} \left\|\frac{\partial^k}{\partial t^k}E\right\|^2_{L^\infty(I,(H^2(\Omega))^2)}$$

for sufficiently small h.

Remark 13.6. For further study of approximation of time dependent Maxwell equations see [Leis], [Martensen], [Monk, 1992b,c], [Picard, 1980] and [Saranen, 1980] for the horizontal line method (Rothe method) and for [Adam, Gourdin-Serveniere, Nedelec, Raviart] for an implicit scheme for integrating Maxwell equations.

13.2. FE-approximation of the time-harmonic problem

Consider the time-harmonic ($e^{-it\omega}$-harmonic) system of Maxwell equations, which we obtain from (13.1)

$$\begin{cases} \operatorname{rot} E - i\omega\mu H = J \\ \operatorname{curl} H + i\omega\varepsilon E = K \end{cases} \quad \text{in } \Omega \tag{13.16}$$

with the homogeneous boundary condition

$$n \wedge E\Big|_{\partial\Omega} = 0. \tag{13.17}$$

In these equations, E and H are vector and scalar functions, respectively, ε is a function that takes positive definite bounded matrix values, μ is a positive bounded real-valued function on $\overline{\Omega}$, J and K are given force densities and ω is a complex number. Moreover, Ω is a bounded smooth domain in the plane $\mathbf{R}^2$.

It is well-known that the Maxwell equations can be generalized to any number of dimensions. In particular, if the solution of the Maxwell boundary value problem in the three-dimensional case is known to be independent of one space variable, we are led to the problem (13.16)–(13.17).

We shall write

$$M = i\begin{pmatrix} 0 & \varepsilon^{-1}\operatorname{curl} \\ -\mu^{-1}\operatorname{rot} & 0 \end{pmatrix}, \qquad U = \begin{pmatrix} u_1 \\ u_2 \end{pmatrix} = \begin{pmatrix} E \\ H \end{pmatrix},$$
$$F = \begin{pmatrix} F_1 \\ F_2 \end{pmatrix} = i\begin{pmatrix} \varepsilon^{-1}K \\ -\mu^{-1}J \end{pmatrix}.$$

Thus (13.16) can formally be rewritten as

$$(M - \omega)U = F.$$

In preparation for treating the boundary value problem we first observe that M can be extended to become a self-adjoint operator $\mathcal{M}$ on a certain Hilbert space $\mathcal{H}_s$, depending on a parameter $s > 0$, and an appropriate domain of definition $D(\mathcal{M})$ "containing" the boundary condition (13.17) in the weak sense. Thus, (13.16), (13.17) can be transformed into the operator equation

$$(\mathcal{M} - \omega)U = F \in \mathcal{H}_s, \quad U \in D(\mathcal{M}). \tag{13.18}$$

Using the projection method of [Weyl], as the main tool, we shall follow [Neittaanmäki, Picard, 1980] and show that for an appropriate choice of $s > 0$ and for ω from the resolvent set $\rho(\mathcal{M})$ of $\mathcal{M}$, we can replace (13.18) by an equivalent sesquilinear form equation

$$\begin{aligned}(\eta, U)_{*,s} - (1+\bar{\omega}^2)(\eta, U)_0 \\ = ((M+\bar{\omega})\eta, F)_0 - \frac{s}{\bar{\omega}}(\operatorname{div}\varepsilon\eta_1, \operatorname{div}\varepsilon F_1)_0,\end{aligned} \tag{13.19}$$

where $\eta = (\eta_1, \eta_2)$ and $F = (F_1, F_2)$.

In order to use variational methods we shall formulate the Maxwell equations in the "weak" sense. Let

$$\mathcal{H} = L^2(\Omega) \times L^2(\Omega).$$

Using the formal differential operator M, we now define the two-dimensional Maxwell operator

$$\mathcal{M} : D(\mathcal{M}) \subset \mathcal{H}_s \to \mathcal{H}_s, \quad U \mapsto \mathcal{M}U,$$

where $D(\mathcal{M}) = H_0(\mathrm{rot}) \cap H(\operatorname{div}\varepsilon) \times H^1(\Omega)$ and $\mathcal{H}_s = H(\operatorname{div}\varepsilon) \times L^2(\Omega)$ with the scalar product

$$(U, V)_s = (U, V)_0 + s(\operatorname{div}\varepsilon U_1, \operatorname{div}\varepsilon V_1)_0. \tag{13.20}$$

with usual complex inner product

$$(U, V)_0 = \int_\Omega U \cdot \overline{V}\, dx.$$

Because $\mathcal{M}$ is a closed operator, $D(\mathcal{M})$ is a Hilbert space with respect to the scalar product

$$(U, V)_{*,s} = (U, V)_s + (MU, MV)_0, \quad s \geq 0, \tag{13.21}$$

and is, in fact, a subspace of the Hilbert space

$$\mathcal{H}_{*,s} = (H(\mathrm{rot}) \cap H(\operatorname{div}\varepsilon)) \times H^1(\Omega)$$

with the same scalar product.

Using the projection theorem we obtain, as in [Picard, 1973] and [Picard, 1977], the following lemma.

Lemma 13.7. *In the inner product $(\cdot\,,\cdot)_{*,s}$ the space $\mathcal{H}_{*,s}$ has the orthogonal decomposition*

$$\mathcal{H}_{*,s} = D(\mathcal{M}) \oplus (\operatorname{Ker}(\operatorname{div}\mu^{-1}\operatorname{curl} + \varepsilon) \times \{0\})$$

with corresponding orthogonal projections P_* and Q_*.

According to [Picard, 1973] also the following lemma holds.

Lemma 13.8. *The space $\mathcal{H}$ has the orthogonal decomposition*

$$\mathcal{H} = (\operatorname{grad} H_0^1(\Omega) \times \{0\}) \oplus (H(\operatorname{div}^0 \varepsilon) \times L^2(\Omega)) \tag{13.22}$$

with the corresponding orthogonal projections P and Q.

We denote the components of P_*, Q_* by P_*^1, P_*^2 and Q_*^1, Q_*^2 and of P, Q by P_1, P_2 and Q_1, Q_2, respectively.

Next we shall deal with the solvability of the Maxwell equations. The Maxwell boundary value problem was introduced in its classical form. To derive the "weak" formulation, equations (13.16), (13.17) are first replaced by (13.18) for some fixed positive number s.

Since $\operatorname{div}\operatorname{curl} = 0$, it follows from (13.18) that

$$-\omega \operatorname{div} \varepsilon U_1 = \operatorname{div} \varepsilon F_1, \tag{13.23}$$

where the index denotes the natural projection on the first factor in the definition of $\mathcal{H}_s$ (as usual we consider these factors as subspaces of the product).

To get further insight into the properties of $\mathcal{M}$ we prove the following lemma.

Lemma 13.9. *The operator $\mathcal{M}$ is selfadjoint in $\mathcal{H}_s$.*

Proof. Since $\mathcal{M}$ is symmetric, we have to prove only that $D(\mathcal{M}^*) \subset D(\mathcal{M})$ where $\mathcal{M}^*$ is the adjoint operator. Let $V \in D(\mathcal{M}^*)$, so that $U \in \mathcal{H}_s$ and there exists $Z \in \mathcal{H}_s$ such that

$$(U, MV)_s = (Z, V)_s \quad \forall V \in D(\mathcal{M}). \tag{13.24}$$

Because $\operatorname{div}\operatorname{curl} = 0$ it follows that

$$(U, MV)_s = (U, MV)_0 \quad \forall V \in D(\mathcal{M})$$

and we have only to show that

$$(Z, V)_s = (Z, V)_0 \quad \forall V \in D(\mathcal{M}). \tag{13.25}$$

From this and (13.24) it will then follow that

$$(U, MV)_0 = (Z, V)_0 \quad \forall V \in D(\mathcal{M}),$$

which implies $U \in D(\mathcal{M})$.

To verify (13.25) we observe that $P\mathcal{H}_s \subset D(\mathcal{M})$, $\mathcal{M}P = 0$ and $\operatorname{div} \varepsilon P_1 V = \operatorname{div} \varepsilon V_1$ so that we obtain from (13.24)

$$0 = (Z, PV)_s = (PZ, V)_0 + s(\operatorname{div} \varepsilon P_1 Z, \operatorname{div} \varepsilon V_1)_0$$

for all $V \in \mathcal{H}_s$. This implies in turn that $PZ = 0$, $QZ = Z$, and finally that (13.25) holds. □

From Lemma 13.9 we see that the spectrum of $\mathcal{M}$ is real, and it is well known that the spectrum of $\mathcal{M}$ consists of isolated points because $(\mathcal{M} - \omega)^{-1} Q\mathcal{H}_s$ is compact, if ω is in the resolvent set $\rho(\mathcal{M})$.

Since (13.18) is not suitable for our method of approximation, we derive instead an equivalent sesquilinear form equation for (13.18). From now on we restrict ω to be in $\rho(\mathcal{M})$.

Lemma 13.10. *Every solution U of (13.18) satisfies the equation*

$$(\eta, U)_{*,s} - (1+\bar{\omega}^2)(\eta, U)_0 = ((\mathcal{M}+\bar{\omega})\eta, F) - (s/\bar{\omega})(\operatorname{div}\varepsilon\eta_1, \operatorname{div}\varepsilon F_1)_0 \quad \forall \eta \in D(\mathcal{M}). \tag{13.26}$$

Proof. Since for η, $U \in D(\mathcal{M})$,

$$(\eta, U)_{*,s} - (1+\bar{\omega}^2)(\eta, U)_0 = ((\mathcal{M}+\bar{\omega})\eta, (\mathcal{M}-\omega)U)_0 + s(\operatorname{div}\varepsilon\eta_1, \operatorname{div}\varepsilon U_1)_0,$$

(13.26) follows from (13.18) and (13.23). □

For an appropriate choice of the parameter s the converse is also valid. To show this the following result is needed.

Lemma 13.11. *There exists $\kappa > 0$ such that*

$$(P_1\phi, \varepsilon P_1\phi)_0 \le \kappa^2 \|\operatorname{div}\varepsilon\phi\|_0^2$$

for all $\phi \in H(\operatorname{div}\varepsilon)$.

Proof. This estimate can be shown by contradiction. Let $\{\phi_j\}_{j=1}^{\infty} \subset H(\operatorname{div}\varepsilon)$ be such that

$$(P_1\phi_j, \varepsilon P_1\phi_j)_0 = 1 \quad \text{and } \operatorname{div}\varepsilon\phi_j = \operatorname{div}\varepsilon P_1\phi_j \to 0, \; j \to \infty.$$

Using Poincaré's inequality (3.32), we can find a sequence $\{\psi_j\}_{j=1}^{\infty} \subset H_0^1(\Omega)$ such that

$$P_1\phi_j = \operatorname{grad}\psi_j \subset H(\operatorname{div}\varepsilon) \quad \text{and } \|\psi_j\|_0 < C.$$

Then it follows that

$$\begin{aligned} 1 = (\operatorname{grad}\psi_j, \varepsilon\operatorname{grad}\psi_j)_0 &= -(\psi_j, \operatorname{div}\varepsilon\operatorname{grad}\psi_j)_0 \\ &\le C\|\operatorname{div}\varepsilon P_1\phi_j\|_0, \end{aligned}$$

in contradiction to $\operatorname{div}\varepsilon P_1\phi_j \to 0$. □

From now on, we will select s to be any fixed number larger than $\kappa|\omega|^2$. Using Lemma 13.10, we can now show the equivalence of (13.18), (13.23) and the sesquilinear form equation (13.26).

Lemma 13.12. *Every solution of (13.26) satisfies equations (13.18) and (13.23).*

Proof. Using the orthogonal projections P and Q, equation (13.26) can be decomposed into two independent equations

$$s\Big(\operatorname{div}\varepsilon\eta_1, \operatorname{div}\varepsilon\Big(U_1 + \frac{1}{\omega}F_1\Big)\Big)_0 - \bar{\omega}^2\Big(\eta_1, \varepsilon P_1\Big(U_1 + \frac{1}{\omega}F_1\Big)\Big)_0 = 0 \tag{13.27}$$

and

$$(\mathcal{M}\eta, \mathcal{M}QU)_0 - \bar{\omega}^2(\eta, QU)_0 = ((\mathcal{M} + \bar{\omega})\eta, QF)_0 \tag{13.28}$$

for all $\eta \in D(\mathcal{M})$.

Since $P_1\mathcal{H}_s \subset H_0(\mathrm{rot}) \cap H(\operatorname{div}\varepsilon)$, (13.27) implies

$$s\Big\|\operatorname{div}\varepsilon\Big(U_1 + \frac{1}{\omega}F_1\Big)\Big\|_0^2 = \bar{\omega}^2\Big(P_1\Big(U_1 + \frac{1}{\omega}F_1\Big), \varepsilon P_1\Big(U_1 + \frac{1}{\omega}F_1\Big)\Big)_0.$$

According to the choice of the parameter s and Lemma 13.11, it follows

$$P_1\Big(U_1 + \frac{1}{\omega}F_1\Big) = 0,$$

and consequently equation (13.23). We note that $(\mathcal{M} - \omega)U = F$ implies $(-\mathcal{M} - \bar{\omega})\bar{U} = \bar{F}$, and so $-\bar{\omega}$ is in $\rho(\mathcal{M})$, too. Equation (13.28) then gives

$$\mathcal{M}QU - \omega QU = QF$$

and so together with $-\omega P_1U_1 = P_1F_1$ we obtain (13.18). □

Equation (13.26) provides a basis for the formulation of the approximate problem.

We approximate the solution of equation (13.26) using piecewise linear vector-valued functions. To construct the finite element spaces let $h \in (0, h_0]$ be a discretization parameter and let $\mathcal{T}_h$ be finite subdivisions of $\overline{\Omega}$ into triangles $K \in \mathcal{T}_h$ in a usual sense (Chapter 4).

The finite element spaces, we need to use, will be of the following type

$$\begin{aligned} S_1^h = \{E^h \in (C(\overline{\Omega}))^2 \mid E^h\big|_K \in (P_1(K))^2\ \forall K \in \mathcal{T}_h,\\ n \wedge E^h = 0 \text{ at the node points on } \partial\Omega\}\\ S_2^h = \{H^h \in C(\overline{\Omega}) \mid H^h\big|_K \in (P_1(K))^2\ \forall K \in \mathcal{T}_h\}. \end{aligned}$$

We define $\mathcal{H}^h = S_1^h \times S_2^h$ as a subspace of $\mathcal{H}_{*,s}$. Because $\mathcal{H}^h \not\subset D(\mathcal{M})$ we will have to deal with non-conforming elements with respect to the boundary conditions. The finite dimensional analogues of our Maxwell problem are:

Let $F \in \mathcal{H}_s$. Find $U^h \in \mathcal{H}^h$ such that

$$\begin{aligned} &(\eta^h, U^h)_{*,s} - (1 + \bar{\omega}^2)(\eta^h, U^h)_0\\ &= ((M + \bar{\omega})\eta^h, F)_0 - \frac{s}{\bar{\omega}}(\operatorname{div}\varepsilon\eta_1^h, \operatorname{div}\varepsilon F_1)_0 \end{aligned} \tag{13.29}$$

for all $\eta^h \in \mathcal{H}^h$.

The solvability of equation (13.29) and the approximation property of its solution with respect to the solution of (13.18) are formulated in the following theorem.

Theorem 13.13. *Under the given assumptions there exists $h_1 > 0$ such that (13.29) is uniquely solvable for all $F \in \mathcal{H}_s$ and $h \in (0, h_1]$. Denoting the solution operator by $R_\omega^h \colon \mathcal{H}_s \to \mathcal{H}^h$ the following asymptotic convergence estimates hold*

$$\|R_\omega F - R_\omega^h F\|_{*,s} \le C_1 h(\|F\|_{*,s} + \|\operatorname{div} \varepsilon F_1\|_1), \tag{13.30}$$

$$\|R_\omega F - R_\omega^h F\|_0 \le C_2 h^2(\|F\|_{*,s} + \|\operatorname{div} \varepsilon F_1\|_1) \tag{13.31}$$

for all $F \in \mathcal{H}_{*,s}$ with $\operatorname{div} \varepsilon F_1 \in H^1(\Omega)$.

Proof. ([Neittaanmäki, Picard, 1980]) Let P^h be the orthogonal projection of $\mathcal{H}_{*,s}$ on $\mathcal{H}^h$. Since $(\mathcal{M} - \omega)U = F$ for $U = R_\omega F$, equation (13.29) can be written

$$\begin{aligned}&(\eta, U^h)_{*,s} - (1 + \bar{\omega}^2)(P^h\eta, U^h)_0\\ &= (\eta, P^hU)_{*,s} - (1 + \bar{\omega}^2)(P^h\eta, U)_0 + \bar{\omega}[(P^h\eta, MU)_0 - (MP^h\eta, U)_0]\end{aligned}$$

or

$$\begin{aligned}&(\eta, U^h - P^hU)_{*,s} - (1 + \bar{\omega}^2)(P^h\eta, U^h - U)_0\\ &= \bar{\omega}((P^h\eta, MU)_0 - (MP^h\eta, U)_0) \quad \forall \eta \in \mathcal{H}_{*,s}.\end{aligned} \tag{13.32}$$

To treat (13.32) by means of perturbation theory we derive an equivalent operator equation in $\mathcal{H}_{*,s}$. To do this we introduce operators

$$\begin{aligned}S \colon \mathcal{H} &\to D(\mathcal{M}), & T \colon \mathcal{H} &\to \mathcal{H}_{*,s}, & B \colon D(\mathcal{M}) &\to \mathcal{H}_{*,s}\\ \phi &\mapsto \psi^{(1)}, & \phi &\mapsto \psi^{(2)}, & \phi &\mapsto \psi^{(3)}\end{aligned}$$

defined by equations

$$(\eta, \psi^{(1)})_{*,s} = (\eta, \phi)_0 \qquad \forall \eta \in D(\mathcal{M}), \tag{13.33}$$

$$(\varphi, \psi^{(2)})_{*,s} = ((I - P_*)\varphi, \phi)_0 \qquad \forall \varphi \in \mathcal{H}_{*,s}, \tag{13.34}$$

and

$$(\theta, \psi^{(3)})_{*,s} = (M\theta, \phi)_0 - (\theta, M\phi)_0 \quad \forall \theta \in \mathcal{H}_{*,s}. \tag{13.35}$$

Then (13.32) can be written in the form ($Q^h = I - P^h$)

$$[I - (1 + \omega^2)P^h(S + T)](U^h - U) = \omega P^h BU - Q^h. \tag{13.36}$$

If $I - (1 + \omega^2)P^h(S + T)$ has a bounded inverse on $X = D(\mathcal{M}) \oplus Q_*\mathcal{H}^h \subset \mathcal{H}_{*,s}$, then (13.36) implies the existence of the solution operator $R_\omega^h \colon \mathcal{H}_s \to \mathcal{H}^h$.

We now proceed to establish the existence of this bounded inverse. First we have to prove

$$I - (1 + \omega^2)S \colon X \to X \quad \text{is injective for } \omega \in \rho(\mathcal{M}) \tag{a}$$

and

$$I - (1+\omega^2)S \quad \text{is surjective.} \tag{b}$$

The statement (a) follows because $(I-(1+\omega^2)S)\phi = 0$ is equivalent to $(\mathcal{M} - \omega)\phi = 0$ (see the proof of Lemma 13.12). The statement (b) is true because even $(I-(1+\omega^2)S)\phi = f \in \mathcal{H}$ is solvable, which can be seen by the decomposition technique used to proved Lemma 13.12. Since $f \in X$ and $S\phi \in D(\mathcal{M})$, we have $\phi \in X$. From (a) and (b) we see that $I-(1+\omega^2)S$ has a bounded inverse.

According to [Neittaanmäki, Picard, 1980], the following estimates hold for sufficiently small h

$$\|U - P^hU\|_1 \le ch\|U\|_2, \tag{13.37}$$

$$\|U - P^hU\|_0 \le ch^2\|U\|_2, \tag{13.38}$$

and, moreover,

$$\|P^hTU\|_{*,s} \le ch^{3/2}\|U\|_{*,s}, \tag{13.39}$$

$$\|P^hTU\|_0 \le ch^2\|U\|_{*,s}, \tag{13.40}$$

as well as

$$\|P^hBU\|_{*,s} \le ch^{3/2}\|U\|_{*,s}, \tag{13.41}$$

$$\|P^hBU\|_0 \le ch^2\|U\|_{*,s}. \tag{13.42}$$

Using perturbation-theoretical arguments and the results of (13.37) and (13.39), the invertibility of $I-(1+\omega^2)P^h(S+T)$ follows from that of $I-(1+\omega^2)S$.

Since, furthermore $((I-(1+\omega^2)P^h(S+T))^{-1}$ is uniformly bounded as $h \to 0$, the asymptotic error estimate (13.30) follows immediately from (13.36) and from the estimates (13.37), (13.39) and (13.41).

The error estimate (13.31) follows similarly from (13.36), (13.38), (13.40) and (13.42) since the uniform boundedness of the inverse of $I-(1+\omega^2)P^h(S+T)\colon \mathcal{H} \to \mathcal{H}$ can be shown in a analogous way. □

Remark 13.14. For the finite element approximation of eigenfrequencies and eigenvectors to the Maxwell boundary value problem (13.16), (13.17) we refer to [Neittaanmäki, Picard, 1981].

13.3. Maxwell problem in polygonal domains

Now we consider the case where Ω is a bounded polygonal domain, whose boundary Γ consists of the smooth part Γ_0 and of the vertices $\{a_j\}_{j=1}^m$.

In this section we discuss the approximation of the pair $U = (E, H)$ in polygonal domains and we shall follow [Neittaanmäki, Saranen, 1981d]. For simplicity we handle only the (physically interesting) case, where the frequence $\omega \neq 0$ is a real number. We also restrict our discussion to a homogeneous isotropic medium and accordingly choose $\varepsilon = 1$ and $\mu = 1$. The weak form of (13.16) reads

$$(M - \omega)U = F, \quad U \in X_0, \ F \in H \tag{13.43}$$

where

$$M = \mathrm{i}\begin{pmatrix} 0 & \mathrm{curl} \\ -\mathrm{rot} & 0 \end{pmatrix}, \quad F = \mathrm{i}\begin{pmatrix} K \\ -J \end{pmatrix}$$

and

$$X_0 = (H_0(\mathrm{rot}) \cap H(\mathrm{div})) \times H^1(\Omega),$$
$$H = (L^2(\Omega))^2 \times L^2(\Omega).$$

Let, moreover,

$$X = (H(\mathrm{rot}) \cap H(\mathrm{div})) \times H^1(\Omega).$$

As in the previous section the variational formulation of (13.43) reads: Find $U \in X_0$ such that

$$(\phi, U)_{*,s} - (1 - \omega^2)(\phi, U)_0 = (\phi, G)_0 \quad \text{for } \phi \in X_0, \tag{13.44}$$

where

$$(\phi, U)_{*,s} = (\phi, U)_0 + s(\mathrm{div}\,\phi_1, \mathrm{div}\,U_1)_0 + (M\phi, MU)_0, \quad s > 0,$$

and

$$G = (M + \omega)F + s\omega^{-1}(\mathrm{grad\,div}\,F_1, 0).$$

Our trial subspaces $\widetilde{S}^h$ consist of piecewise linear continuous fields (resp. functions for the second argument) which are augmented by singular terms near the nonregular boundary points. A crucial point is the a priori knowledge of the singularities of the solution in neighbourhoods of the vertices (see [Saranen 1980]). The finite element solution $U^h \in S^h$ is defined by

$$(\phi, U^h)_{*,s} - (1 + \omega^2)(\phi, U^h)_0 = (\phi, G)_0 \tag{13.45}$$

for every $\phi \in \widetilde{S}^h$. Since in the case Ω is polygonal we have $S^h \subset X_0$, we obtain a conforming model for solving the problem (13.44).

In order to describe the behaviour of the solution $U = (E, H)$ in neighbourhood of the vertices some auxiliary notations must be introduced. Let α_j be the inner angle in the corner a_j, $0 < \alpha_j < 2\pi$. Take $R > 0$ so small that the neighbourhoods $U_j(2R) = \{x \mid |x - a_j| < 2R\}$ are mutually disjoint. Further, take $R_0 \in (0, R)$, and let $\xi \in C_0^\infty(\mathbf{R})$ be a test function such that $\xi(r) \equiv 1$, $0 < r < R_0$, $\xi(r) \equiv 0$,

$r \geq 2R_0$. Let ϕ_j be the polar angle in the corner a_j and let $\{\bar{e}^j_r, \bar{e}^j_\phi\}$ be the corresponding obvious orthonormal system. Define the fields

$$S^r_{\ell,j}(x) = |x - a_j|^{-1+\ell\pi/\alpha_j}\xi(|x - a_j|)v^j_\ell(\phi_j)\bar{e}^j_r, \quad \ell \geq 1,$$
$$S^\phi_{\ell,j}(x) = |x - a_j|^{-1+\ell\pi/\alpha_j}\xi(|x - a_j|)w^j_\ell(\phi_j)\bar{e}^j_\phi, \quad \ell \geq 1,$$
$$S^\phi_{0,j}(x) = |x - a_j|\xi(|x - a_j|)\bar{e}^j_\phi,$$

where

$$v^j_\ell(\phi_j) = \sqrt{2/\alpha_j}\sin\left(\ell\frac{\pi}{\alpha_j}\phi_j\right), \quad w^j_\ell(\phi_j) = \sqrt{2/\alpha_j}\cos\left(\ell\frac{\pi}{\alpha_j}\phi_j\right).$$

Theorem 13.15. *Let Ω be a polygonal domain such that $\alpha_j \neq k(\pi/2)$, $k = 1,2,3$, for every $j \in \{1, ..., m\}$. Assume that $F \in X_0$, $\operatorname{div} F_1 \in H^1_0(\Omega)$, $G = (M+\omega)F + s\omega^{-1}(\operatorname{grad}\operatorname{div} F_1, 0) \in X$, and that ω is not an eigenvalue of (13.43), $0 \neq \omega \in \mathbf{R}$. If $U \in X_0$ is a solution of (13.44) with $s > \omega^2\lambda_0^{-1}$, we have the decomposition*

$$U = (E, H) = (S_E, S_H) + \widetilde{U}, \tag{13.46}$$

with the singular components of the form

$$S_E = \sum_{j=1}^m \left(\sum_{0<\ell\pi/\alpha_j<2} (\delta^1_{\ell,j}S^r_{\ell,j} + \delta^2_{\ell,j}S^\phi_{\ell,j}) + \delta^2_{0,j}S^\phi_{0,j}\right), \tag{13.47}$$

$$S_H = \sum_{\pi/\alpha_j<1} \delta_j|x - a_j|^{\pi/\alpha_j}\xi(|x - a_j|)w^j_1(\phi_j) \tag{13.48}$$

and with the regular part $\widetilde{U} = (\widetilde{E}, \widetilde{H}) \in (H^2(\Omega))^3$ such that

$$\|\widetilde{U}\|_2 \leq c(\|F\|_{*,s} + \|\Delta \operatorname{div} F_1\|_0). \tag{13.49}$$

Proof. ([Neittaanmäki, Saranen, 1981d]) Since $(\operatorname{curl}\Phi_2, \operatorname{curl} H)_0 = (\operatorname{grad}\Phi_2, \operatorname{grad} H)_0$ and since $\operatorname{div} E \in H^1_0(\Omega)$, equation (13.44) can be written as

$$\begin{aligned}(\operatorname{curl}\Phi_1, \operatorname{curl} E)_0 - \omega^2(\Phi_1, E)_0 + (\operatorname{grad}\Phi_2, \operatorname{grad} H)_0 - \omega^2(\Phi_2, H)_0 \\ = (\Phi_1, \widetilde{G}_1)_0 + (\Phi_2, G_2)_0\end{aligned} \tag{13.50}$$

for all $\Phi = (\Phi_1, \Phi_2) \in X_0$, where $\widetilde{G}_1 = G_1 + s\operatorname{grad}\operatorname{div} E$. As $\widetilde{G}_1 \in (L^2(\Omega))^2$, $\operatorname{curl}\widetilde{G}_1 = \operatorname{curl} G_1 \in L^2(\Omega)$ and

$$\operatorname{div}\widetilde{G}_1 = -\omega^2 \operatorname{div} E \in H^1_0(\Omega), \tag{13.51}$$

Theorem 1.4 of [Saranen, 1980] is applicable to the component E. Moreover, equation (13.50) contains Neumann's boundary value problem for the second component H. Decomposition (13.46) is therefore a consequence of [Kondratiev, Theorem 3] and [Saranen, 1980, Theorem 1.4]. Further, these results imply

$$\|\widetilde{U}\|_2 \leq c(\|U\|_0 + \|\widetilde{G}_1\|_0 + \|\operatorname{curl}\widetilde{G}_1\|_1 + \|G_2\|_0). \tag{13.52}$$

To estimate the right-hand side of (13.52) we first note that

$$\|U\|_0 \leq c\|G\|_0, \tag{13.53}$$

because ω^2 is not an eigenvalue of (13.45). From (13.50), (13.51) it further follows that

$$(-s\Delta - \omega^2)\operatorname{div} E = \operatorname{div} G_1, \quad \operatorname{div} E \in H_0^1(\Omega). \tag{13.54}$$

The choice of s then yields

$$\|\operatorname{grad}\operatorname{div} E\|_0 \leq \omega^{-2}\|\operatorname{div} G_1\|_0.$$

Hence, we have using the Poincaré inequality and (13.51)

$$\|\widetilde{G}_1\|_0 + \|\operatorname{div}\widetilde{G}_1\|_1 \leq c(\|G\|_0 + \|\operatorname{div} G_1\|_0). \tag{13.55}$$

The estimates (13.53) and (13.55) imply (13.49). □

In the following another version of Theorem 13.15 is also needed. For a given $Y \in \mathcal{H}$, and $s > 0$, let $V \in X_0$ be the unique solution of

$$(\Phi, V)_{*,s} = (\Phi, Y)_0, \quad \text{for every } \Phi \in X_0. \tag{13.56}$$

Theorem 13.16. *Let the assumptions of Theorem 13.15 for the domain Ω be valid and let $s > 0$. If $Y \in X$ then the solution V of (13.56) has the decomposition $V = S + \widetilde{V}$, where $\widetilde{V} \in (H^2(\Omega))^3$ and where*

$$\|\widetilde{V}\|_2 \leq c\|Y\|_{*,s}. \tag{13.57}$$

The components of $S = (S_1, S_2)$ are of the form (13.47), (13.48).

Proof. ([Neittaanmäki, Saranen, 1981d]) The proof follows along the lines of the proof of Theorem 13.15 if we first show that

$$\operatorname{div} V_1 \in H_0^1(\Omega). \tag{13.58}$$

This can be seen by using the projection P_1 introduced after Lemma 13.8. Since $\Phi = (P_1\eta, 0) \in X_0$, if $\eta \in H(\operatorname{div})$, the equation

$$s(\operatorname{div}\eta, \operatorname{div} V_1)_0 = (\eta, P_1(Y_1 - V_1))_0$$

holds for all $\eta \in H(\text{div})$ due to (13.56). This mean the relation (13.58) ([Weber, Lemma 3.2]). □

Let $\mathcal{F} = \{\mathcal{T}_h\}$, $0 < h \le h_0$ be a regular family of triangulations of $\overline{\Omega}$ in the usual sense (Chapter 4). Let $\mathcal{N}_h$ be the set of all nodes of $\mathcal{T}_h$ lying on the boundary Γ. Especially the vertices a_j belong to $\mathcal{N}_h$. We define finite element spaces

$$\begin{aligned} S_1^h &= \{\Phi \in (C(\overline{\Omega}))^2 \mid \Phi|_K \in (P_1(K))^2 \quad \forall K \in \mathcal{T}_h,\ n \wedge \Phi(x) = 0 \\ &\qquad \forall x \in \mathcal{N}_h \cap \Gamma_0,\ \Phi(a_j) = 0,\ j = 1, ..., m\}, \\ S_2^h &= \{\eta \in C(\overline{\Omega}) \mid \eta|_K \in P_1(K) \quad \forall K \in \mathcal{T}_h\}. \end{aligned}$$

The trial subspaces $\widetilde{S}^h$ which will be used are given by $\widetilde{S}^h = \widetilde{S}_1^h \times \widetilde{S}_2^h$ with

$$\begin{aligned} \widetilde{S}_1^h = \Bigg\{ \Phi = \sum_{j=1}^{m} \Bigg(\sum_{0<\ell\pi/\alpha_j<2} (\alpha_{\ell,j} S_{\ell,j}^r + \beta_{\ell,j} S_{\ell,j}^\phi) + \beta_{0,j} S_{0,j}^\phi \Bigg) + \widetilde{\Phi} \mid \\ \widetilde{\Phi} \in S_1^h,\ \alpha_{\ell,j}, \beta_{\ell,j} \in \mathbb{C} \Bigg\} \end{aligned}$$

and with

$$\widetilde{S}_2^h = \Bigg\{ \eta = \sum_{\pi/\alpha_j<1} \gamma_j |x - a_j|^{\pi/\alpha_j} \xi(|x - a_j|) w_1^j(\Phi_j) + \tilde{\eta} \mid \tilde{\eta} \in S_2^h,\ \gamma_j \in \mathbb{C} \Bigg\}.$$

Since $\widetilde{S}^h \subset X_0$, we are using a conforming method. The finite element solution $U^h \in \widetilde{S}^h$ is now defined by (13.45). We shall obtain the solvability of (13.45) for small h as a by-product of the convergence result which can be proved by analogous arguments as Theorem 13.13 (for further details see [Neittaanmäki, Saranen, 1981d]).

Theorem 13.17. *Let the assumptions of Theorem 13.15 be valid. For sufficiently small h, $0 < h \le h_1$, the equation (13.45) has a unique solution $U^h \in \widetilde{S}^h$ with*

$$\|U - U^h\|_{*,s} \le ch\|\widetilde{U}\|_2, \tag{13.59}$$

$$\|U - U^h\|_0 \le ch^{3/2}\|\widetilde{U}\|_2, \tag{13.60}$$

as $h \to 0$.

Note that the rate $\mathcal{O}(h^{3/2})$ of convergence in (13.60) is a consequence of the estimate (13.49).

14. Methods for optimal shape design of electrical devices

14.1. Background

Often the primary problem facing designers of structural systems is determining the shape of the structure. In spite of graphical work stations and modern software for analyzing the structure, finding the best geometry for the structure by "trial and error" is still a very tedious and timeconsuming task. The goal in optimal shape design (structural optimization, or redesign) is to computerize the design process and therefore shorten the time it takes to design new products or improve the existing design. Structural optimization is already used in many applications in industry. In general, however, structural optimization is just beginning to penetrate the industrial community. Integrating FEM (Finite Element Method) and CAD (Computer Aided Design) technologies within the optimization loop will (hopefully quite soon) fully computerize the design loop.

In Figure 14.1(a) we see a traditional CAD/FEM-system and in Figure 14.1(b) the new generation of a CAD/FEM-system.

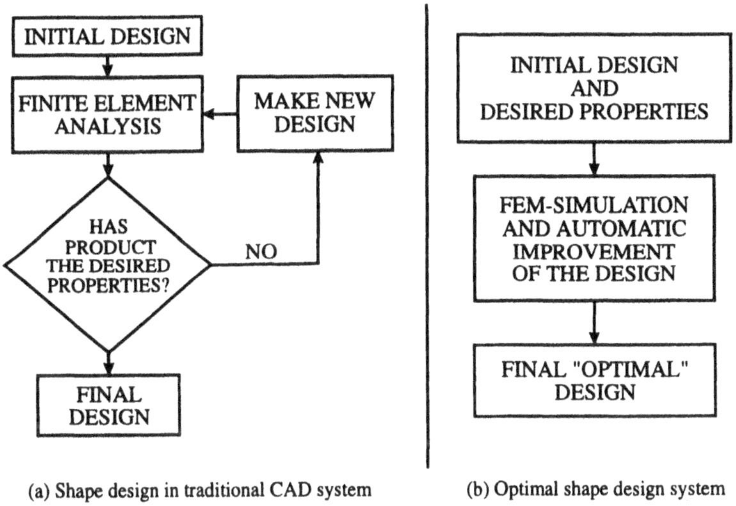

Figure 14.1.

Parameters chosen to describe the design (geometry) of the system are called *design variables*. The design parameters can be either finite dimensional (vector) or distributed parameters. Optimal structural design problems can be divided roughly into two classes: geometry optimization and topology optimization.

The third class would be optimal sizing: We assume that the layout of the structure is given and we try to find optimal sizes of the structural members. The sizes of the members are chosen as the design parameters that can be of a vector or distributed type. Typical sizing problems in structural optimization are optimal sizing of a beam (distributed parameter) and of a frame (vector parameter), [Arora], [Brandt], [Chleboun], [Haslinger, Neittaanmäki, 1988, 1996], [Haug and Céa].

In domain optimization (or variable boundary optimization) the shape of the domain is sought. Usually the problem is reduced to finding a function which defines the unknown boundary [Chleboun, Mäkinen], [Haslinger, Neittaanmäki, 1996], [Hlaváček, Křížek, 1992], [Pironneau].

Topology optimization deals with the search of optimal layout of the system. In topology optimization, the mechanical body can be considered as a domain in space with a high density of material, that is, the body is described by the global density function that assigns the material to points that are part of the body. Topology optimization problems have an on-off nature and are therefore extremely difficult to solve in the distributed case. For some topology optimization type of problems see [Bendsøe, 1989, 1995], [Haslinger, Neittaanmäki, 1996], [Kohn, Strang].

In this chapter we consider the domain optimization problem. We shall first formulate the problem in an abstract setting. We shall outline the main points in a numerical implementation (sensitivity analysis, etc.) and give some examples which are interesting from an industrial point of view.

14.2. Formulation of the problem

Let $\Omega \in \mathcal{O}$ ($\mathcal{O}$ is a nonempty set of admissible domains) be a domain for which we want to find an optimal design (an optimal geometrical layout). We suppose that $\mathcal{O}$ is a subset of some larger family $\tilde{\mathcal{O}}$; $\mathcal{O} \subseteq \tilde{\mathcal{O}}$.

For any $\Omega \in \tilde{\mathcal{O}}$ we associate a Hilbert space $V(\Omega)$ of functions, defined on Ω. In order to handle the situation mathematically, we introduce topology in $\tilde{\mathcal{O}}$ and in $\{V(\Omega) \mid \Omega \in \tilde{\mathcal{O}}\}$. If $\Omega_j, \Omega \in \tilde{\mathcal{O}}$, we have to define the limit

$$\Omega_j \overset{\tilde{\mathcal{O}}}{\to} \Omega \quad \text{as } j \to \infty. \tag{14.1}$$

Analogously, if $u_j \in V(\Omega_j)$, $u \in V(\Omega)$, $\Omega_j, \Omega \in \tilde{\mathcal{O}}$, then we specify the convergence

$$u_j \to u, \tag{14.2}$$

(see the example illustrated in Fig. 14.3). Let

$$\Omega \in \mathcal{O} \mapsto u(\Omega) \in V(\Omega) \tag{14.3}$$

be a mapping which associates with any domain $\Omega \in \mathcal{O}$ the unique solution u of a state problem (given by equations, inequalities etc. in Ω) and let

$$G = \{(\Omega, u(\Omega)) \mid \Omega \in \mathcal{O}\} \tag{14.4}$$

be its graph.

Finally, let $I = I(\Omega, u)$ (with $\Omega \in \mathcal{O}$ and $u \in V(\Omega)$) be a real cost function (criterion function), whose restriction on G will be denoted by $J(\Omega)$, i.e.,

$$J(\Omega) = I(\Omega, u(\Omega)), \quad (\Omega, u(\Omega)) \in G. \tag{14.5}$$

The *abstract optimal shape design problem* is stated as follows:

$$\begin{cases} \text{Find } \Omega^* \in \mathcal{O} \text{ such that} \\ J(\Omega^*) \leq J(\Omega) \text{ for all } \Omega \in \mathcal{O}. \end{cases} \tag{14.6}$$

We will say that $(\Omega^*, u(\Omega^*))$ is an optimal pair for (14.6).

Theorem 14.1. *Assume that G is compact in the following sense: If $\{\Omega_j\}_{j=1}^{\infty}$, $\Omega_j \in \mathcal{O}$, is an arbitrary sequence, there exist a subsequence $\{(\Omega_{j_k}, u(\Omega_{j_k}))\} \subset \{(\Omega_j, u(\Omega_j))\}$ and an element $(\Omega, u(\Omega)) \in G$ such that $\Omega_{j_k} \overset{\tilde{\mathcal{O}}}{\to} \Omega$ (the topology in $\mathcal{O}$ is induced by the topology in $\tilde{\mathcal{O}}$) and $u(\Omega_{j_k}) \to u(\Omega)$. Let I be lower semicontinuous: If $\Omega_j, \Omega \in \mathcal{O}$ with $\Omega_j \overset{\tilde{\mathcal{O}}}{\to} \Omega$ and if $u_j \in V(\Omega_j)$, $u \in V(\Omega)$ with $u_j \to u$, then*

$$\liminf_{j\to\infty} I(\Omega_j, u_j) \geq I(\Omega, u). \tag{14.7}$$

Then there exists at least one solution $\Omega^* \in \mathcal{O}$ of (14.6).

The proof of Theorem 14.1 is given in [Haslinger, Neittaanmäki, 1988, p. 29].

A large range of important optimal shape design problems which arise in structural mechanics, acoustics, electric fields, fluid flow and other areas of engineering and applied sciences can be included in the above formulation. Typically,

$$J(\Omega) = \int_\Omega dx \qquad \text{(minimization of the weight)},$$

$$J(\Omega) = \int_\Omega u^2(x)\, dx \qquad \text{(minimization of displacements)},$$

$$J(\Omega) = \int_\Omega \|\operatorname{grad} u(x)\|^2\, dx \qquad \text{(minimization of stresses)},$$

$$J(\Omega) = \int_{\partial\Omega} \left(\frac{\partial}{\partial n} u(x)\right)^2 ds \qquad \text{(minimization of the boundary flux)}.$$

It is well-known that, in general, problem (14.6) has no solution (see an example in [Dal Maso, Mosco]). The reason is that if we try to apply the direct method of calculus of variations we find that every minimizing sequence Ω_j has a subsequence such that the corresponding solutions u_j of (14.6) converge weakly to a function u. But in general, we cannot find an open set $\Omega \in \mathcal{O}$ such that u is the solution

$u(\Omega)$ of (14.6). On the other hand it can be proved (see [Dal Maso, Mosco]) that u is the solution of a certain relaxed problem.

Example 14.2. ([Céa, 1981]) Let $\Omega(u)$ be an interval of the real line

$$\Omega(\alpha) = \{x \in \mathbf{R} \mid 0 < x < \alpha\}.$$

Let the governing state equation be

$$\begin{cases} -u''(\alpha; x) = 2, & x \in (0, \alpha), \\ u'(\alpha; 0) = 0, \\ u(\alpha; \alpha) = 0. \end{cases} \tag{14.8}$$

The cost functional to be minimized is

$$J(\alpha) = \int_{\Omega(\alpha)} (u(\alpha; x) - 1)^2 \, dx = \int_0^\alpha (u(\alpha; x) - 1)^2 \, dx. \tag{14.9}$$

Since the solution of (14.8) is

$$u(\alpha; x) = \alpha^2 - x^2,$$

the cost function reads, by (14.9),

$$J(\alpha) = \frac{8}{15}\alpha^5 - \frac{4}{3}\alpha^3 + \alpha.$$

The graph of J is given in Figure 14.2.

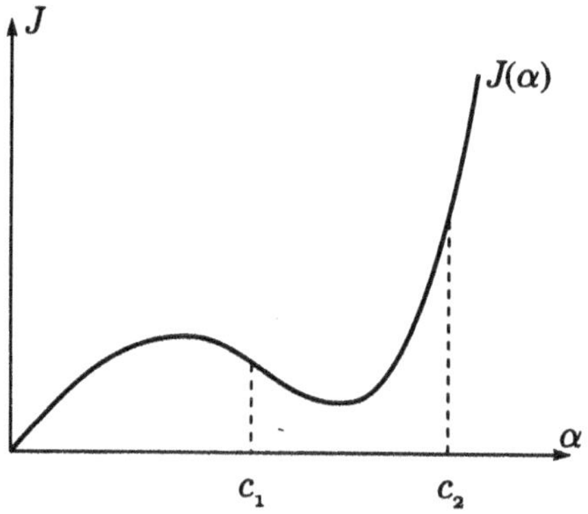

Figure 14.2.

We see in Figure 14.2 that in general the mapping $\alpha \to J(\alpha)$ is not convex and usual optimization methods need not converge to the optimal solution. However, if one puts box constraints of the type $c_1 \leq \alpha \leq c_2$ in an appropriate way one can get a convex problem.

Approximation of the optimal shape design problem. Let $h > 0$ be a parameter of discretization (finite element mesh parameter for example). With any $h > 0$ we associate a family $\mathcal{O}_h \subset \tilde{\mathcal{O}}$. If $\mathcal{O}_h \subset \mathcal{O}$, we say that $\mathcal{O}_h$ is an *internal approximation* of $\mathcal{O}$. Otherwise we use the term *external approximation.*

With any $\Omega_h \in \mathcal{O}_h$ we associate a finite dimensional space $V_h(\Omega_h) \subset V(\Omega_h)$ of functions defined on Ω_h. The convergence

$$\Omega_h \xrightarrow{\tilde{\mathcal{O}}} \Omega \quad \text{as } h \to 0+ \quad (\Omega_h \in \mathcal{O}_h,\ \Omega \in \tilde{\mathcal{O}}), \tag{14.10}$$

and the convergence

$$u_h \to u \quad \text{as } h \to 0+ \quad (u_h \in V_h(\Omega_h),\ u \in V(\Omega)), \tag{14.11}$$

are defined in the same way as in the continuous case.

In Ω_h we define the discrete state problem with the solution $u_h \in V_h(\Omega_h)$. Analogously to (14.3), we can define a mapping

$$u_h \colon \Omega_h \mapsto u_h(\Omega_h) \in V_h(\Omega_h) \tag{14.12}$$

with the graph

$$G_h = \{(\Omega_h, u_h(\Omega_h)) \mid \Omega_h \in \mathcal{O}_h\}. \tag{14.13}$$

Finally, we introduce the discretization I_h of the cost functional I and set

$$J_h(\Omega_h) = I_h(\Omega_h, u_h(\Omega_h)) \tag{14.14}$$

for $\Omega_h \in \mathcal{O}_h$.

The *discretization of the optimal shape design problem* for h fixed can be stated as follows:

$$\begin{cases} \text{Find } \Omega_h^* \in \mathcal{O}_h \text{ such that} \\ J_h(\Omega_h^*) \le J_h(\Omega_h) \text{ for all } \Omega_h \in \mathcal{O}_h. \end{cases} \tag{14.15}$$

Again $\{\Omega_h^*, u_h(\Omega_h^*)\} \in G_h$ will be called an optimal pair for (14.15).

Next we shall introduce assumptions which guarantee that solutions of (14.15) (at least some of them) are in certain sense close to those of (14.6).

(A1) For any $\Omega \in \mathcal{O}$ there exists $\Omega_h \in \mathcal{O}_h$ such that

$$\Omega_h \to \Omega \quad \text{as } h \to 0+.$$

(A2) The family $\{G_h\}, h \to 0+$, has the following property: There exist a subsequence $\{(\Omega_{h_j}, u_{h_j}(\Omega_{h_j}))\} \subset \{\Omega_h, u_h(\Omega_h)\}$, $(\Omega_h, u_h(\Omega_h)) \in G_h$ and an element $(\Omega, u(\Omega)) \in G$ such that

$$\begin{aligned} \Omega_{h_j} &\xrightarrow{\tilde{\mathcal{O}}} \Omega, && \text{as } j \to \infty, \\ u_{h_j}(\Omega_{h_j}) &\to u(\Omega), && \text{as } j \to \infty. \end{aligned}$$

(A3) If $\{\Omega_h\}$ where $\Omega_h \in \mathcal{O}_h$ and $\Omega \in \mathcal{O}$ are such that

$$\Omega_h \xrightarrow{\tilde{\mathcal{O}}} \Omega \quad \text{as } h \to 0+$$

and if $u_h(\Omega_h) \in V_h(\Omega_h)$, $u(\Omega) \in V(\Omega)$ are solutions of (4.12) and (4.3), respectively, with

$$u_h(\Omega_h) \to u(\Omega),$$

then

$$\lim_{h\to 0+} J_h(\Omega_h) = J(\Omega).$$

Theorem 14.3. *Let the assumptions (A1)–(A3) be satisfied and suppose that for any $h > 0$ there exists a solution $(\Omega_h^*, u_h(\Omega_h^*))$ of (14.15). Then there exists a subsequence $\{(\Omega_{h_j}^*, u_{h_j}(\Omega_{h_j}^*))\} \subset \{(\Omega_h^*, u_h(\Omega_h^*))\}$ such that*

$$\Omega_{h_j}^* \xrightarrow{\mathcal{O}} \Omega^* \quad \text{as } j \to \infty \tag{14.16}$$

and

$$u_{h_j}(\Omega_{h_j}^*) \to u(\Omega^*), \quad \text{as } j \to \infty, \tag{14.17}$$

where $(\Omega^*, u(\Omega^*)) \in G$ is an optimal pair for (14.6). Moreover, any such cluster point $(\Omega^*, u(\Omega^*))$ of a sequence $\{(\Omega_h^*, u_h(\Omega_h^*))\}$ satisfying (14.16) and (14.17) is an optimal pair for (14.6).

The proof is presented in [Haslinger, Neittaanmäki, 1996].

Trouble may arise in applying a gradient method to minimize $J_h(\Omega_h)$. Namely the mapping $\Omega_h \to J(\Omega_h)$ is not necessarily differentiable and thus the numerical implementation of the problem becomes troublesome (see [Mäkelä, Neittaanmäki]). Then one can either regularize the problem or apply a subgradient method [Haslinger, Neittaanmäki, 1996], [Mäkelä, Neittaanmäki].

14.3. Design sensitivity analysis

A substantial literature has been developed in the field of shape design sensitivity analysis, see for example [Haslinger, Neittaanmäki, 1988, 1996]. Contributions to this field have been made using two fundamentally different approaches. The first approach uses the discretized model, based on finite element analysis, and proceeds to carry out shape design sensitivity analysis by controlling finite element node movement and differentiating the algebraic finite element equations (algebraic approach). The second approach to sensitivity analysis usually proceeds by applying material derivative approach to the continuous optimal shape design problem, see [Céa, 1986], [Zolésio, 1981a,b]. The sensitivity information is expressed as integrals (domain or boundary integrals) of the solution of the state and adjoint state and these expressions are discretized to obtain ∇J. For the interrelation of the different methods and their use in practice see [Haslinger, Neittaanmäki, 1996] and [Neittaanmäki, Salmenjoki].

In this chapter we shall outline various methods for doing this technically tedious step. We shall first consider sensitivity analysis for optimal shape design problems with linear state problem. Moreover, sensitivity analysis for nonlinear state problems will be discussed.

Let $\partial\Omega = \overline{\Gamma}_1 \cup \overline{\Gamma(\alpha)}$ be the boundary of the domain $\Omega \subset \mathbf{R}^2$. Let the boundary $\Gamma(\alpha)$ of the domain $\Omega(\alpha)$ be controlled by a function $\alpha \in U_{\mathrm{ad}}$,

$$U_{\mathrm{ad}} = \left\{\alpha \in C^{(0),1}([0,1]) \mid 0 < C_0 \le \alpha \le C_1,\ |\alpha'| \le C_2,\ \int_0^1 \alpha(x_2)\,dx_2 = C_3\right\}$$

(set of admissible controls, see Figure 14.3). Here $C^{(0),1}([0,1])$ stands for the space of Lipschitz continuous functions on $[0,1]$.

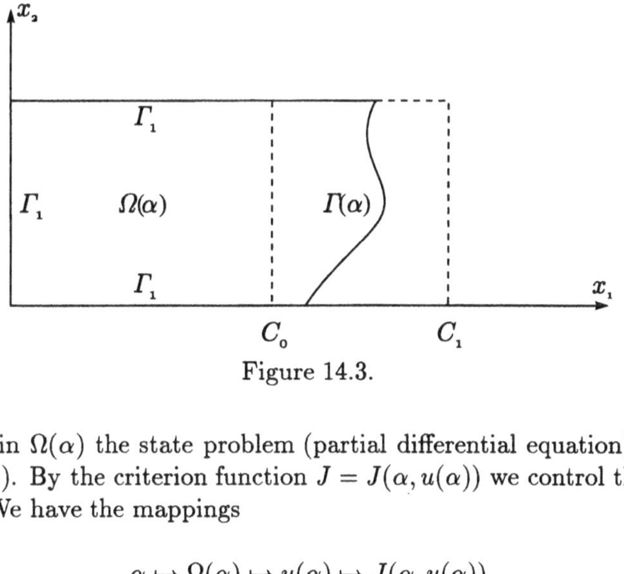

Figure 14.3.

We solve in $\Omega(\alpha)$ the state problem (partial differential equation) and obtain the state $u(\alpha)$. By the criterion function $J = J(\alpha, u(\alpha))$ we control the quality of the design. We have the mappings

$$\alpha \mapsto \Omega(\alpha) \mapsto u(\alpha) \mapsto J(\alpha, u(\alpha)) \tag{14.18}$$

and have the optimal shape design problem: Find $\alpha^* \in U_{\mathrm{ad}}$ such that

$$J(\alpha^*, u(\alpha^*)) \le J(\alpha, u(\alpha)) \quad \text{for all } \alpha \in U_{\mathrm{ad}}. \tag{14.19}$$

In this concrete case we can specify convergence (14.1) "$\Omega_j \to \Omega$" as $\alpha_j \rightrightarrows \alpha$ (uniformly) on $[0,1]$, where $\Omega_j = \Omega(\alpha_j)$ and $\Omega = \Omega(\alpha)$. Moreover, if $u_j \in V(\alpha_j)$, $u \in V(\alpha)$, $\alpha, \alpha_j \in U_{\mathrm{ad}}$, the convergence "$u_j \to u$" in (14.2) is defined by $\tilde{u}_j \rightharpoonup \tilde{u}$ (weakly) in $H^1(\widehat{\Omega})$, where $\tilde{u}_j$, $\tilde{u}$ denote the Calderon extension of u_j, u from Ω_j, Ω, respectively, to $\widehat{\Omega} = (0,1) \times (0,1)$.

In practice one solves problem (14.19) in an iterative manner: One seeks a sequence $\{\alpha^k\}$ such that $J(\alpha^{k+1}, u(\alpha^{k+1})) < J(\alpha^k, u(\alpha^k))$ usually by the gradient method

$$\alpha^{k+1} = \alpha^k - \varrho P\left(\nabla_{\alpha^k} J(\alpha^k, u(\alpha^k))\right), \quad k = 0, 1, 2, \ldots, \tag{14.20}$$

where the projection P and real parameter ϱ are chosen in such way that $\alpha^{k+1} \in U_{\mathrm{ad}}$. A crucial point in the procedure (14.20) is to compute the gradient of J with respect to the design variable α. This step is called design sensitivity analysis.

Consider the optimal shape design problem

$$\underset{\alpha \in U_{\mathrm{ad}}}{\text{minimize}} \left\{ J(\alpha, u(\alpha)) = \frac{1}{2} \int_{\Omega(\alpha)} (u(\alpha))^2 \, dx \right\} \tag{14.21}$$

subject to the state problem:

$$\begin{cases} -\Delta u(\alpha) = f & \text{in } \Omega(\alpha), \\ u(\alpha) = 0 & \text{on } \Gamma_1, \\ \dfrac{\partial}{\partial n} u(\alpha) = 0, & \text{on } \Gamma(\alpha), \end{cases} \tag{14.22}$$

where $f \in L^2(\widehat{\Omega})$, $\Omega(\alpha) = \{(x_1, x_2) \in \mathbf{R}^2 \mid 0 < x_1 < \alpha(x_2), x_2 \in (0,1)\}$, $\Gamma(\alpha) = \{(x_1, x_2) \mid x_1 = \alpha(x_2),\ x_2 \in (0,1)\}$, $\Gamma_1 = \partial\Omega \setminus \overline{\Gamma(\alpha)}$, and $\alpha \in U_{\mathrm{ad}}$ (see Figure 14.3).

The discrete analogue of (14.21) and (14.22) reads

$$\underset{\mathbf{x} \in U_{\mathrm{ad}}}{\text{minimize}} \left\{ J(\mathbf{x}, \mathbf{u}(\mathbf{x})) = \frac{1}{2} (\mathbf{u}(\mathbf{x}), M(\mathbf{x}) \mathbf{u}(\mathbf{x})) \right\} \tag{14.23}$$

subject to variational equality

$$K(\mathbf{x}) \mathbf{u}(\mathbf{x}) = \mathbf{f}(\mathbf{x}) \tag{14.24}$$

where $\mathbf{u}$ is the solution vector with components $\mathbf{u}_\ell$ which give an approximation to u at the nodal point X_ℓ of the triangulation $\mathcal{T}_h$ of $\Omega(u)$, i.e., $\mathbf{u}_\ell \cong u(X_\ell)$, $\ell = 1, \ldots, n$. The symbol $\mathcal{U}_{\mathrm{ad}} \subset \mathbf{R}^{D+1}$ stands for the set of admissible design variables and $\mathbf{x} = (\mathbf{x}_0, \ldots, \mathbf{x}_D)$ is the vector of design variables ($\mathbf{x}_i = \alpha(a_i)$, $a_i = ih$, $i = 0, \ldots, D(h)$) and $X_i^D = (\mathbf{x}_i, a_i)$ are the design nodes, $X_{i,j}^M$ are associated design nodes, see Figure 14.4. Moreover, $M(\mathbf{x})$ is the mass matrix, which is formally defined as follows $M_{ij}|_K = \int_K \varphi_i \varphi_j \, dx$, $K(\mathbf{x})$ is the stiffness matrix, $K_{ij}|_K = \int_K \operatorname{grad} \varphi_i \operatorname{grad} \varphi_j \, dx$, $\mathbf{f}(\mathbf{x})$ the force vector, $\mathbf{f}_j|_K = \int_K f \varphi_j \, dx$ and φ_j are basis functions.

In the method given by (14.20) one needs the gradient of J with respect to the design vector $\mathbf{x}$. We shall outline different methods for doing this important step.

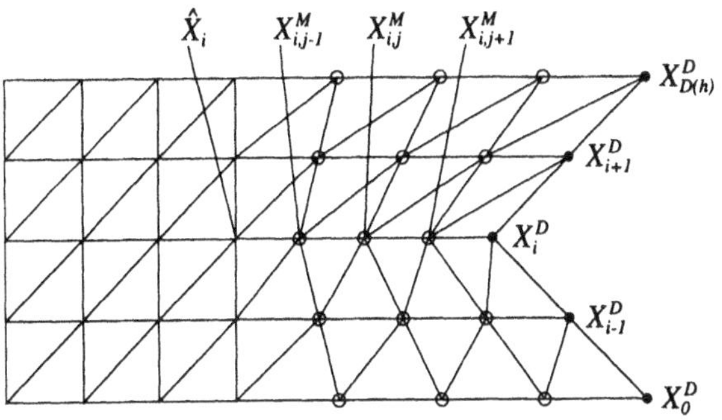

Figure 14.4.

Finite difference. We can define an approximation to each component of the gradient as follows

$$\frac{\partial J}{\partial \mathbf{x}_k} \cong \frac{J(\mathbf{x}+\varepsilon e_k, \mathbf{u}(\mathbf{x}+\varepsilon e_k)) - J(\mathbf{x}, \mathbf{u}(\mathbf{x}))}{\varepsilon}, \qquad k = 0, ..., D, \tag{14.25}$$

where e_k denotes the standard basis vector and $\varepsilon > 0$ is some small parameter. We note that in this method we have to solve the state system (14.24) once for each component of $\mathbf{x}$ in order to obtain $\mathbf{u}(\mathbf{x}+\varepsilon e_k)$ for all $k = 0, ..., D$.

Derivation of the gradient of J with respect to the design variable $\mathbf{x}$. We derive an expression for the directional derivative of J with respect to the design vector $\mathbf{x}$ in the direction $\mathbf{v} \in \mathbb{R}^{D+1}$. Using (14.23), we have through analytic differentiation

$$\begin{aligned} J'(\mathbf{x}) = J'(\mathbf{x}, \mathbf{u}(\mathbf{x}); \mathbf{v}) &= \lim_{\varepsilon \to 0} \frac{J(\mathbf{x}+\varepsilon \mathbf{v}, \mathbf{u}(\mathbf{x}+\varepsilon \mathbf{v})) - J(\mathbf{x}, \mathbf{u})}{\varepsilon} \\ &= (\mathbf{u}'(\mathbf{x}), M(\mathbf{x})\mathbf{u}(\mathbf{x})) + \frac{1}{2}(\mathbf{u}(\mathbf{x}), M'(\mathbf{x})\mathbf{u}(\mathbf{x})), \end{aligned} \tag{14.26}$$

where $\mathbf{u}'(\mathbf{x})$ and $M'(\mathbf{x})$ denote the directional derivatives of $\mathbf{u}$ and M (see [Haslinger, Neittaanmäki, 1996]).

In order to eliminate $\mathbf{u}'(\mathbf{x})$ from (14.26) we introduce the adjoint state equation

$$K(\mathbf{x})\mathbf{p} = M(\mathbf{x})\mathbf{u}(\mathbf{x}). \tag{14.27}$$

Differentiating (14.24), we obtain

$$K(\mathbf{x})\mathbf{u}'(\mathbf{x}) = \mathbf{f}'(\mathbf{x}) - K'(\mathbf{x})\mathbf{u}(\mathbf{x}).$$

From here, (14.26) and (14.27) we arrive at

$$\begin{aligned} J'(\mathbf{x}) &= (\mathbf{u}'(\mathbf{x}), K(\mathbf{x})\mathbf{p}) + \frac{1}{2}(\mathbf{u}(\mathbf{x}), M'(\mathbf{x})\mathbf{u}(\mathbf{x})) \\ &= (K(\mathbf{x})\mathbf{u}'(\mathbf{x}), \mathbf{p}) + \frac{1}{2}(\mathbf{u}(\mathbf{x}), M'(\mathbf{x})\mathbf{u}(\mathbf{x})) \\ &= (\mathbf{f}'(\mathbf{x}) - K'(\mathbf{x})\mathbf{u}(\mathbf{x}), \mathbf{p}) + \frac{1}{2}(\mathbf{u}(\mathbf{x}), M'(\mathbf{x})\mathbf{u}(\mathbf{x})). \end{aligned} \tag{14.28}$$

Now we shall present various methods for computing K' and $\mathbf{f}'$.

Finite difference applied to $K'(\mathbf{x})$ and $\mathbf{f}'(\mathbf{x})$. The usefulness of the finite difference approach can be somewhat extended. The computation of the gradient of J can be done using equation (14.26) where $K'(\mathbf{x})$ and $\mathbf{f}'(\mathbf{x})$ are computed by the difference method,

$$\frac{\partial K_{ij}|_K}{\partial \mathbf{x}_k} \cong \frac{K_{ij}(\mathbf{x}+\varepsilon e_k)|_K - K_{ij}(\mathbf{x})|_K}{\varepsilon}. \tag{14.29}$$

Here ε is the difference parameter, $k = 0, ..., D$ and $\mathbf{f}'(\mathbf{x})$ is computed correspondingly. Computation time is saved as $\mathbf{u}(\mathbf{x})$ is calculated only once for $J'(\mathbf{x})$.

Algebraic method by applying analytical derivation in local matrices. The algebraic method of sensitivity analysis is very often used in engineering applications. In [Haslinger, Neittaanmäki, 1996] a material derivative concept is presented for design sensitivity analysis for stiffness matrices, mass matrices and force vectors. In this connection we shall apply algebraic sensitivity analysis based on the isoparametric technique ([Brockman, 1985], [Mäkinen]). We follow the approach of [Mäkinen] and introduce the method in the case of nonlinear state problem.

Suppose that the state problem is the nonlinear potential problem with mixed boundary conditions

$$\left.\begin{aligned} -\operatorname{div}(\nu(x, \|\operatorname{grad} u\|^2)\operatorname{grad} u) &= f \quad \text{in } \Omega \subset \mathbf{R}^d,\ d=2,\\ u &= 0 \quad \text{on } \Gamma_1,\\ \nu(x, \|\operatorname{grad} u\|^2)\operatorname{grad} u \cdot n &= g \quad \text{on } \Gamma_2. \end{aligned}\right\} \tag{14.30}$$

Here $\partial\Omega = \Gamma_1 \cup \Gamma_2$, $\Gamma_1 \cap \Gamma_2 = \emptyset$, $f \in L^2(\Omega)$, $g \in L^2(\Gamma_2)$ and $\nu\colon \mathbf{R}^d \times \mathbf{R}^1 \to \mathbf{R}^1$ is a given positive smooth function. We assume that for given data the problem (14.30) is elliptic and has a unique weak solution.

The matrix form of the discretized problem (14.30) is the system of nonlinear equations

$$K(\mathbf{u})\mathbf{u} = \mathbf{f}, \tag{14.31}$$

where $K(\mathbf{u})$ is the stiffness matrix and $\mathbf{f}$ is the load vector. The unknown vector $\mathbf{u}$ contains the nodal values of u_h (a finite-element approximation of u).

We present the computation of the sensitivities for K and $\mathbf{f}$ via an element-by-element technique with a reference element. Suppose now that in (14.31) the nodes of the finite-element mesh depend on a real parameter α. Our aim is to find the sensitivity of the solution vector $\mathbf{u}$ with respect to α, that is, to find $\partial\mathbf{u}/\partial\alpha$. In what follows we will use $(\cdot)'$ to denote $\partial(\cdot)/\partial\alpha$, etc.

If the nodes of the finite-element mesh depend smoothly on α, we can differentiate (14.31) to obtain

$$[K(\mathbf{u})]'\mathbf{u} + K(\mathbf{u})\mathbf{u}' = \mathbf{f}'. \tag{14.32}$$

The terms $[K(\mathbf{u})]'\mathbf{u}$ and $\mathbf{f}$ can be computed element by element using the relations

$$K(\mathbf{u})\mathbf{u} = \sum_e P^e K^e(\mathbf{u}^e)\mathbf{u}^e, \qquad \mathbf{f} = \sum_e P^e \mathbf{f}^e. \tag{14.33}$$

Here P^e is the "local-to-global" expanding matrix, P^{eT} is the "global-to-local" gathering matrix and $\mathbf{u}^e = P^{eT}\mathbf{u}$ is the vector of the nodal values of u_h associated with the eth element.

In the case of isoparametric elements each element K_e is obtained from the parent reference element $\widehat{K} = [-1,1]^d$, for example, by the mapping $\widehat{K} \to K_e$ such that $\xi \mapsto x(\xi)$. Let

$$v = \begin{bmatrix} \varphi_1 \\ \vdots \\ \varphi_m \end{bmatrix}, \qquad L = \begin{bmatrix} \partial\varphi_1/\partial\xi_1 & \dots & \partial\varphi_m/\partial\xi_1 \\ \vdots & \ddots & \vdots \\ \partial\varphi_1/\partial\xi_d & \dots & \partial\varphi_m/\partial\xi_d \end{bmatrix} \tag{14.34}$$

be the matrices containing the shape functions and their derivatives for the parent element. Denote by $J = [\partial x_j/\partial\xi_i]_{i,j=1}^d$ the Jacobian of the mapping $\xi \mapsto x(\xi)$. Finally, let

$$X^e = \begin{bmatrix} X_1^1 & \dots & X_d^1 \\ \vdots & \ddots & \vdots \\ X_1^m & \dots & X_d^m \end{bmatrix} \tag{14.35}$$

be the matrix containing the nodal coordinates of the eth element. In what follows, we will omit the superscript e because we are now working only with the eth element. At a point $x(\xi)$ the Cartesian derivatives of the shape functions are now given by $B = J^{-1}L$ and the Jacobian is given by $J = LX$.

Gaussian quadrature with integration points and weights (ξ^k, W_k) $(k = 1, ..., q)$ is then used to perform the numerical integration needed for computing the element stiffness matrix, yielding

$$K(\mathbf{u}) = \sum_{k=1}^{q} W_k \nu(x^k, s_k) B_k^T B_k |J_k|, \tag{14.36}$$

where $s_k = \|\operatorname{grad} u_h(x^k)\|^2$, $x^k = x(\xi^k)$, $B_k = B(\xi^k)$, $J_k = J(\xi^k)$ and $|J_k| = \det J_k$.

The following results of [Brockman, 1985] and [Mäkinen] give the formulae for computing the sensitivities:

$$B_k' = -B_k X' B_k, \tag{14.37}$$

$$|J_k|' = |J_k| \sum_{j=1}^{m} [\operatorname{grad} \varphi_j(x^k)]^T (X^j)', \tag{14.38}$$

$$(x^k)' = (X')^T \varphi_k, \tag{14.39}$$

$$s_k' = 2(B_k \mathbf{u}^e)^T B_k' \mathbf{u}^e + 2(B_k \mathbf{u}^e)^T B_k (\mathbf{u}^e)', \tag{14.40}$$

and

$$\begin{aligned} \nu(x^k, s_k)' = {} & 2\frac{\partial \nu(x^k, s_k)}{\partial s} (B_k \mathbf{u}^e)^T B_k' \mathbf{u}^e \\ & + [\operatorname{grad}_x \nu(x^k, s_k)]^T (X')^T \varphi_k \\ & + 2\frac{\partial \nu(x^k, s_k)}{\partial s} (B_k \mathbf{u}^e)^T B_k (\mathbf{u}^e)'. \end{aligned} \tag{14.41}$$

Moreover,

$$[K(\mathbf{u}^e)]'\mathbf{u}^e = S^e(\mathbf{u}^e)(\mathbf{u}^e)' + T^e(\mathbf{u}^e)\mathbf{u}^e, \tag{14.42}$$

where

$$S^e(\mathbf{u}^e) = \sum_{k=1}^{q} C_k B_k^T B_k \mathbf{u}^e (\mathbf{u}^e)^T B_k^T B_k, \tag{14.43}$$

$$\begin{aligned} T^e(\mathbf{u}^e) = \sum_{k=1}^{q} [&C_k B_k^T B_k \mathbf{u}^e (\mathbf{u}^e)^T B_k^T B_k' + D_k (B_k')^T B_k \\ &+ D_k B_k^T B_k' + E_k B_k^T B_k + F_k B_k^T B_k] \end{aligned} \tag{14.44}$$

and

$$\begin{aligned} C_k &= 2W_k |J_k| \partial\nu(x^k, s_k)/\partial s, & D_k &= W_k |J_k| \nu(x^k, s_k), \\ E_k &= W_k |J_k| [\operatorname{grad}_x \nu(x^k, s_k)]^T (\mathbf{X}')^T \varphi_k, & F_k &= W_k |J_k|' \nu(x^k, s_k). \end{aligned} \tag{14.45}$$

In the absence of surface terms (that is, $g \equiv 0$) the element force vector is given by

$$\mathbf{f} = \sum_k W_k f(x^k) \varphi_k |J_k|. \tag{14.46}$$

Differentiating (14.46), we get

$$(\mathbf{f})' = \sum_k W_k [\operatorname{grad}_x f(x^k)^T (x^k)' \varphi_k |J_k| + f(x^k) \varphi_k |J_k|']. \tag{14.47}$$

Performing the assembly process, we get the following expression for the sensitivity of the solution vector.

The sensitivity of $\mathbf{u}$ is given as the solution of the linear system of equations (see (14.32) and (14.42))

$$[K(\mathbf{u}) + S(\mathbf{u})]\mathbf{u}' = \mathbf{f}' - T(\mathbf{u})\mathbf{u}. \tag{14.48}$$

We remark that in (14.37)–(14.48) the only matrix depending on a specific application (mesh topology, design parametrization, etc.) is X'. All the other matrices are available from the assembly of the system (14.31). To get accurate numerical values for the sensitivities it is recommended that system (14.31) is solved as accurately as possible. It is beneficial to use the Newton–Raphson method in solving (14.31), because after the convergence of the Newton–Raphson iteration the coefficient matrix in (14.48) is (approximately) that used in the last iteration.

The adjoint state technique for the sensitivity of a functional. Let $\boldsymbol{\alpha} = (\alpha_1, ..., \alpha_D)^T \in \mathbf{R}^D$ ($D = D(h)$) be a parameter vector and let the functional $F: \mathbf{R}^D \times \mathbf{R}^d \to \mathbf{R}$, that is, $(\boldsymbol{\alpha}, \mathbf{u}) \mapsto F(\boldsymbol{\alpha}, \mathbf{u})$, be a cost or a constraint functional. The sensitivity of F with respect to α_s ($s = 1, ..., D$) is given by

$$\frac{dF}{d\alpha_s} = \frac{\partial F}{\partial \alpha_s} + (\text{grad}_u F)^T \frac{\partial \mathbf{u}}{\partial \alpha_s}. \tag{14.49}$$

The form of (14.49) is not suitable when the gradient of F with respect to $\boldsymbol{\alpha}$ is needed as it requires D solutions of the linear system (14.48). Employing the standard adjoint equation technique of optimal control theory to eliminate $\partial \mathbf{u}/\partial \alpha_s$, we obtain

$$\frac{dF}{d\alpha_s} = \frac{\partial F}{\partial \alpha_s} + \mathbf{p}^T \left(\frac{\partial \mathbf{f}}{\partial \alpha_s} - T^{(s)}(\mathbf{u})\mathbf{u} \right), \tag{14.50}$$

where $\mathbf{p}$ is the solution of the adjoint equation

$$[K(\mathbf{u}) + S(\mathbf{u})]\mathbf{p} = \nabla_u F. \tag{14.51}$$

Now the computation of $\nabla_{\boldsymbol{\alpha}} F$ requires only one solution of the linear system (14.51).

Applications. Sensitivity analysis for the magnetic field. As explained in Section 7.2 the nonlinear magnetostatic field in a two-dimensional simply connected domain Ω expressed in terms of the potential function A, is governed by the equation

$$\text{rot}(\nu(x, \|\text{curl}\, A\|^2)\, \text{curl}\, A) = j, \tag{14.52}$$

where j is the current density and ν is the magnetic reluctivity. Let the domain under consideration be given by $\Omega = \Omega_{\text{air}} \cup \Omega_{\text{copper}} \cup \Omega_{\text{iron}}$. In this case the function ν is of the form

$$\nu(x, \|\mathbf{B}\|^2) = \begin{cases} 1/\mu_0, & x \in \Omega_{\text{air}} \cup \Omega_{\text{copper}}, \\ \nu(\|\mathbf{B}\|^2), & x \in \Omega_{\text{iron}}. \end{cases} \tag{14.53}$$

where $\mathbf{B} = \text{curl}\, A$.

As in $\mathbf{R}^2$ $\text{curl}\, A = (\partial_2 A, -\partial_1 A)$, $\text{rot}\, v = \partial_1 v_2 - \partial_2 v_1$ and $\|\text{curl}\, A\| = \|\text{grad}\, A\|$, (14.52) reduces into the nonlinear potential problem

$$-\text{div}(\nu(x, \|\text{grad}\, u\|^2)\, \text{grad}\, u) = j. \tag{14.54}$$

Although the mapping $x \mapsto \nu(x, s)$ is not continuous, no fundamental problems arise if the finite-element boundaries coincide with the material boundaries. The results of (14.50)–(14.51) are now directly applicable.

Remark 14.4. The sensitivity formulae presented here are both simple to program correctly and efficient because basic-linear-algebra-subroutine (BLAS) packages can be utilized. Our approach is general because it applies to all isoparametric Lagrangian finite elements. General-purpose programs can be easily developed because the dependence on the specific application can be isolated into separate modules. The same approach can clearly be applied to different state problems.

14.4. Implementation of optimal shape design procedures

Consider the problem

$$\underset{\mathbf{x}\in U_{\mathrm{ad}}}{\text{minimize}}\, J(\mathbf{x}, \mathbf{u}(\mathbf{x})) \tag{14.55}$$

subject to the constraints

$$K(\mathbf{x}, \mathbf{u}(\mathbf{x}))\mathbf{u}(\mathbf{x}) = \mathbf{f}(\mathbf{x}). \tag{14.56}$$

As seen in the previous section, the partial derivative of J with respect to x_k can be written in the form

$$\frac{dJ}{dx_k}(\mathbf{x}, \mathbf{u}(\mathbf{x})) = \frac{\partial J}{\partial x_k}(\mathbf{x}, \mathbf{u}) + \mathbf{p}^T\left(\frac{\partial \mathbf{f}(\mathbf{x})}{\partial x_k} - \frac{\partial K(\mathbf{x}, \mathbf{u})}{\partial x_k}\mathbf{u}(\mathbf{x})\right), \tag{14.57}$$

where p solves the adjoint state problem

$$[K(\mathbf{x}, \mathbf{u}(\mathbf{x})) + S(\mathbf{x}, \mathbf{u}(\mathbf{x}))]p = \nabla_u J(\mathbf{x}, \mathbf{u}(\mathbf{x})). \tag{14.58}$$

To solve the discrete problem (14.55) on a computer the use of existing optimization routines is preferred. One can employ, for example, the sequential quadratic programming subroutine E04VCF from the NAG library, for example, which calls a subroutine provided by the user to calculate the values of the cost function and its gradient at a given point.

The basic structure of the optimization program is as follows.

1. Read the data necessary for the optimization problem.
2. Set up the constraints for the optimization problem.
3. Call the optimization routine. (The optimization routine is called until the optimum is found).
4. Do the necessary postprocessing.

14.5. Industrial applications

We shall present a collection of optimal shape design problems which are interesting from an industrial point of view. For the technical background of these problems as well as the numerical solution we refer to [Neittaanmäki, Rudnicki, Savini], [Haslinger, Neittaanmäki, 1996].

Example 14.5. Optimal shape design of an electromagnet.
We will consider a real-life electromagnet belonging to the starter system of a vehicle, a model of which is shown in Figure 14.5 ([Di Barba, Navarra, Savini, Sikora], [Neittaanmäki, Rudnicki, Savini]).

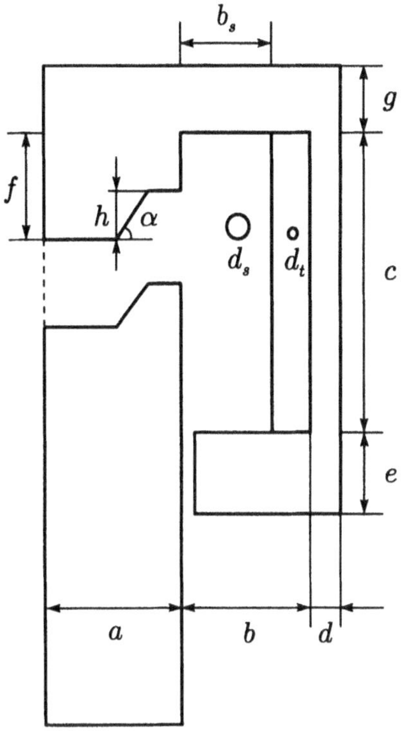

Figure 14.5. A two-dimensional model of an iron-core electromagnet.

The device is investigated in a steady state with a constant supply voltage and the nonlinear characteristics of iron are accounted for. The optimization problem essentially consists of minimizing the volume of the electromagnet in order to have a prescribed force on the movable part at a given air gap. In this case the field region is discretized into a rectangular finite-element mesh of 280 nodes subdivided into 500 triangular elements. Design variables can be selected to represent the device performance. They are: a, b, c, d, e, f, g, h, α, d_s, d_t, b_s where d_t and d_s are the diameters of the wires in the windings (Figure 14.5). Linear constraints corresponding to admissible values of the variables are introduced as well as nonlinear constraints on the number of turns and on the free space between windings; the prescribed force for two assigned positions of the movable part.

The optimization process (Figure 14.6), starting from the initial set of design parameters, first performs a check of the linear and nonlinear constraints. The vector of the actual variables $\mathbf{x}$ is then passed to the optimization code which is based on the search method by Rosenbrock and Storey ([Jacoby, Kowalik, Pizzo]);

according to the results of this code, the regeneration of the finite-element mesh is performed (the new mesh exhibits the same topology as the initial mesh). After a check of the constraints the current values of the design variables are updated by the optimizer. The whole process is automatically repeated until the optimum of the objective function is reasonably well approximated.

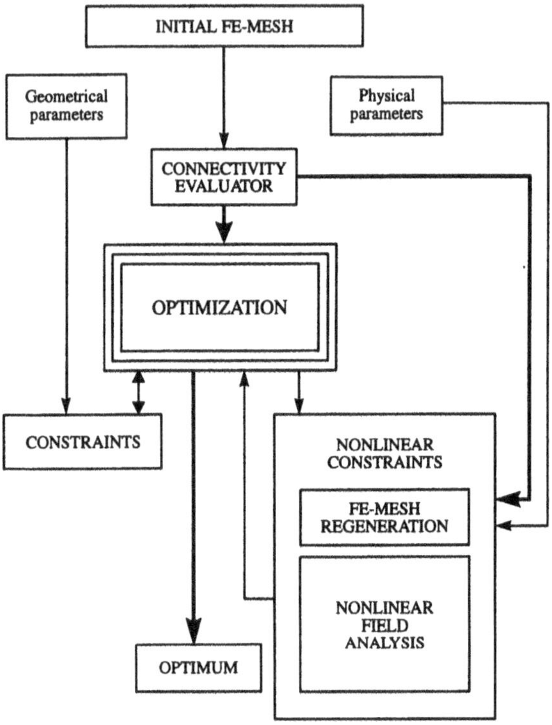

Figure 14.6. A flow chart of the optimization algorithm.

The starting configuration of the electromagnet is shown in Figure 14.7(a). The corresponding value of the objective function is $8.3 \times 10^{-5}\,\mathrm{m}^3$. The evolution of the design variables during computation is summarized in Table 14.1 (all the parameters are expressed in centimetres, except α which is measured in degrees).

a	b	c	d	e	f	g	h	α	d_s	d_t	b_s
12.50	11.87	25.00	2.62	7.00	9.00	6.00	3.12	52.04	1.13	0.64	8.00
12.48	11.85	25.95	2.62	7.00	8.97	6.00	3.13	51.88	1.12	0.65	8.00
12.37	11.89	23.82	2.69	6.86	7.60	5.89	3.49	48.02	1.17	0.69	7.87
12.68	12.31	22.64	1.96	6.45	6.90	5.66	3.28	48.05	1.23	0.71	7.68

Table 14.1. Evolution of the design variables.

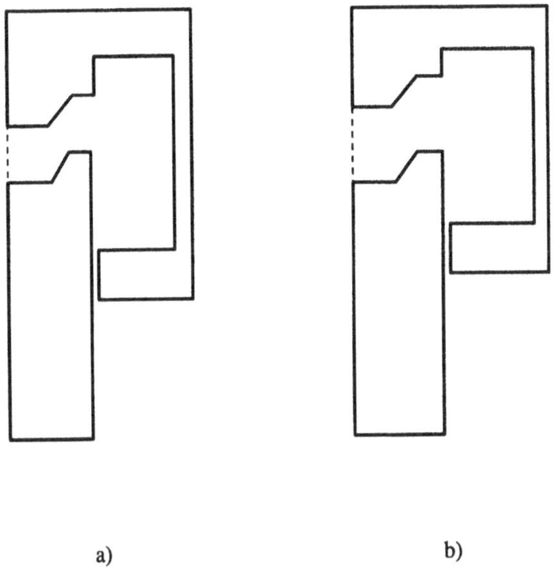

Figure 14.7. (a) The initial and (b) the final configuration of an iron-core electromagnet.

The objective function on exit is $7.93 \times 10^{-5}\,\mathrm{m}^3$. The final configuration of the electromagnet is shown in Figure 14.7 (b). From Table 14.1 and Figure 14.7 we can see that the final configuration is quite different from the initial one, the saving in the volume being nearly 11%.

Example 14.6. Optimal shape design of an air-filled capacitor.
Let us consider the cross section of an air-filled capacitor with parallel plates of finite extension at a distance equal to 2 m (Figure 14.8). The problem is described in [Rudnicki, Savini, Di Barba, Pawluk]. The state equation is Laplace's equation with Dirichlet boundary conditions at the upper S_2 and lower S_1 plates, and with homogeneous Neumann boundary conditions along the symmetry axis S_3 and along S_4 which roughly represents a field line truncating the end region of the field (a fictitious boundary); that is,

$$\Delta u = 0 \quad \text{in } \Omega, \tag{14.59}$$

$$u = 0 \quad \text{on } S_1, \tag{14.60}$$

$$u = 1 \quad \text{on } S_2, \tag{14.61}$$

$$\frac{\partial u}{\partial n} = 0 \quad \text{on } S_3 \cup S_4. \tag{14.62}$$

The optimal shape design problem consists of modifying the shape of the plates in order to have, in the end region Ω_0, the same uniform distribution of electric field as exists in the region adjacent to the symmetry axis Ω_c. The problem of synthesis can be formulated as follows:

$$\underset{\Omega_0 \in \mathcal{O}}{\text{minimize}} \int_{\Omega_0} |\operatorname{grad} u(\Omega) - E_c|^2 \, dx, \tag{14.63}$$

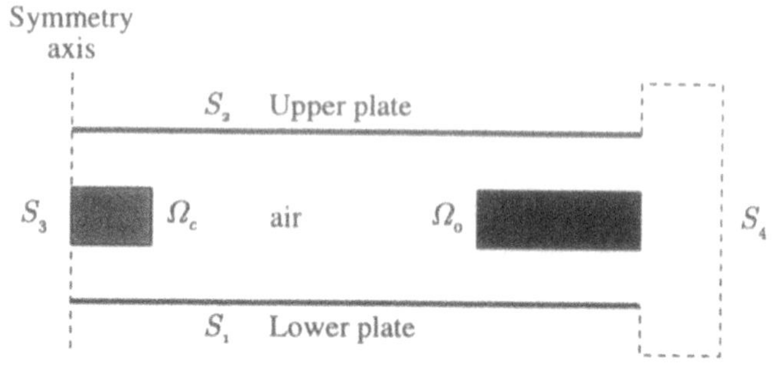

Figure 14.8. A cross-section of a parallel-plate capacitor.

where $u(\Omega)$ is a solution of the state problem (14.59)–(14.62), E_c is the normal component of the electric field in Ω_c and $\mathcal{O}$ denotes a family of admissible domains (determined by the constraints).

Introducing a finite-element model (Figure 14.9), Ω_c is covered by two elements, Ω_0 is given by four elements, while four moving nodes are used: two nodes on the upper plate and two nodes on the lower plate. These nodes are allowed to move along the vertical axis. For reasons of symmetry, extra symmetry constraints are imposed on the movement of the design nodes.

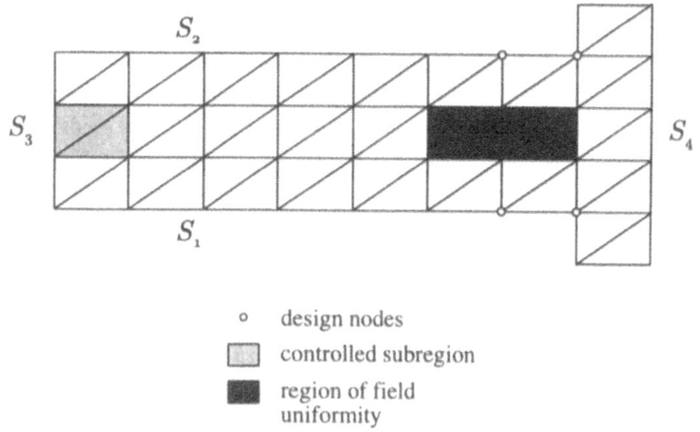

Figure 14.9. A finite-element mesh of a parallel-plate capacitor.

In Figure 14.10 the final shapes of the capacitor plates are shown. Results of the computation using a sequential-linear-programming optimizer with adaptive move limits are summarized in Table 14.2 (x_2 refers to the moving coordinate to design nodes). In [Neittaanmäki, Rudnicki, Savini] boundary element method is applied for solving the state problem.

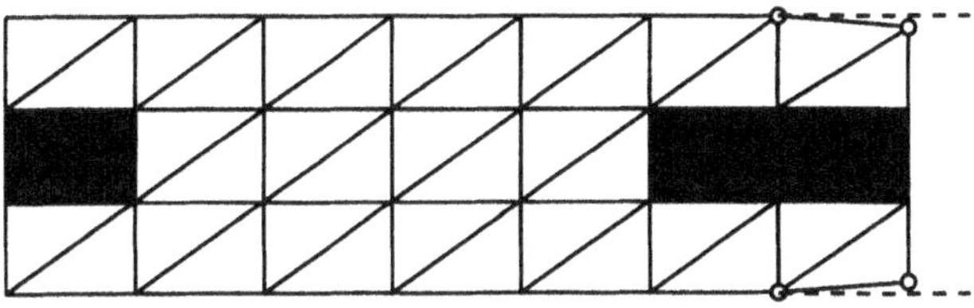

Figure 14.10. The final shape.

Initial x_2	Final x_2	Initial field $[\mathrm{V\,m^{-1}}]$	Final field $[\mathrm{V\,m^{-1}}]$
0.0	0.001	0.4772	0.5000
0.0	0.045	0.4982	0.5001
2.0	1.999	0.4982	0.5001
2.0	1.995	0.4998	0.5000

Table 14.2. Results of a finite-element simulation.

Example 14.7. Optimal shape design of a pole profile in a linear H-shaped magnet. A classical problem in magnet design is that of finding the pole profile of, for example, an H-shaped magnet, in order to have a prescribed uniform field in a subregion of the air gap. As a first approach, the iron is assumed to behave linearly. Figure 14.11 shows a cross section of one quarter of the magnet, composed of the iron core Ω_f, the coil Ω_c, the air gap Ω_a with the controlled subregion Ω_0 and the pole profile.

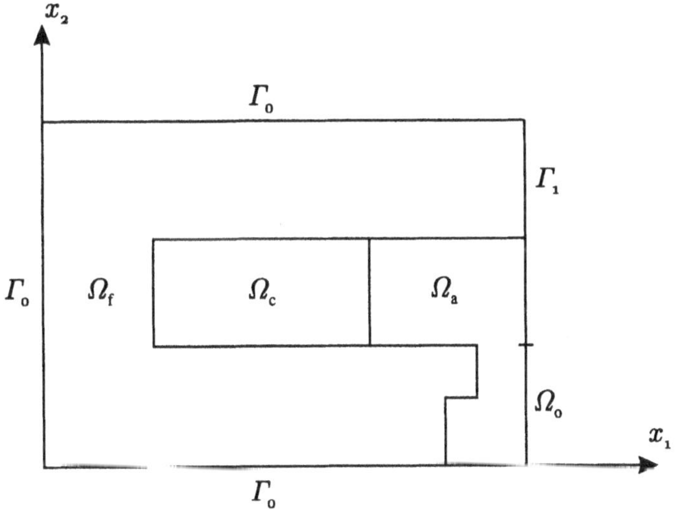

Figure 14.11. A model of the H-shaped magnet.

It is known that the magnetic vector potential $\mathbf{A}$ and the current density $\mathbf{J}$ are in $\mathbf{R}^3$ directed perpendicularly to the field region Ω, i.e., $\mathbf{A} = (0, 0, A)$, $\mathbf{J} = (0, 0, J)$.

The equation in each homogeneous part of Ω, in terms of the nonzero component of the magnetic vector potential $\mathbf{A}$ [Wb m^{-1}], is

$$-\Delta A = \mu_0 \mu_r J \quad \text{in } \Omega, \tag{14.64}$$

$$A = 0 \quad \text{on } \Gamma_0, \tag{14.65}$$

$$\frac{\partial A}{\partial n} = 0 \quad \text{on } \Gamma_1, \tag{14.66}$$

where μ_r is the relative permeability and μ_0 is the vacuum permeability.

The analysis of the magnetic field can be performed by means of the finite-element method. The field region $\Omega = \Omega_f \cup \Omega_c \cup \Omega_a$, represented by a rectangle $(0, 1.2) \times (0, 1)$, is discretized into the mesh shown in Figure 14.12, with 65 nodes and 104 triangular linear elements.

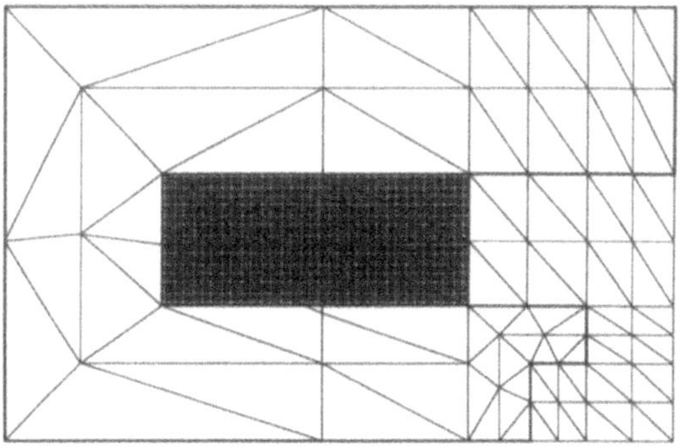

Figure 14.12. Finite-element mesh.

The profile of the pole shoe is defined by means of five moving nodes whose coordinates $(x_1(i), x_2(i))$ constitute the design vector (cf. Figure 14.13). To keep the mesh regular some side constraints are imposed on the movement of the design nodes. Because of the symmetry the x_2-component of the design node 1 is fixed: $x_2(1) = 0$.

The optimal values of the design variables are determined by minimizing the objective function F of the relative mean square error measure in the controlled region, consisting of four finite elements located close to the symmetry axis (see Figure 14.13), where

$$F(\mathbf{x}) = \sum_{i=1}^{N} \left(\frac{B_i - B_0}{B_0} \right)^2 . \tag{14.67}$$

In (14.67) B_i ($i = 1, ..., 4$) is the amplitude of the calculated flux density and $B_0 = 0.4$ [T] is the amplitude of the prescribed flux density.

Let $\mu_r = 1000$ and $J = 5 \times 10^6$ [A m^{-2}]. Using the sequential-linear-programming technique, the results shown in Table 14.3 were obtained for two different cases. In the first case tight constraints were imposed on the movement of the design nodes; in the second case side constraints on the movement of design nodes 1 and 2 were a little wider. The evolution from the initial to the final shape is shown in Figure 14.13. The initial and final values of the flux densities in the controlled subregion are given in Table 14.4.

Design node number		1	2	3	4	5	Objective
Case 1	Initial coordinates	(1,0)	(1,0.08)	(1,0.16)	(1.04,0.16)	(1.08, 0.16)	0.49×10^{-2}
	Final coordinates	(1.0403,0)	(1.0401, 0.03)	(1.0409, 0.1268)	(1.04,0.11)	(1.03,0.21)	0.80×10^{-4}
Case 2	Initial coordinates	(1,0)	(1,0.08)	(1,0.16)	(1.04,0.16)	(1.08,0.16)	0.49×10^{-2}
	Final coordinates	(1.0465,0)	(1.03867, 0.064)	(1.0452, 0.087)	(1.0734, 0.109)	(1,0.1379)	0.84×10^{-4}

Table 14.3. Results of an optimization.

Case 1	Flux density values			
initial	0.4290	0.4500	0.3980	0.3580
final	0.4036	0.4067	0.3957	0.3982
Case 1	Flux density values			
initial	0.4270	0.4480	0.3960	0.3570
final	0.4037	0.4068	0.3960	0.3970

Table 14.4. Evolution of the flux densities B [T] in the controlled subregion.

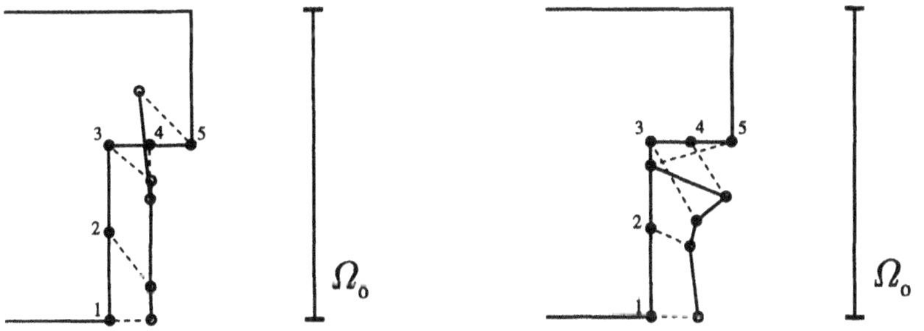

Figure 14.13. Pole profile for the initial and the final configuration for two cases. (o) Moving nodes.

Example 14.8. Optimal shape design of a pole profile in a nonlinear H-shaped magnet.
The analysis of both very large electromagnets and of tape recorder heads has already become a benchmark in computational electromagnetics. It has been extensively studied by [Pironneau] and by [Haslinger, Neittaanmäki, 1988].

In Figure 14.14 a cross section of an H-shaped magnet is shown, exhibiting the controlled subregion Ω_0, where it is desired that the magnetic field is constant. Moreover, Ω_f denotes the iron, Ω_c the copper, and Ω_a the air region. Because of symmetry we can restrict the analysis to one quarter of the cross section of the magnet (Figure 14.15).

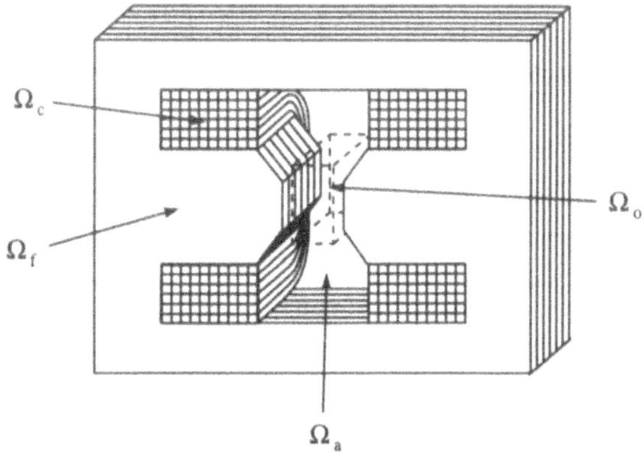

Figure 14.14. A cross-sectional view of the H-shaped magnet.

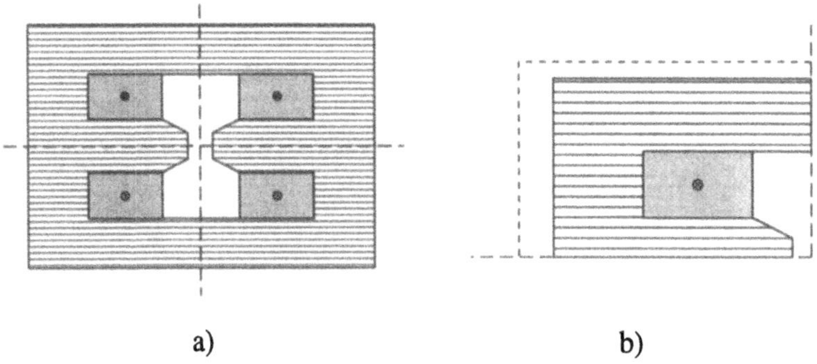

Figure 14.15. One quarter of the cross section.

In the field region $\Omega = \Omega_f \cup \Omega_a \cup \Omega_c$ shown in Figure 14.16 the magnetic potential satisfies the nonlinear Poisson equation with homogeneous mixed boundary conditions:

$$
\begin{aligned}
-\operatorname{div}\left(\nu_r(x, \|\operatorname{grad} A\|^2)\operatorname{grad} A\right) &= \mu_0 J, && \text{in } \Omega,\\
A &= 0, && \text{on } \Gamma_0\\
\frac{\partial A}{\partial n} &= 0, && \text{on } \Gamma_1,
\end{aligned}
$$

where $\mu_0 = 4\pi \cdot 10^{-7}$ $[\mathrm{m\,kg\,s^{-2}A^{-2}}]$, J is the x_3-component of the current density vector and

$$
\nu_r(x, \|\operatorname{grad} A\|^2) = \begin{cases} \nu_A = 1 & \text{for all } x \in \Omega_a,\\ \nu_C = 1 & \text{for all } x \in \Omega_c,\\ \nu_F = e + (b-e)\dfrac{\|\operatorname{grad} A\|^{2\alpha}}{\|\operatorname{grad} A\|^{2\alpha} + g} & \text{for all } x \in \Omega_f, \end{cases}
$$

(see Figure 14.16). The constants α, e, b and g above can be fixed by experiments. For example, we choose $e = 5\cdot 10^{-4}$, $b = 0.175775$, $g = 8758.756$ and $\alpha = 5.419241$.

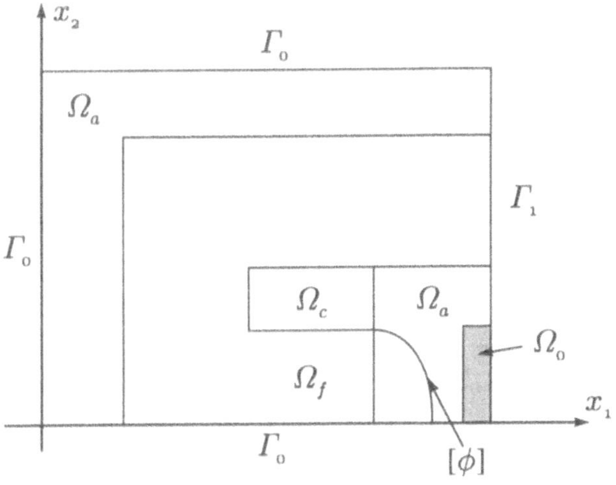

Figure 14.16. The field region.

The optimal shape design problem can be formulated as follows: Find the shape of the pole, that is, design a part of the boundary of Ω_f (separating Ω_a and Ω_f and described by a curve ϕ) in such a way that the magnet produces a constant magnetic field B_d in the controlled subregion $\Omega_0 \subset \Omega_a$ (Figure 14.16).

The optimization problem can, therefore, be set as the minimization problem

$$
\operatorname*{minimize}_{\Omega_f \in \mathcal{O}} F(\Omega_f), \quad \mathcal{O} = \left\{\Omega_f \mid [\phi] \in U\right\}. \tag{14.68}
$$

Here F is the objective function given by

$$
F(\Omega_f) = \int_{\Omega_0} \|\operatorname{curl} A - B_d\|^2 \, dx \tag{14.69}
$$

and we suppose that the graph $[\phi]$ of the curve ϕ lies inside a given set U where $U \subset \Omega_a$ is a subregion (see (14.72)).

In the discretization, triangular elements with a linear approximation have been applied (see Figure 14.17). In numerical experiments we take $\Omega = (0, 1.2) \times (0, 1)$.

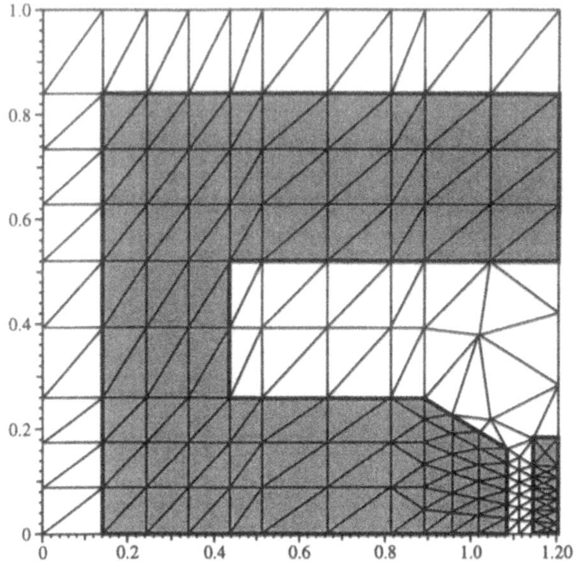

Figure 14.17. Triangulation for Ex1 and Ex2.

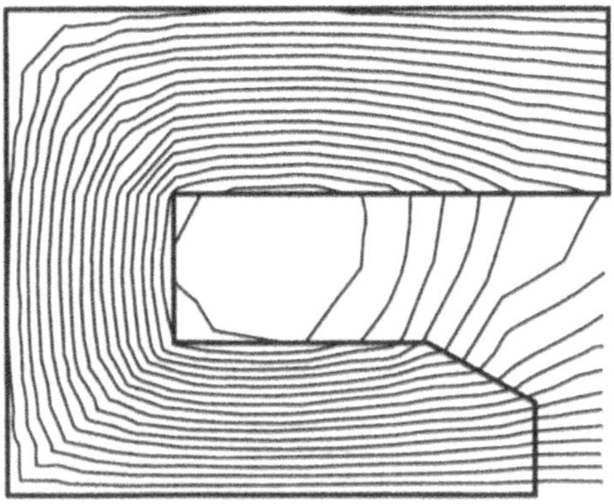

Figure 14.18. Initial shape.

The design vector is composed of the x_1- and x_2-coordinates of N design nodes along the pole boundary. When moving the design nodes, one has to move as well the neighbouring nodes in order to have the triangulation regular. Let $(x_1(i), x_2(i))$ be the coordinates of the design node X_i^D and let $(x_1(i,j), x_2(i,j))$ be the coordinates of the associated moving nodes $X_{i,j}^M$ along the line joining $(x_1(i), x_2(i))$ and the corresponding fixed node $\widehat{X}_i = (\hat{x}_1(i), \hat{x}_2(i))$. We suppose that

$$x_1(i,j) = \hat{x}_1(i) + \frac{j(\hat{x}_1(i) - x_1(i))}{M}, \qquad j = 0, ..., M-1, \tag{14.70}$$

$$x_2(i,j) = \hat{x}_2(i) + k_i(x_1(i,j) - \hat{x}_1(i)), \quad j = 0, ..., M-1 \tag{14.71}$$

and that

$$0.921 \leq x_1(i) \leq 1.131, \quad x_2(i) = \hat{x}_2(i) + k_i(x_1(i) - \hat{x}_1(i)), \tag{14.72}$$

where M is the number of associated moving nodes and k_i is the slope of the line joining X_i^D and $\widehat{X}_i$.

In numerical tests two different examples Ex1 and Ex2 are considered. These correspond to the cases when Ω_0 depends on the graph $[\phi]$ (moving Ω_0) or when it does not (fixed Ω_0). For the fixed domain Ω_0, we take (in metres)

$$\Omega_{0F} = (1.14, 1.2) \times (0, 0.18), \tag{14.73}$$

while, for the moving Ω_0 we take

$$\Omega_{0M} = \Omega_{0F} \cup Q_M, \tag{14.74}$$

where Q_M is the quadrilateral with vertices $(1.14, 0)$, $(1.14, 0.18)$, X_N^D and X_1^D.

Problem (14.68) is discretized by the FEM and a design sensitivity analysis is performed according to the method explained in Section 14.3. The state problem is solved using the nonlinear SOR method. For optimization we use the sequential quadratic programming (SQP).

We consider two examples with the data given in Table 14.5. In Ex1 and Ex2 we use a triangulation of 267 elements and 158 nodes. The corresponding mesh is shown in Figure 14.17.

Example	B_d [T]	J [Am^{-2}]	N
Ex1	(0,1.3)	5×10^6	7
Ex2	(0,1.6)	5×10^6	7

Table 14.5. Data for the two examples considered.

In Figures 14.19 and 14.20 we present results for Ex1 and Ex2 showing the final configurations with the corresponding contour plots of the equipotentials to be compared to the initial configuration of Figure 14.18.

Table 14.6 gives the values of the cost function for the initial and final configurations obtained after six SQP iterations.

Example	Initial F	Final F
Ex1	3.2767×10^{-3}	2.1750×10^{-5}
Ex2	8.3845×10^{-4}	6.0020×10^{-5}

Table 14.6. Initial and final values of the cost F.

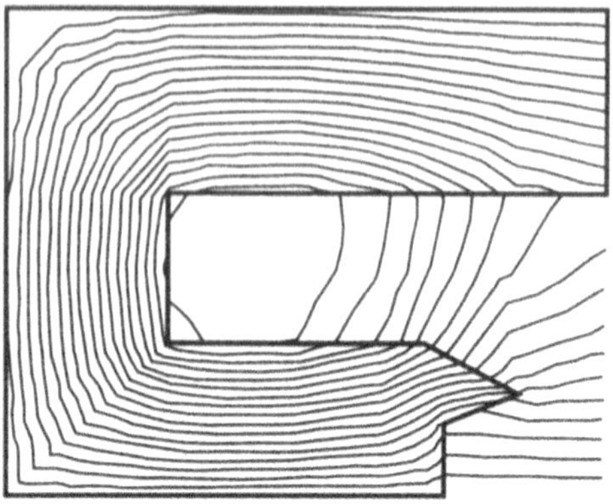

Figure 14.19. Final shape and equipotentials for Ex1.

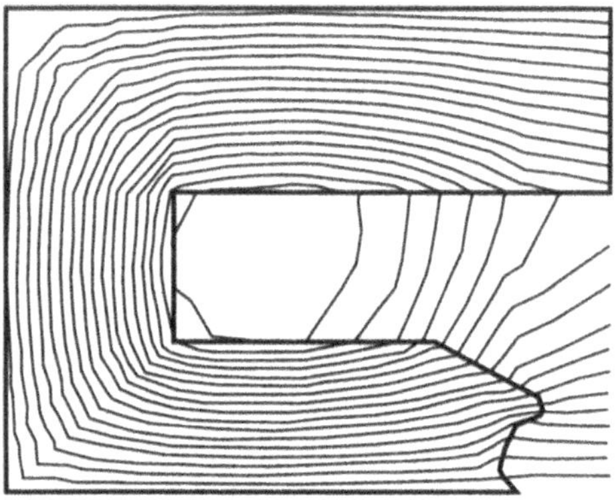

Figure 14.20. Final shape and equipotentials for Ex2.

References

Adam, J. C., Gourdin-Serveniere, A., Nedelec, J. C. and Raviart, P. A., *Study of an implicit scheme for integrating Maxwell's equations*, Comput. Methods Appl. Mech. Engrg. **22** (1980), 327–346.

Adams, R. A., "Sobolev spaces," Academic Press, New York, 1975.

Agmon, S., "Lectures on elliptic boundary value problems," Van Nostrand Math. Studies 2, New York, 1965.

Akin, J. E., "Application and implementation of finite element methods," Academic Press, London, 1982.

Amick, C. J., *Some remarks on Rellich's theorem and Poincaré inequality*, J. London Math. Soc. (2), **18** (1978), 81–93.

Apel, T. and Dobrowolski, M., *Anisotropic interpolation with applications to the finite element method*, Computing **47** (1992), 277–293.

Arora, J., "Introduction to optimum design," McGraw-Hill, New York, 1989.

Axelsson, O., "Iterative solution methods," Cambridge University Press, Cambridge, 1994.

Axelsson, O. and Barker, V. A., "Finite element solution of boundary value problems. Theory and computation," Academic Press, New York, 1984.

Axelsson, O. and Kolotilina, L. Y. (eds.), "Preconditioned conjugate gradient methods," (Proc. Conf. Nijmegen, 1989), Lecture Notes in Mathematics 1457, Springer-Verlag, Berlin, 1990.

Axelsson, O. and Lindskog, G., *On the eigenvalue distribution of a class of preconditioning methods*, Numer. Math. **48** (1986), 479–498.

Babuška, I., *The finite element method for elliptic equations with discontinuous coefficients*, Computing **5** (1970), 207–213.

Babuška, I., *Uncertainties in engineering design: mathematical theory and numerical experience*, in "The Optimal Shape," (J. Bennet and M. M. Botkin, eds.), Plenum Press, 1986; also in Technical Note BN-1044, Univ. of Maryland, 1985, 1–35.

Babuška, I. and Aziz, A. K., *Survey lectures on the mathematical foundations of the finite element method*, in "The Mathematical Foundations of the Finite Element Method with Applications to Partial Differential Equations," (A. K. Aziz, ed.), Academic Press, New York, 1972, pp. 3–359.

Babuška, I. and Aziz, A. K., *On the angle condition in the finite element method*, SIAM J. Numer. Anal. **13** (1976), 214–226.

Babuška, I. and Miller, A., *A feedback finite element method with a posteriori error estimation. Part I*, Comput. Methods Appl. Mech. Engrg. **61** (1987), 1–40.

Baker, G. A., *Error estimates for finite element methods for second order hyperbolic equations*, SIAM J. Numer. Anal. **13** (1976), 564–576.

Bank, R. E. and Weiser, A., *Some a posteriori error estimators for elliptic partial differential equations*, Math. Comp. **44** (1985), 283–301.

Barnhill, R. E. and Gregory, J. A., *Sard kernel theorems on triangular domains with application to finite element error bounds*, Numer. Math. **25** (1976), 215–229.

Barrett, J. W. and Elliott, C. M., *Fitted and unfitted finite-element methods for elliptic equations with smooth interfaces*, IMA J. Numer. Anal. **7** (1987), 283–300.

Beckman, F. S., *The solution of linear equations by the conjugate gradient method*, in "Mathematical Methods for Digital Computers," (A. Ralston and H.S. Wilf, eds.), John Wiley & Sons, New York, 1960, pp. 62–72.

Bendsøe, M. P., *Shape design as a material distribution problem*, Mat. Report No. 11 (1989), Danmarks Technishe Højskola.

Bendsøe, M. P., "Optimization of structural topology, shape and material," Springer-Verlag, New York, 1995.

Berger, M. S., "Nonlinearity and functional analysis," Academic Press, New York, 1977.

Bernardi, C., *Optimal finite-element interpolation on curved domains*, SIAM J. Numer. Anal. **26** (1989), 1212–1240.

Bossavit, A., *Mixed elements and a two-phase free-boundary problem in magnetostatics*, in "Numerical Methods for Free Boundary Problems," Proc. of the Conference, Jyväskylä, 1990, (P. Neittaanmäki, ed.), International Series in Numerical Mathematics 99, Birkhäuser, Basel, 1991, pp. 104–113.

Brandt, A. M., "Criteria and methods of structural optimization," Martinus Nijhoff Publishers, Dordrecht, 1986.

Brebbia, C. A. (ed.), "Topics in boundary element research," vol. 6– Electromagnetic applications, Springer-Verlag, Berlin, 1989.

Brezzi, F., *Theoretical and numerical problems in semiconductor devices*, in "Proc. Conf. Numerical Analysis, Dundee, 1985," (D. F. Griffiths and G. A. Watson, eds.), Pitman Research Notes in Mathematics Series, Longman Scientific & Technical, Harlow, 1986, pp. 24–31.

Brockman, R. A., *Geometric sensitivity analysis with isoparametric finite elements*, Comm. Appl. Numer. Methods **3** (1987), 495–499.

Brokate, M., "Optimale Steuerung von gewöhnlichen Differentialgleichungen mit Nichtlinearitäten vom Hysteresis–Typ," Verlag Peter Lang, Frankfurt am Main, 1987.

Bruaset, A. M., "A survey of preconditioned iterative methods," Longman Scientific & Technical, Harlow, 1995.

Céa, J., "Optimisation, theórie et algorithmes," Dunod, Paris, 1971; "Optimization, theory and algorithms," Springer-Verlag, 1987.

Céa, J., *Problems of shape optimal design*, in "Optimization of distributed parameter structures, Part II," (E. J. Haug and J. Céa, eds.), 1981, pp. 1005–1048.

Céa, J., *Conception optimale ou identification de formes, calcul rapide de la dérivée directionelle de la fonction coût*, RAIRO Modèl Math. Anal. Numér. **20** (1986), 371–402.

Čermák, L. and Zlámal, M., *Transformation of dependent variables and the finite element solution of nonlinear evolution equations*, Internat. J. Numer. Methods Engrg. **15** (1980), 31–40.

Červ, V. and Segeth, K., *A comparison of the accuracy of the finite-difference solution to boundary-value problems for the Helmholtz equation obtained by direct and iterative methods*, Apl. Mat. **27** (1982), 375–390.

Chen, C. M., *Superconvergence of finite element approximations to nonlinear elliptic problems*, in "Proc. China–France Sympos. on Finite Element Methods," Beijing, 1982, Gordon and Breach, New York, 1983, pp. 622–640.

Chen, C. M. and Huang, Y. Q., "High accuracy theory of finite element methods," Hunan Science and Technology Press, Changsha, 1995. (in Chinese)

Chleboun, J., *Optimal design of an elastic beam on an elastic basis*, Apl. Mat. **31** (1986), 118–140.

Chleboun, J. and Mäkinen, R., *Primal hybrid formulation of an elliptic equation in smooth optimal shape problems*, Adv. Math. Sci. Appl. **5** (1995), 139–162.

Ciarlet, P. G., "The finite element method for elliptic problems," North-Holland, Amsterdam, 1978.

Ciarlet, P. G. and Lions, J. L. (eds.), "Handbook of numerical analysis," Vol. II Finite element methods, North-Holland, Amsterdam, 1991.

Ciarlet, P. G. and Raviart, P. A., *The combined effect of curved boundaries and numerical integration in isoparametric finite element method*, in "The Mathematical Foundations of the Finite Element Method with Applications to Partial Differential Equations," (A. K. Aziz, ed.), Academic Press, New York, 1972, pp. 409–474.

Coddington, E. A. and Levinson, N., "Theory of ordinary differential equations," McGraw-Hill, New York, 1955.

Coppersmith, D. and Winograd, S., *Matrix multiplication via arithmetic progression*, in "Proc. 19th Ann. ACM Sympos. Theory of Computing," 1987, pp. 1–6.

Courant, R., *Variational methods for problems of equilibrium and vibration*, Bull. Amer. Math. Soc. **49** (1943), 1–23.

Cullum, J. and Willoughby, R. A., "Lanczos algorithms for large symmetric eigenvalue computations," Birkhäuser, Boston, 1985.

Dal Maso, R. and Mosco, V., *Wiener's criterion and Γ-convergence*, Appl. Math. Optim. **15** (1987), 15–63.

Dauge, M., "Elliptic boundary value problems on corner domains," Lecture Notes in Mathematics 1341, Springer-Verlag, Berlin, 1988.

Delfour, M. C., Payre, G. and Zolésio, J. P., *Approximation of nonlinear problems associated with radiating bodies in the space*, SIAM J. Numer. Anal. **24** (1987), 1077–1094.

Di Barba, P., Navarra, P., Savini, A. and Sikora, R., *Optimum design of iron-core electromagnets*, IEEE Trans. Magnetics **26** (1990), 646–649.

Doktor, P., *On the density of smooth functions in certain subspaces of Sobolev spaces*, Comment. Math. Univ. Carolin. **14** (1973), 609–622.

Doležel, I., *Numerical calculation of the leakage field in the window of a transformer with magnetic shielding*, Acta Tech. ČSAV (1981), 563–588.

Douglas, J. and Dupont, T., *A Galerkin method for a nonlinear Dirichlet problem*, Math. Comp. **29** (1975), 689–696.

Douglas, J., Dupont, T. and Serrin, J., *Uniqueness and comparison theorems for nonlinear elliptic equations in divergence form*, Arch. Rational Mech. Anal. **42** (1971), 157–168.

Douglis, A. and Nirenberg, L., *Interior estimates for elliptic systems of partial differential equations*, Comm. Pure Appl. Math. **8** (1955), 503–538.

Drkošová, J., Rozložník, M., Strakoš, Z. and Greenbaum, A., *Numerical stability of GMRES*, BIT **3** (1995), 309–330.

Duff, I. S., Erisman, A. M. and Reid, J. K., "Direct methods for sparse matrices," Clarendon Press, Oxford, 1987.

Dunavant, D. A., *High degree efficient symmetrical Gaussian quadrature rules for the triangle*, Internat. J. Numer. Methods Engrg. **21** (1985a), 1129–1148.

Dunavant, D. A., *Economical symmetrical quadrature rules for complete polynomials over a square domain*, Internat. J. Numer. Methods Engrg. **21** (1985b), 1777–1784.

Dupont, T., *L^2-estimates for Galerkin methods for second order hyperbolic equations*, SIAM J. Numer. Anal. **10** (1973), 880–889.

Ekeland, I. and Temam, R., "Convex analysis and variational problems," North-Holland, Amsterdam, 1976.

Engels, H., "Numerical quadrature and cubature," Academic Press, London, 1980.

Feistauer, M., *On the finite element approximation of a cascade flow problem*, Numer. Math. **50** (1987), 655–684.

Feistauer, M., *On the finite element approximation of functions with noninteger derivatives*, Numer. Funct. Anal. Optim. **10** (1989), 91–110.

Feistauer, M., "Mathematical methods in fluid dynamics," Longman, Harlow, 1993.

Feistauer, M., Kalis, H., and Rokyta, M., *Mathematical modelling of an electrolysis process*, Comment. Math. Univ. Carolin. **30** (1989), 465–477.

Feistauer, M., Křížek, M. and Sobotíková, V., *An analysis of finite element variational crimes for a nonlinear elliptic problem of a nonmonotone type*, East-West J. Numer. Math. **1** (1993), 267–285.

Feistauer, M. and Sobotíková, V., *Finite element approximation of nonlinear elliptic problems with discontinuous coefficients*, RAIRO Modèl. Math. Anal. Numér. **24** (1990), 457–500.

Feistauer, M. and Ženíšek, A., *Finite element solution of nonlinear elliptic problems*, Numer. Math. **50** (1987), 451–475.

Feistauer, M. and Ženíšek, A., *Compactness method in finite element theory of nonlinear elliptic problems*, Numer. Math. **52** (1988), 147–163.

Feynman, R. P., Leighton, R. B. and Sands, M., "The Feynman lectures on physics, vol. 3," Addison-Wesley Publ. Company, 1966.

Fletcher, R., *Conjugate gradient methods for indefinite systems*, in "Proceedings of Dundee Conference on Numerical Analysis," Lecture Notes in Mathematics 506, Springer-Verlag, New York, 1975, pp. 73–89.

Franců, J., *Monotone operators. A survey directed to applications to differential equations*, Apl. Math. **35** (1990), 257–301.

Freund, R. W., *Conjugate gradient type methods for linear systems with complex symmetric coefficient matrices*, SIAM J. Sci. Stat. Comput. **13** (1992), 425–448.

Freund, R. W. and Nachtigal, N. M., *QMR: A quasi-minimal residual method for non-Hermitian linear systems*, Numer. Math. **60** (1991), 315–339.

Fučík, S., Kratochvíl, A. and Nečas, J., *Kačanov-Galerkin method*, Comment. Math. Univ. Carolin. **14** (1973), 651–659.

Fučík, S. and Kufner, A., "Nonlinear differential equations," Elsevier, Amsterdam, 1980.

Gajewski, H., *On uniqueness and stability of steady-state carrier distributions in semiconductors*, in "Proc. Conf. EQUADIFF 6," (J. Vosmanský and M. Zlámal, ed.), Springer-Verlag, Berlin, 1986, pp. 209–214.

Gajewski, H., Gröger, K. and Zacharias, K., "Nichtlineare Operatorgleichungen und Operatordifferentialgleichungen," Akademie-Verlag, Berlin, 1974.

Gelbaum, B. R. and Olmsted, J. M. H., "Counterexamples in analysis," Holden-day, San Francisco, 1964.

George, A. and Liu, J. W.-H., "Computer solution of large sparse positive definite systems," Prentice-Hall, Englewood Cliffs, NJ, 1981.

George, P. L., "Automatic mesh generation. Application to finite element methods," John Wiley & Sons, Chichester, 1991.

Gilbarg, D. and Trudinger, N. S., "Elliptic partial differential equations of second order," Springer-Verlag, Berlin, 1977.

Girault, V. and Raviart, P. A., "Finite element methods for Navier–Stokes equations," Springer-Verlag, Berlin, 1986.

Glowinski, R., "Numerical methods for nonlinear variational problems," Springer-Verlag, New York, 1984.

Glowinski, R. and Marrocco, A., *Analyse numérique du champ magnetique d'un alternateur par elements finis et sur-relaxation ponctuelle non lineaire*, Comput. Methods Appl. Mech. Engrg. **3** (1974), 55–85.

Goldberg, M., *Three infinite families of tetrahedral space-fillers*, J. Combin. Theory Ser. A **16** (1974), 348–354.

Golub, G. H. and Van Loan, C. F., "Matrix computations," North Oxford Academic, London, 1986.

Gregory, J. A., *Error bounds for linear interpolation on triangles*, in "Proc. MAFELAP II," (J. R. Whiteman, ed.), Academic Press, London, 1976, pp. 163–170.

Grisvard, P., "Elliptic problems in nonsmooth domains," Pitman Monographs and Surveys in Pure and Applied Mathematics 24, Harlow, 1985.

Gröger, K., *On steady-state carrier distributions in semiconductor devices*, Apl. Mat. **32** (1987), 49–56.

Haslinger, J. and Hlaváček, I., *Convergence of a finite element method based on the dual variational formulation*, Apl. Mat. **21** (1976), 43–65.

Haslinger, J. and Neittaanmäki, P., *On different finite element methods for approximating the gradient of the solution to the Helmholtz equation*, Comput. Methods Appl. Mech. Engrg. **42** (1984), 131-148.

Haslinger, J. and Neittaanmäki, P., "Finite element approximation for optimal shape design: theory and applications," John Wiley & Sons, Chichester, 1988.

Haslinger, J. and Neittaanmäki, P., "Finite element approximation for optimal shape, material and topology design," John Wiley & Sons, Chichester, 1996.

Haug, E. J. and Céa, J. (eds.), "Optimization of distributed parameter structures," Parts I & II. Nato Advances Study Institute Series, Series E, Sijthoff & Noordhoff, Alphen aan den Rijn, 1981.

Haus, H. A. and Melcher, J. R., "Electromagnetic fields and energy," Englewood Cliffs, Prentice Hall, 1989.

Heise, B., *Analysis of a fully discrete finite element method for a nonlinear magnetic field problem*, SIAM J. Numer. Anal. **31** (1994), 745–759.

Hewitt, E. and Stromberg, K., "Real and abstract analysis," Springer-Verlag, Berlin, 1975.

Hirzebruch, F. and Scharlau, W., "Einführung in die Funktionalanalysis," Bibliographisches Institut, Mannheim, 1971.

Hlaváček, I. and Křížek, M., *On a superconvergent finite element scheme for elliptic systems*, Parts I–III, Apl. Mat. **32** (1987), 131–154, 200–213, 276–289.

Hlaváček, I. and Křížek, M., *Weight minimization of elastic bodies weakly supporting tension*, Parts I–II, Appl. Math. **37** (1992), 201–240, 289–312.

Hlaváček, I. and Křížek, M., *On a nonpotential and nonmonotone second order elliptic problem with mixed boundary conditions*, Stability Appl. Anal. Contin. Media **3** (1993a), 85–97.

Hlaváček, I. and Křížek, M., *Dual finite element analysis of three-dimensional axisymmetric elliptic problems*, Parts I–II, Numer. Methods Partial Differential Equations **9** (1993b), 507–526, 527–550.

Hlaváček, I., Křížek, M. and Malý, J., *On Galerkin approximations of quasilinear nonpotential elliptic problem of a nonmonotone type*, J. Math. Anal. Appl. **184** (1994), 168–189.

Hlaváček, I., Křížek, M. and Pištora, V., *How to recover the gradient of linear elements on nonuniform triangulations?*, Appl. Math. **41** (1996), No. 3.

Hlaváček, I. and Nečas, J., *On inequalities of Korn's type*, Part I, Arch. Rational Mech. Anal. **36** (1970), 305–311.

Hottel, H. C. and Sarofim, A. F., "Radiative transport," McGraw-Hill, New York, 1965.

Irons, B., *A frontal solution program for finite element analysis*, Internat. J. Numer. Methods Engrg. **2** (1970), 5–32.

Jacobs, D. A. H., *A generalization of the conjugate-gradient method to solve complex systems*, IMA J. Numer. Anal. **6** (1986), 447–452.

Jacoby, S. I. S., Kowalik, J. S. and Pizzo, J. T., "Iterative methods for nonlinear optimization problems," Prentice-Hall, Englewood Cliffs, NJ, 1972.

Jamet, P., *Estimations de l'erreur pour des éléments finis droits presque dégénérés*, RAIRO Anal. Numér. **10** (1976), 43–60.

Jennings, A. and Malik, G. M., *The solution of sparse linear equations by the conjugate gradient method*, Internat. J. Numer. Methods Engrg. **12** (1978), 141–158.

Johnson, C., "Numerical solutions of partial differential equations by the finite element method," Cambridge University Press, Cambridge, 1988.

Kačúr, J., Nečas, J., Polák, J. and Souček, J., *Convergence of a method for solving the magnetostatic field in nonlinear media*, Apl. Mat. **13** (1968), 456–465.

Kadlec, J., *On the regularity of the solution of the Poisson problem on a domain with boundary locally similar to the boundary of a convex open set*, Czechoslovak Math. J. **14** (1964), 386–393.

Kohn, R. V. and Strang, G., *Optimal design and relaxation of variational problems*, Comm. Pure Appl. Math. **XXXIX** (1986), Part I 113–137, Part II 139–182, Part III 353–378.

Kondratiev, V. A., *Boundary problems for elliptic equations in domains with conical or angular points*, Trans. Moscow Math. Soc. **16** (1967), 277–313.

Koukal, S., *Piecewise polynomial interpolations in the finite element method*, Apl. Mat **18** (1973a), 146–160.

Koukal, S., *Piecewise rational interpolation in a domain whose boundary consists of straight line segments and parts of conic sections*, in "Proc. of the 2nd Seminary on the Finite Element Method," Plzeň, 1973b, pp. 191–196.

Koukal, S. and Křížek, M., *Curved affine quadratic finite elements*, J. Comput. Appl. Math. **63** (1995), 333–339.

Krasnoselskij, M. A. and Pokrovskij, A. V., "Systems with hysteresis," Springer-Verlag, Berlin, 1989.

Kreisinger, V. and Adam, J., *Magnetic fields in nonlinear anisotropic ferromagnetics*, Acta Tech. ČSAV (1984), 209–241.

Krejčí, P., "Hysteresis, convexity and dissipation in hyperbolic equations," Gakuto Int. Series, Gakkotosho, Tokyo, 1996.

Křížek, M., *An equilibrium finite element method in three-dimensional elasticity*, Apl. Mat. **27** (1982), 46–75.

Křížek, M., *Conforming equilibrium finite element methods for some elliptic plane problems*, RAIRO Anal. Numér. **17** (1983), 35–65.

Křížek, M., *On semiregular families of triangulations and linear interpolation*, Appl. Math. **36** (1991), 223–232.

Křížek, M., *On the maximum angle condition for linear tetrahedral elements*, SIAM J. Numer. Anal. **29** (1992), 513–520.

Křížek, M., *On approximation of the Neumann problem by the penalty method*, Appl. Math. **38** (1993), 459–469.

Křížek, M. and Lin, Q., *On diagonal dominance of stiffness matrices in 3D*, East-West J. Numer. Math. **3** (1995), 59–69.

Křížek, M. and Litvinov, W. G., *On the methods of penalty functions and Lagrange's multipliers in the abstract Neumann problem*, Z. Angew. Math. Mech. **74** (1994), 216–218.

Křížek, M. and Mlýnek, J., *On the preconditioned biconjugate gradients for solving linear complex equations arising from finite elements*, Banach Center

Publ. **29** (1994), 195–205.

Křížek, M. and Neittaanmäki, P., *On the validity of Friedrichs' inequalities*, Math. Scand. **54** (1984a), 17–26.

Křížek, M. and Neittaanmäki, P., *Finite element approximation for a div-rot system with mixed boundary conditions in non-smooth plane domains*, Apl. Mat. **29** (1984b), 272–285.

Křížek, M. and Neittaanmäki, P., *Solvability of a first order system in three-dimensional non-smooth domains*, Apl. Mat. **30** (1985), 307–315.

Křížek, M. and Neittaanmäki, P., *Internat. FE-approximation of spaces of divergence-free functions in three-dimensional domains*, Internat. J. Numer. Methods Fluids **6** (1986), 811–817.

Křížek, M. and Neittaanmäki, P., *On superconvergence techniques*, Acta Appl. Math. **9** (1987), 175–198.

Křížek, M. and Neittaanmäki, P., *On time-harmonic Maxwell equations with nonhomogeneous conductivities: solvability and FE-approximation*, Apl. Mat **34** (1989), 480–499.

Křížek, M. and Neittaanmäki, P., "Finite element approximation of variational problems and applications," Longman Scientific & Technical, Harlow, 1990.

Křížek, M., Neittaanmäki, P. and Vondrák, M., *A nontraditional approach for solving the Neumann problem by the finite element method*, Mat. Apl. Comput. **11** (1992), 31–40.

Křížek, M. and Preiningerová, V., *Calculation of the 3d temperature field of synchronous and of induction machines by the finite element method*, Elektrotechn. obzor **80** (1991), 78–84. (in Czech)

Křížek, M. and Strouboulis, T., *How to generate local refinements of unstructured tetrahedral meshes satisfying a regularity ball condition?*, Preprint, Math. Inst., Prague (1995), 1–13.

Kufner, A., John, O. and Fučík, S., "Function spaces," Academia, Prague, 1977.

Larsson, S., Thomée, V. and Zhang, N. Y., *Interpolation of coefficients and transformation of the dependent variable in the finite element methods for the nonlinear heat equation*, Math. Methods Appl. Sci. **11** (1989), 105–124.

Leis, R., "Initial boundary value problems in mathematical physics," John Wiley & Sons, New York, 1986.

Lenoir, M., *Optimal isoparametric finite elements and error estimates for domains involving curved boundaries*, SIAM J. Numer. Anal. **23** (1986), 562–580.

Lewis, R. W. and Morgan, K. (eds.), "Numerical methods in thermal problems," Proc. of the Sixth Int. Conf. held at Swansea, Pineridge Press, 1989.

Lin, Q. and Zhu, Q. D., "Preprocessing and postprocessing for the finite element method," Shanghai Scientific and Technical Publishers, 1994. (in Chinese)

Lions, J. L., "Quelques méthodes de résolution des problémes aux limites non linéaires," Dunod, Paris, 1969.

Lions, J. L. and Magenes, E., "Problèmes aux limites non homogènes et applications," Dunod, Paris, 1968.

Liu, L. and Křížek, M., *Finite element analysis of a radiation heat transfer problem*, Preprint, Math. Inst., Prague (1995), 1–11.

Luenberger, D. G., *Hyperbolic pairs in the method of conjugate gradients*, SIAM J. Appl. Math. **17** (1969), 1263–1267.

Mackerle, J., "Finite element methods. A guide to information sources," Elsevier, Amsterdam, 1991.

Mäkelä, M. and Neittaanmäki, P., "Nonsmooth optimization: Analysis and algorithms with applications to the optimal control," World Scientific Publishing Company, Singapore, 1992.

Mäkinen, R., *Finite element design sensitivity analysis for nonlinear potential problems*, Comm. Appl. Numer. Methods **6** (1990), 343–350.

Markowich, P. A., "The stationary semiconductor device equations," Springer-Verlag, Wien, 1986.

Markowich, P. A., Ringhofer, C. A. and Schmeiser, C., "Semiconductor equations," Springer-Verlag, Berlin, 1990.

Martensen, E., *The convergence of the horizontal line method for Maxwell's equations*, Math. Methods Appl. Sci. **1** (1979), 101–113.

Matyska, C., *Variational principles for the momentum equation of mantle convection with Newtonian and power-law rheologies*, Geophys. J. Int. (to appear).

Mayergoyz, I. D., "Mathematical models of hysteresis," Springer-Verlag, Berlin, 1991.

Mazja, V. G., "Spaces of S.L. Sobolev," (Russian), Izd. Leningrad Univ., St. Petersburg, 1985.

Melkes, F. and Zlámal, M., *Numerical solution of nonlinear quasistationary magnetic fields*, Internat. J. Numer. Methods Engrg. **19** (1983), 1053–1062.

Meyers, N. G., *An example of non-uniqueness in the theory of quasi-linear elliptic equations of second order*, Arch. Rational Mech. Anal. **14** (1963), 177–179.

Milka, Z., *Finite element solution of a stationary heat conduction equation with the radiation boundary condition*, Appl. Math. **38** (1993), 67–79.

Mitchell, A. R. and Wait, R., "The finite element method in partial differential equations," John Wiley & Sons, London, 1975.

Moan, T., *Experiences with orthogonal polynomials and "best" numerical integration formulas on a triangle; with particular reference to finite element approximations*, Z. Angew. Math. Mech. **54** (1971), 501–508.

Mock, M. S., *On equations describing steady-state carrier distributions in a semiconductor device*, Comm. Pure Appl. Math. **25** (1972), 781–792.

Mock, M. S., *An example of nonuniqueness of stationary solutions in semiconductor device models*, COMPEL **1** (1982), 165–174.

Mock, M. S., "Analysis of mathematical models of semiconductor devices," Boole Press, Dublin, 1983.

Moise, E. E., "Geometric topology in dimensions 2 and 3," Springer-Verlag, Berlin, 1977.

Monk, P., *A finite element method for approximating the time-harmonic Maxwell equations*, Numer. Math. **63** (1992a), 243–261.

Monk, P., *Analysis of a finite element method for Maxwell's equations*, SIAM J. Numer. Anal. **29** (1992b), 714–729.

Monk, P., *A comparison of three mixed methods for the time dependent Maxwell equations*, SIAM J. Sci. Statist. Comput. **13** (1992c), 1097–1122.

Nečas, J., "Les méthodes directes en théorie des équations elliptiques," Academia, Prague, 1967.

Nečas, J., "Introduction to the theory of nonlinear elliptic equations," Teubner, Leipzig, 1983.

Nečas, J. and Hlaváček, I., "Mathematical theory of elastic and elasto-plastic bodies: an introduction," Elsevier, Amsterdam, Oxford, New York, 1981.

Neittaanmäki, P. and Křížek, M., *Conforming FE-method for obtaining the gradient of a solution to the Poisson equation*, in "Proc. of a GAMM–Seminar: Efficient Solutions of Elliptic Systems," (W. Hackbusch, ed.), Kiel, 1984, Vieweg & Sons, Wiesbaden, 1984, pp. 74–86.

Neittaanmäki, P. and Picard, R., *Error estimates for the finite element approximation to a Maxwell type boundary value problems*, Numer. Funct. Anal. Optim. **2** (1980), 263–285.

Neittaanmäki, P. and Picard, R., *On the convergence of the finite element approximation of eigenfrequences and eigenvectors to Maxwell's boundary value problem*, Ann. Acad. Sci. Fenn. Ser AI Math. **6** (1981), 255–271.

Neittaanmäki, P., M. Rudnicki and A. Savini, "Inverse problems and optimal design in electricity and magnetism," Oxford University Press, 1995.

Neittaanmäki, P. and Salmenjoki, K., *Sensitivity analysis for optimal shape design problems*, Structural Optimization **1** (1989), 241–251.

Neittaanmäki, P. and Saranen, J., *Finite element approximation of vector fields given by curl and divergence*, Math. Methods Appl. Sci. **3** (1981a), 328–335.

Neittaanmäki, P. and Saranen, J., *Finite element approximation of electromagnetic fields in the three dimensional space*, Numer. Funct. Anal. Optim. **2** (1981b), 487–506.

Neittaanmäki, P. and Saranen, J., *Semi-discrete Galerkin approximation method applied to initial boundary value problems for Maxwell's equations in anisotropic, inhomogeneous media*, Proc. Roy. Soc. Edinburgh Sect. A **89** (1981c), 125–133.

Neittaanmäki, P. and Saranen, J., *On the finite element approximation for Maxwell's problem in polynomial domains of the plane*, Appl. Anal. **12** (1981d), 73–83.

Nitche, J. A., *On L_∞-convergence of finite element approximations to the solution of a nonlinear boundary value problem*, in "Proc. Conf. Numer. Analysis," Dublin, 1976, pp. 1–17.

Noor, A. K., *Bibliography of books and monographs on finite element technology*, Appl. Mech. Rev. **44** (1991), 307–311.

Ohayon, R. and Gorge, Y., *Variational analysis of non-linear non-homogeneous heat conduction problem*, in "Proc. Conf. Numer. Methods for Non-linear Problems, Swansea," Pineridge Press, Swansea, 1980, pp. 673–681.

Ortega, J. M. and Voigt, R. G., *Solution of partial differential equations on vector and parallel computers*, SIAM Rev. **27** (1985), 149–240.

Picard, R., *Zur Theorie der zeitunabhängigen Maxwell'schen Gleichungen mit der Randbedingung* $n(\nabla \times E) = n(\nabla \times H) = 0$ *im inhomogenen anisotropen Medium*, Bonner Math. Schriften **65** (1973).

Picard, R., *Zur Existenz des Wellenoperators bei Anfangsrandwertproblemen vom Maxwell-Typ*, Math. Z. **156** (1977), 175–185.

Picard, R., *On convergence and error estimates for the horizontal line method applied to a Maxwell's boundary value problem in anisotropic, inhomogeneous media*, Proc. Roy. Soc. Edinburgh Sect. A **86** (1980), 53–59.

Picard, R., "An elementary proof for a compact imbedding result in generalized electromagnetic theory," SFB 72, Universität Bonn, preprint 624, 1983.

Pironneau, O., "Optimal shape design for elliptic systems," Springer series in Computational Physics, Springer-Verlag, New York, 1984.

Polak, S. J., den Heijer, C., Schilders, W. H. A. and Markowich, P., *Semiconductor device modelling from the numerical point of view*, Internat. J. Numer. Methods Engrg. **24** (1987), 763–838.

Pospíšek, M., *Nonlinear boundary value problems with application to the semiconductor device equations*, Appl. Math. **39** (1994), 241–258.

Pospíšek, M., Segeth, K. and Šilhán, P., *Numerical modeling of semiconductor devices*, Colloq. Math. Soc. János Bolyai **59** (1990), 319–328.

Preiningerová, V., Křížek, M. and Kahoun, V., *Temperature distribution in large transformer cores*, in "Proc. of GANZ Conf.," (M. Franyó, ed.), Budapest, 1985, pp. 254–261.

Rannacher, R. and Scott, R., *Some optimal error estimates for piecewise linear finite element approximations*, Math. Comp. **38** (1982), 437–445.

Rektorys, K., "Survey of applicable mathematics," Iliffe Books Ltd, London, 1969; Kluwer, Amsterdam, 1994.

Rektorys, K., "The method of discretization in time and partial differential equations," D. Reidel, Dordrecht, 1982.

Rice, J. R., "Matrix computations and mathematical software," McGraw-Hill, New York, 1981.

Rudin, W., "Real and complex analysis," McGraw-Hill, New York, 1966.

Rudnicki, M., Savini, A., Di Barba, P. and Pawluk, K., *Optimal synthesis of electric devices using the boundary element method*, in "Computational modelling of free and moving boundary problems," Vol. 2, Heat transfer (L. C. Wrobel and C. A. Brebbia, eds.), Walter de Gruyter, Berlin, 1991, pp. 269–279.

Ryžik, I. M. and Gradštejn, I. S., "Tables of integrals, sums, series and products," Izd. tech. lit., Moscow, 1951.

Saranen, J., *Über die Approximation der Lsungen der Maxwellschen Randwertaufgabe mit der Methode der finiten Elemente*, Appl. Anal. **10** (1980), 15–30.

Saranen, J., *The finite element method for the electric field of Maxwell's boundary value problem in polygonal domains of plane*, Ann. Acad. Sci. Fenn. Ser. A I Math. **6** (1981), 63–75.

Saranen, J., *On an inequality of Friedrichs*, Math. Scand. **51** (1982), 310–322.

Sarkar, T. K., "Application of conjugate gradient method to electromagnetics and signal analysis," Elsevier, Amsterdam, 1991.

Schulte, E., *Tilings*, in "Handbook of convex geometry," Vol. B, North-Holland, Amsterdam, 1993.

Schwartz, L., "Analyse mathématique," Vol. 1, Hermann, Paris, 1967.

Selberherr, S., "Analysis and simulation of semiconductor devices," Springer-Verlag, Wien, 1984.

Silvester, P. P. and Ferrari, R. L., "Finite elements for electrical engineers," Cambridge Univ. Press, Cambridge, 1983.

Solymar, L., "Lectures on electromagnetic theory," Oxford Univ. Press, Oxford, 1987.

Steele, C. W., "Numerical computation of electric and magnetic fields," Van Nostrand Reinhold Co., New York, 1987.

Strang. G., *Variational crimes in the finite element method*, in "The Mathematical Foundations of the Finite Element Method with Applications to Partial Differential Equations," (A. K. Aziz, ed.), Academic Press, New York, 1972, pp. 689–710.

Strang, G. and Fix, G., "An analysis of the finite element method," Prentice-Hall, Inc., Englewood Cliffs, New Jersey, 1973.

Strassen, V., *Gaussian elimination is not optimal*, Numer. Math. **13** (1969), 354–356.

Strassen, V., *Algebraic Complexity Theory*, in "Handbook of Theoretical Computer Science, Vol. A, Algorithms and Complexity," (J. van Leeuwen, ed.), Elsevier, Amsterdam, 1990, pp. 634–672.

Synge, J. L., "The hypercircle in mathematical physics," Cambridge Univ. Press, Cambridge, 1957.

Szabó, B. A., *Solution of electrostatic problems by the p-version of the finite element method*, Comm. Appl. Numer. Methods **4** (1988), 565–572.

Szabó, B. A. and Babuška, I., "Finite element analysis," John Wiley & Sons, New York, 1991.

Sze, M. S., "Physics of semiconductor devices," John Wiley & Sons, New York, 1981.

Taylor, A. E., "Introduction to functional analysis," John Wiley & Sons, New York, 1958.

Temam, R., "Navier–Stokes equations," North-Holland, Amsterdam, 3rd revised edn, 1984.

Temam, R., "Infinite-dimensional dynamical systems in mechanics and physics," Springer-Verlag, New York, 1988.

Thomée, V., *Polygonal domain approximation in Dirichlet's problem*, Inst. Math. Appl. Conf. Ser. **11** (1973), 33–44.

Thomée, V., "Galerkin finite element methods for parabolic problems," Lecture Notes in Mathematics 1054, Springer-Verlag, Berlin, 1984.

Tichonov, A. N., *Ein Fixpunktsatz*, Math. Ann. **111** (1935), 767–776.

Van der Sluis, A., *Condition numbers and equilibration of matrices*, Numer. Math. **14** (1969), 14–23.

Van der Vorst, H. A. and Dekker, K., *Conjugate gradient type methods and preconditioning*, J. Comput. Appl. Math. **24** (1988), 73–87.

Vanmaele, M., Morton, K. W., Süli, E. and Borzì, A., *Analysis of the cell vertex finite volume method for the Cauchy–Riemann equations*, SIAM J. Numer. Anal. (to appear).

Vanmaele, M. and Van Keer, R., *Convergence and error estimates for a finite element method with numerical quadrature for a second order elliptic eigenvalue problem*, Internat. Ser. Numer. Math. **96** (1991), 225–236.

Van Roosbroeck, W., *Theory of the flow of electrons and holes in Germanium and other semiconductors*, Bell Syst. Tech. J. **29** (1950), 560–607.

Varga, R. S., "Matrix iterative analysis," Prentice-Hall, New Jersey, 1962.

Vaško, A., "Infra-red radiation," Iliffe Books LTD, London, 1968.

Verzuh, F. M., *The solution of simultaneous linear equations with the aid of the 602 calculating punch*, Math. Comp. (Math. Tables and other Aids to Computation) **3** (1949), 453–462.

Visintin, A., *Mathematical models of hysteresis*, in "Topics in Nonsmooth Mechanics," (J. J. Moreau, P. D. Panagiotopoulos and G. Strang, eds.), Birkhäuser, Basel, 1987.

Vitásek, E., "Principles of the theory of numerical methods for solving differential equations," Academia, Prague, 1994. (in Czech)

Vujanovič, B. and Strauss, A. M., *Heat transfer with nonlinear boundary conditions via a variational principle*, AIAA (1971), 327–330.

Wahlbin, L. B., "Superconvergence in Galerkin finite element methods," Springer-Verlag, Berlin, 1995.

Weatherill, N. P. and Hassan, O., *Efficient three-dimensional Delaunay triangulations with automatic point creation and imposed boundary constraints*, Internat. J. Numer. Methods Engrg. **37** (1994), 2005–2039.

Weber, Ch., *A local compactness theorem for Maxwell's equations*, Math. Methods Appl. Sci. **2** (1980), 12–25.

Weisz, J., *On a-posteriori error estimate of approximate solutions to a mildly nonlinear nonpotential elliptic boundary value problem*, Math. Nachr. **153** (1991), 231–236.

Weyl, H, *The method of orthogonal projection in potential theory*, Duke Math. J. **7** (1940), 411–444.

Wheeler, M. F., *A priori L_2 error estimates for Galerkin approximations to parabolic partial differential equations*, SIAM J. Numer. Anal. **10** (1973), 723–759.

White, R. E., "An introduction to the finite element method with applications to nonlinear problems," John Wiley & Sons, New York, 1985.

Whiteman, J. R., *Singularities in two- and three-dimensional elliptic problems and finite element methods for their treatment*, in "Proceedings of the EQUADIFF 6 Conference," Brno, 1985, Springer-Verlag, Berlin, 1986, pp. 345–352.

Yosida, K., "Functional analysis," Springer-Verlag, Berlin, 1965.

Zeidler, E., "Nonlinear functional analysis and its applications," Vol. 1 Fixed-point theorem, Vol. 2 Monotone operators, Vol. 3 Variational methods and optimization, Vol. 4 Applications to mathematical physics, Springer-Verlag, New York, 1985, 1989, 1986, 1988.

Ženíšek, A., *The convergence of the finite element method for boundary value problems of a system of elliptic equations*, Apl. Mat. **14** (1969), 355–377.

Ženíšek, A., *Curved triangular finite C^m-elements*, Apl. Mat. **23** (1978), 346–377.

Ženíšek, A., *Discrete forms of Friedrichs' inequalities in the finite element method*, RAIRO Anal. Numér. **15** (1981), 265–286.

Ženíšek, A., *The finite element method for nonlinear elliptic equations with discontinuous coefficients*, Numer. Math. **58** (1990a), 51–77.

Ženíšek, A., "Nonlinear elliptic and evolution problems and their finite element approximations," Academic Press, London, 1990b.

Ženíšek, A., *Variational problems in domains with cusp points*, Appl. Math. **38** (1993), 381–403.

Zienkiewicz, O. C., "The finite element method in engineering science," 3rd. edn., McGraw-Hill, London, 1977.

Zienkiewicz, O. C. and Morgan, K., "Finite elements and approximation," John Wiley & Sons, New York, 1983.

Zlámal, M., *On the finite element method*, Numer. Math. **12** (1968), 394–409.

Zlámal, M., *Curved elements in the finite element method*, SIAM J. Numer. Anal. **10** (1973), 229–240.

Zlámal, M., *A linear scheme for the numerical solution of nonlinear quasistationary magnetic fields*, Math. Comp. **41** (1983), 425–440.

Zlámal, M., *Finite element solution of the fundamental equations of semiconductor devices I.*, Math. Comp. **49** (1986), 27–43.

Zolésio, J. P., *The material derivative (or speed) method for shape optimization*, in "Optimization of distributed parameter structures, Part II," (E. J. Haug and J. Céa, eds.), 1981a, pp. 1089–1151.

Zolésio, J. P., *Domain variational formulation for free boundary problems*, in "Optimization of distributed parameter structures, Part II," (E. J. Haug and J. Céa, eds.), 1981b, pp. 1152–1194.

Author index

Subject index

GPSR Compliance

The European Union's (EU) General Product Safety Regulation (GPSR) is a set of rules that requires consumer products to be safe and our obligations to ensure this.

If you have any concerns about our products, you can contact us on ProductSafety@springernature.com

In case Publisher is established outside the EU, the EU authorized representative is:

Springer Nature Customer Service Center GmbH
Europaplatz 3
69115 Heidelberg, Germany

Batch number: 10761599

Printed by Printforce, the Netherlands